# Meter-Wave Synthetic Aperture Radar for Concealed Object Detection

For a listing of recent titles in the
*Artech House Radar Library*,
turn to the back of this book.

# Meter-Wave Synthetic Aperture Radar for Concealed Object Detection

Hans Hellsten

ARTECH HOUSE

BOSTON | LONDON

artechhouse.com

Library of Congress Cataloging-in-Publication Data
A catalog record for this book is available from the U.S. Library of Congress

British Library Cataloguing in Publication Data
A catalog record for this book is available from the British Library.

ISBN-13: 978-1-63081-025-2

**Cover design by John Gomes**

Cover image is a FOI CARABAS II SAR image, obtained on November 23, 2000, flying at 4.7 km altitude across the city of Uppsala. Processed frequency band was 22–82 MHz, while depression angle interval was 55–71.


**685 Canton St.**
**Norwood, MA**



10 9 8 7 6 5 4 3 2 1

*To my family who endured and to Björn Sten who taught me mathematics*

# Contents

# Foreword

The exciting story in this book and the CARABAS family of airborne synthetic aperture radar (SAR) systems span more than three decades. In the early 1980s, theoretical work by Dr. Hans Hellsten at FOA (today known as FOI, the Swedish Defense Research Agency) led him to apply ultrawideband signals to avoid the ubiquitous speckle problem in coherent radar imaging. Together with a team of engineers at FOA, he took the lead to realize the first airborne SAR with these ideas in the frequency band 20–90 MHz. This led to CARABAS-I in 1992 and myself becoming part of the project team. CARABAS-II made its first flight in 1996 with major system improvements (e.g., antenna pattern and unambiguous Doppler). The FOA team went on and developed LORA (low-frequency radar) with its first flight in 2002, which is based on the same concept as CARABAS but expanding the frequency range to 200 to 800 MHz. Since 2012, LORA can be operated in two bands (i.e., 20 to 90 MHz and 200 to 800 MHz). These three airborne systems have conducted flights in numerous campaigns, mostly in Sweden but also abroad (USA, Panama, Finland, France, Switzerland, and Thailand), providing evidence of the capability and robustness of meter-wave SAR imaging principles. Hellsten left FOI in 2001 to lead the development of an industrial CARABAS prototype at Ericsson Microwave Systems. The effort led to the development of the dual-band CARABAS-III installed in a small helicopter.

Our collaboration also resulted in pioneering research in other areas. One example is the fast factorized backprojection algorithm, which was originally developed for ultrawideband SAR image formation, subsequently adopted by others and further developed for a broad range of imaging conditions. The CARABAS systems also showed significant potential for civilian applications, most notably forest mapping due to the high sensitivity of radar backscatter in the very high frequency (VHF) band to forest stem volume and biomass. A joint research effort with Chalmers University of Technology resulted in electromagnetic models and parameter retrieval algorithms. The work spurred a lot of interest in the community for using VHF-band and ultrahigh frequency–band SAR in large-scale forest inventory and, in fact, became part of a research effort, which eventually led to the selection of the BIOMASS mission as the seventh Earth Explorer by the European Space Agency in 2013. Its payload consists of a SAR instrument, which will be the first of its kind operating at a center frequency of 435 MHz. The satellite mission is scheduled for launch into low-Earth orbit in 2021 and is dedicated to global forest mapping, in particular monitoring above-ground biomass and disturbance of tropical forests. The collaboration with Chalmers has continued in recent years on other research topics, including research of integrating autofocus methods into the fast factorized backprojection algorithm.

Noteworthy is that the airborne systems from an early stage were designed together with engineers from several companies, mainly Ericsson, which later became Saab, a fact that ensured industrial radar experience brought into the projects. Furthermore, the Swedish Armed Forces and the Swedish Defence Materiel Administration (FMV) provided necessary funding, aeronautical expertise, as well as the Sabreliner airborne platform.

*Lars Ulander*
*Research Director, FOI*
*Professor, Chalmers University of Technology*
*IEEE Fellow*

# Acknowledgments

Neither the book nor its contents would have come into being without the skill and devotion of colleagues. The subject matter of the book has grown gradually over three decades. Up to 2000, this development took place under the auspices of the Swedish Defense Research Establishment FOI, thereafter being transferred to the Swedish companies Ericsson and Saab in turn. Out of the several FOI contributors, the author is in particular indebted to Lars Ulander also for subsequent good collaboration up to recent times. Other major contributions came from Gunnar Stenström, while continuous strong support was received from Björn Larsson, Anders Gustavsson, Tommy Jonsson, Per-Olov Frölind, and Mats Pettersson. For the significant recent developments, the important contributions made by Patrik Dammert deserve highlighting. The author has had the pleasure to work closely with a strong and competent team of Saab engineers, where Lars-Gunnar Andersson, Nils Dagås, Torbjörn Elfström, Hannes Illipe, Wolfgang Staberg, Claes Claesson, Anders Åhlander, Jonas Lindgren, Gary Smith-Jonforsen, Hanna Isaksson, Daniel Svensson, Niklas Eriksson, Jonathan Arvidsson, and Fredrik Dicander all have been crucial partners. For this type of development to take place, management creating the time and space necessary is just as important as engineering competence. The author wishes to express his most sincere appreciation to Saab for providing resources to this program in the midst of other pressing business. Appreciation also goes to the Swedish Defence Materiel Administration

FMV and the European Defense Agency SIMCLAIRS program, for providing further support from 2010 to 2013. Finally, the aeronautical engineering skill of Sven Eriksson and Kjell-Ove Gisselson, who have turned odd ideas on antenna concepts into airworthy systems, for the last two decades of development, have been invaluable.

# 1

# This Book: Why and How

Using wavelengths in the meter regime, radar obtains a practically useful capacity for media penetration. Installed in an aircraft and imaging the ground by the principle of synthetic aperture radar (SAR), meter-order resolution is attainable. Several important applications benefit from this type of radar. The foremost is standoff detection of targets screened by tree canopies and the related topic of penetrating artificial camouflage. Other applications are standoff detection of underground objects and building penetration. A further application is interferometric reconstruction of ground topography below a forest layer. The aim of this book is to provide in detail the physics and technology necessary to analyze and understand meter-wave SAR and its applications. Targets screened by forest or underground will be treated in detail up to the level of present understanding and experimental evidence. For building penetration and ground topography reconstruction, experimental data and theoretical development are still not at the level to provide a corresponding in-depth account.

To realize meter-wave SAR with the performance required, a number of significant hurdles have to be surmounted. The traditional microwave frequencies must be transposed into the meter-wave regime, while resolution must remain essentially the same as microwave systems. Resolution depends on bandwidth, so low-frequency SAR must exploit large fractional bandwidth—be *ultrawideband* in popular jargon. The large bandwidth significantly affects the design; in particular, the methods for SAR processing and hardware implementation have to be reworked. Standard design rules such as the radar equation take a different shape. Also entirely new challenges enter; foremost, the issue of cohabitation with communication signals within the fractionally wideband exploited by the radar.

The cohabitation aspects are crucial. Communication interferes with the radar by one-way propagation, while radar signals are received only as a result of the echo from the ground, with response orders of magnitude weaker than the communication signals impacting the radar—these may be extremely strong coming from nearby commercial broadcast transmitters. However, hard experimental evidence, gathered by operating this type of radar for several decades and across the world, shows that designs are possible enabling radar operation in just about any place at any time. Adopting emerging technologies for waveforms and transceiver deign, there will be little or discernable impact from radar operation on communication.

Another significant concern is antenna implementation. High directivity is not a requirement for meter-wave SAR, but there are strict conditions for antenna patterns. These turn out hard to make compliant with the demands for wideband operation and the aerodynamic restrictions imposed by aircraft integration. Satellites would allow more flexibility, but have to be ruled out at least in the lower frequency part of band considered, due to ionospheric interference. Already at low microwave frequencies, the ionosphere disrupts the propagation of electromagnetic signals and requires compensation in satellite navigation applications. Going to lower frequencies, compensatory measures increase in complexity, while at still lower frequencies and also considering large fractional bandwidth, compensation becomes impossible. There is an ESA satellite program BIOMASS with a launch date around 2024. However, with the radar intended for environmental monitoring, the requirement for resolution is modest and transmitted bandwidth is thus narrow. Moreover, it will operate at around 500 MHz; that is, in the high end of the frequency band assumed in this book (see [1]). For the wideband applications, also utilizing frequencies below 100 MHz, airborne platforms are the only solution.

The ambition has been to compose a book, that will be useful at universities, at specialist courses, and for professionals in the radar community. Scientific methodology has been a priority. To achieve a strict logical development, claims are not made unless proved from basic and well-established theory, combined with experimental evidence to the extent possible. In this spirit, the book starts with the physics of scattering from natural terrain as well as artificial; that is, man-made objects, at meter wavelengths. Fresnel's equations for lossy media, Mie scattering from spheres and cylinders, and the small-scale perturbation model for scattering from rough surfaces are developed with the specific application of meter-wave SAR in mind. The results are employed to create generic models for scattering from buried objects and scattering and attenuation caused by vegetation. Some notational simplification has been achieved from the start by already adapting the theoretical development to the current topic (choosing units suitably and avoiding explicit use magnetic permeability—here always unity). Derivations and resulting formulas thereby appear slightly different (and indeed simpler) than found elsewhere. As can be numerically checked, the agreement between the alternative formulations is nonetheless exact.

Based on the scattering theory thus established, achievable resolution when using large fractional bandwidth is investigated and SAR theory for large fractional bandwidth is developed. The exact SAR inversion formulas, which are basic to this theory, were originally derived by Lars-Erik Andersson in collaboration with the author around 1985. The account of SAR inversion principles is also augmented to include moving targets. Again, the formulation is exact.

Next, radar system theory and hardware aspects are developed. A formulation of the radar equation for meter wavelength SAR is provided. Its derivation requires a parallel analysis of the wideband performance of antennas of wavelength order extension, which is also provided. Also, suitable radar waveforms are discussed, as well as the impact of communication radio interference and mitigation methods to suppress the radar signal impacting these.

Thereafter, SAR processing methods for low-frequency SAR images are accounted for. The difficulty in these lies not least in how the navigational data shall be used to compensate for a nonlinear radar aircraft path. It turns out that for meter-wave SAR (just as for high-resolution microwave SAR), very high accuracy is required for these data. Methods relying on inertial measurements, satellite navigation data, as well as autofocusing methods are discussed for achieving the required accuracy.

Finally, methods are derived for meter-wave SAR target detection. Rather than attempting any overview of the many approaches that have been attempted, the account provides in-depth insight into three explicit methods, addressing, respectively, the detection of stationary targets, moving target detection, and subsurface targets. The three methods are all examples of what might be termed multidata detection, that is, target detection achieved by comparing SAR images collected under different circumstances (in time, antenna location on the aircraft or polarization). Accounts of experiments are included, showing how the methods have been successfully adopted.

The book focuses on today's best practices. Writing the book, it soon became clear that reference material more than a few years old could not be used directly. Partly, this is due to the radical advances in microelectronics, digital technology, and computers, which affect the entire outlook on radar design. However, the ambition is also to compose a holistic view on the field of meter-wave SAR, which requires derivations and analytical reasoning to be recast into a unified format for the fairly diverse set of topics that has to be covered. It should also be made clear that it is not the intention (or even within the capacity) of the author to attempt a historical expose or review of the many research efforts that have been ongoing for meter waves (e.g., concerning high-resolution SAR, SAR processing, or target detection). A good overview is already provided in the book by Davis [2], an author whose position with DARPA put him in an excellent position to compose such. Thus, the current list of references is restricted to what constitutes the original or the other work, which directly bears on the book subject matter.

Still, it is considered pertinent to provide some account of meter-wave SAR history from the personal perspective of the author, if for no other reason than to make the reader understand why and how this particular book has come into being.

The author's background is as a doctor in theoretical physics, pursuing basic research and lecturing in the seventies. Personal reasons caused a career change toward radar research at the Swedish Defense Research Agency FOI. The author's involvement with wideband radar and subsequently meter-wave SAR commenced around 1984, and resulted in the foundation of a foliage penetration meter-wave SAR research program around 1985. Initially, modest funds sufficed to form a skeleton research team to address what was understood to be the outstanding issues for the development of such a radar. Setting up the mathematics for SAR image generation from wideband meter-wave data was one on the first priorities. In parallel, high dynamic range radar receivers were studied as a means to deal with the severe communication interference environment. Permits to transmit were granted out of the potentially high military priority for the project. The good progress secured funding from the Swedish Armed Forces for building the 25 to 90-MHz CARABAS I research SAR. The radar was installed on a Sabreliner business jet belonging to the Swedish Air Force. A significant obstacle was the antennas. To ensure good wideband antenna performance for the several meters of wavelengths adopted, the antenna system was configured as two parallel 5m long wideband dipoles, end firing to the broadside (basically the same arrangement is used in the CARABAS II and CARABAS III systems). Not to impose undue mechanical loads, the antennas were made flexible, made as cloth sacks, held to air intakes at the aircraft tail, and stabilized by ram pressure (see Figure 1.1(a)).

Data collections with the CARABAS I system started in 1992, and soon proved successful. Considerable interest from the U.S. came with the first tests. The Swedish team was joined in turn by both participants from Hughes Aircraft Company (now Raytheon) and MIT Lincoln Laboratory, the latter under DARPA sponsorship. In 1993, DARPA funded a U.S. campaign with CARABAS I aircraft participating together with Stanford Research Institute FOLPEN SAR in data collections in Panama (for detecting targets concealed in tropical rainforests), Yuma (investigating the subsurface imaging capability), and Maine (imaging in northern hemisphere coniferous forests). The FOLPEN

**Figure 1.1** (a) and (b) The original 1990 sack antenna configuration for CARABAS I and the 1995 push boom configuration for CARABAS II, installed on the FMV TP861 and TP 862 Sabreliner business jets. The sacks are 5m long and the push booms are 8m long. The antenna separation is about 2m. The 2000 LORA system adopts UHF antenna arrays, which are firm and fit with the CARABAS II push booms. Thus, from the exterior, the two systems look the same.

SAR operated in wideband mode with frequencies from 150 MHz and up. Compared with CARABAS I data, the advantage of the lower CARABAS I frequencies for the purpose of foliage penetration became obvious.

In retrospect, it was a pity that these extensive tests were conducted with CARABAS I still so immature. Throughout the mid-nineties, significant improvements were made on the system. The FOI team was joined by Lars Ulander. Jointly, Lars Ulander and the author developed the fast factorized processing (FFBP) method, which in a modern guise is treated in Chapter 5. Also, the exact treatment of SAR for moving targets was the outcome of such a joint effort (see Chapter 3). The FFBP method opened the way to effectively deal with nonlinear SAR paths thereby considerably improving imaging quality.

Another major effort was the redesign of the antenna arrangement. The sack antennas had proved far from satisfactory, degrading image quality by being slightly unstable in air, and not possible to calibrate accurately. Furthermore, during the U.S. campaign, a sack antenna ripped in flight, actually damaging the aircraft. The Swedish Material Administration FMV, being responsible for the aircraft, demanded that the antenna system must be redesigned. The redesign decided on was to put antenna in front of the aircraft rather than to the rear. With this new placement, the antennas had to be rigid structures—so-called push booms shown in Figure 1.1(b). The design, which had to take into account aspects like aircraft flutter, proved aerodynamically and operationally entirely successful. Also, the original transceiver was modernized. After the modifications, the radar was renamed CARABAS II.

A second set of U.S. meter-wave SAR tests took place in 1997. Data collections at frequencies in the bands 25 to 90 and 200 to 700 MHz were conducted in parallel, using CARABAS II and the NADC P3 ultrawideband SAR. The tests site was Fort Indian Town Gap in Pennsylvania. The test outcome contributed in the DARPA FOPEN concept study, which resulted in the Lockheed Martin FOPEN demonstrator, later marketed as the TRACER ultrahigh-frequency (UHF)/very high frequency SAR. The Swedish input in the U.S. programs essentially ended there, that is, around 1997. In parallel, the Swedish development work continued in adding higher frequencies into the CARABAS system, resulting in the multichannel LORA SAR system, operating in the frequency band 200 to 600 MHz. The main rationale for LORA was to investigate the options for moving target detection. The first data collections with the system were made around 2000.

Up to 2004, meter-wave SAR was scheduled to become an operational asset for the Swedish Armed Forces. To this end, FMV wanted an industrial partner to FOI, and to ensure the author moved 2001 from FOI to take the position of technical manager for the program at what is now Saab Surveillance.

At the time, the company—then known as EMW—was owned by the Ericson Group, internationally well known by their telecommunication products. The choice of EMW as industrial partner was obvious, since right through the nineties, they had provided both technical and economic support to the CARABAS program. However, from 2000 and onward, due to the changing world political climate, a very significant downsizing of the Swedish Defense commenced. The change affected the procurement strategy of largely relying on domestic defense products. Many of the government sponsored technical military programs were abolished. For CARABAS, several initiatives were taken by FMV to share funding for the remaining development in an international arrangement. In the end, under the pressure of the downsizing, such a program never took off. Without a defined customer, Ericson was unwilling to continue to spend internal funds for finalizing the development.

The situation improved in 2006, when EMW was purchased by Saab. The Saab Company had its origins as a Swedish aircraft manufacturer, with car manufacturing since long side tracked into a separate company. At the time, it was diversifying into an international aerospace establishment, with a broad portfolio of defense related products and radars included. Today, Saab Surveillance is a world recognized radar developer, with well-established products such as ERIEYE and GIRAFFE. With the new ownership came a new willingness to internal investments and meter-wave SAR was recognized as one of the areas for future products.

Hence, from 2006, a second era of CARABAS development began. Based on the knowledge acquired from CARABAS I and II, initial work focused on system modeling in order to derive proper requirements for the basic radar components, such as the transmitter, receiver, and antenna system. The ambition was to obtain designs, which could be adapted into any aircraft, ranging from medium-sized unmanned aerial vehicle (UAVs) to helicopters and fighters. One major complication was how antennas could be integrated on these platforms, with acceptable wideband performance below 100 MHz. Though well functioning, the CARABAS II push boom arrangement was not suitable outside a quite narrow class of candidate radar platforms. It also turned out that the antenna diagram inherent in a push boom antenna arrangement was not ideal and negatively affected achievable resolution and performance. On candidate aircraft for a CARABAS type of radar was the Saab Gripen fighter. The radar was conceived as a pod complete with its antenna system. In the Gripen study, the effect on antenna performance by having the low-frequency wideband antenna just beside the aircraft fuselage was given close scrutiny. Another request was for the Saab Skeldar, which is a medium-sized rotor UAV with a payload capacity of around 20 kg. For SAR operating below 100 MHz, a possible antenna solution was to utilize the Skeldar landing skids. Antennas could, however, be no more than 1.5m long, which meant that they were electrically small and with antenna transfer characteristics significantly varying across (say) a 25 to 90-MHz radar frequency band. It turned out that the actual effects of using this type of small antennas all could be well handled. They would not degrade the performance, given sufficient transmit power, which anyway would be quite low for small SAR systems. The theory developed is accounted for in Chapter 4, and was put in practice in building the CARABAS III demonstrator system, on which work commenced around 2008.

While originally intended for Skeldar, a wise decision was taken to exploit a small manned helicopter—the Schweitzer 300C—for the demonstrator. Making the front end fully digital, a lightweight though high-performance transceiver was designed—the CARABAS II and LORA transceivers weigh around 150 kg, while the CARABAS III receiver weight is 11.5 kg. Increasing the transmit power to that of CARABAS II, the transceiver weight would still be less than 15 kg. The concept of electrically small antennas was exploited with an antenna length of 1.5m for the chosen radar band 27.5 to 82.5 MHz. The low band was complemented with a high band 137.5 to 357.5 MHz. As described in Chapter 4, CARABAS III allows change of polarization and also detection of moving targets, receiving data from two along track antenna positions.

Use of a helicopter platform for CARABAS III stressed the need of SAR processing algorithms remaining fully functional in situations where the aircraft trajectory is far from straight. The original 1985 solution to the SAR inversion problem assumed just an ideal straight aircraft path. With the Sabreliner platform, deviations from straight track remained small, and even when neglected in the early development, a SAR image of reasonable quality was obtained. In contrast, for the Schweitzer helicopter only flying at speeds of 25 m/s, wind gusts and turbulence cause major deviations from a straight and linear trajectory, and efficient means of handling nonlinear aircraft paths are necessary. The FFBP method is well suited to deal with this situation for real-time imaging of large areas. However, the accuracy requirements for motion demands logging of not only the aircraft motion as a whole, but also actually the phase center motion of the antennas. The equipment achieving this logging for CARABAS III is precision GPS, complemented with a low-grade INS system, integrated on the CARABAS III transceiver. A Kalman filter fuses the data from the two systems. Motion measurement methods are treated in Chapter 5. Parallel to the CARABAS III hardware development, an FFBP-based real-time processing structure was established, following the FFBP principles of Chapter 5.

For CARABAS III, as well as for the previous CARABAS systems, so-called change detection is the preferred method for finding targets covered by foliage. As it turns out, the ground response (in particular below 100 MHz) is temporally very stable and is only affected by sizeable changes, for example, a tree falling or a vehicle entering or vanishing out of the scene. In this way, a scene may remain unchanged for months. The entering or vanishing of vehicle is not affected by this being screened or camouflaged. Moreover, change detection can be performed by efficient algorithms allowing vehicle targets to be detected automatically, while imaging vast areas. At FOI, development of the change detection method started already in the mid-nineties. As most detection methods, it adopted so-called CFAR thresholding. To assess the confidence of the detections (i.e., the false alarm rate), the method thereby required extensive tests allowing true targets to be repeatedly and independently encountered. Such tests were conducted by FOI. For CARABAS III, the economy allowing extensive testing was no longer around. The author happened to come across how probabilistic reasoning in artificial intelligence used the Bayes theorem, and understood that Bayesian methods could also be suited to change detection and then computing false alarm probabilities within a single scene. Hence, a tool was obtained to objectively judge from a single image the confidence of the detections made therein. The method seemed to fit the case of meter-wave SAR exceptionally well, since it properly addressed the issue scene dependent clutter statistics, which is a characteristic of meter-wave SAR imagery. In Chapter 6, the Bayesian method is described and exemplified for actual data.

Data collections with CARABAS III commenced in 2012. In 2013, Bayesian change detection was successfully applied to data indicating an improvement in performance compared to CARABAS II (concealed targets down to the size of personal cars were reliably detected, while for CARABAS II, the minimum target size was that of a small truck).

From 2012 to 2013, further work went into CARABAS III fine-tuning and calibration. One particular objective was to enhance the sensitivity for imaging subsurface targets such as mines, cables, pipelines, and so forth. In parallel, the author carried out work on modeling the phenomenology for such targets, using generic shapes (a sphere for a mine and a cylinder for linearly extended objects) embedded in a lossy dielectric half space. Competing surface clutter was dealt with, assuming the small-scale perturbation model. The approach allowed a purely analytical treatment, with the physical insight such provides. In particular, it was possible to conclude that detectability of subsurface targets in competing surface clutter was highest at near grazing angle and using polarimetric measurements. A test with CARABAS III was duly conducted and confirmed what had been theoretically concluded. The development is accounted for in Chapters 2 and 6.

With the CARABAS III radar calibrated and tuned, the detectability of moving targets was investigated next. Initial GMTI tests were made in the summer 2014. They showed detectability of slowly moving vehicle targets but not targets as small as humans. The tests were followed up by what aimed to be more thorough campaign in October 2014, which, however, turned out largely a failure. Afterward, a malfunctioning connector could be identified as the reason. With improved processing methods and the connector attended to, a new test series was carried out in October 2016. It proved fully successful, also in detecting humans at walking speed in a forest. The test is described in Chapter 6.

All work done over the years—not least from 2006 and onward—has amounted to quite a complete and in-depth analysis of meter-wave SAR. In particular, the analysis provides the information required for system optimization regarding frequency, bandwidth, resolution, required transmit power, and so forth. The analysis is useful for properly understanding, specifying, and procuring meter-wave SAR, and thus deserves wide publication. The Saab policy has been to allow and even promote this, with proprietary rights (as far as they apply) first secured by patents. Academic cooperation has been pursued throughout the development, with a large number of articles already published on its different aspects.

This book project was initiated in late 2014. A publication providing a cohesive account of the work had been lacking. The suggestion of the author to compose a book was well met within Saab. In late 2014, Artech House was contacted and agreed to its publication, with February 2017 set as the target date for a finished manuscript. Within the two years of writing, the phenomenological modeling regarding subsurface targets has been augmented also to include surface objects, in particular considering backscattering and signal attenuation for vegetation. This late work forms part of Chapter 2, while a comparison with measurements is provided in Chapter 3.

## References

[1] Le Toan, T., S. Quegan, M.W.J. Davidson, et al. "The BIOMASS Mission: Mapping Global Forest Biomass to Better Understand the Terrestrial Carbon Cycle," *Remote Sensing of Environment*, Vol. 115, No. 11, November 15, 2011.

[2] Davis, M.E. *Foliage Penetration Radar—Detection and Characterization of Objects under Trees*. New Jersey: SciTech Publishing, 2011.

# 2

# Meter Wavelength Scattering from Natural Ground and Targets

## 2.1 Introduction

When considering radar signal interaction with the ground and targets, relevant theory will stress different aspects for microwave and meter-wave radars. At the scale of centimeter wavelengths, nature is manifestly rough and radar signals backscattered from natural objects are best considered as arising from random occurrences involving a vast number of elementary scatterers. Manmade objects are smooth at centimeter scale and will therefore to a large extent scatter according to laws of optics, with their response well described by ray optics and diffraction theory. In contrast, for meter wavelengths bare ground appears as nearly smooth, providing only weak backscattering, which (as will be seen) has a direct dependence on ground topography. Both manmade and natural objects (such as trees) scatter according to common principles, with their scattering characteristics depending on object gross features, but entirely independent on fine details. The statistical procedures characterizing analysis of microwave radar signals—typically giving rise to the standard clutter distributions—will have little applicability. Rather low frequency radar backscatter statistics will be directly expressible in the gross ground and object parameters. As an example, the intensity distributions in low-frequency radar imagery of forest stand in a direct and surprisingly accurate relation to the forest stem volume distribution (from which forest biomass follow by well-known forest models).

The objective of this chapter is to provide scattering theory demonstrating these features of low-frequency radar ground scattering. Development will be analytical, thus deriving parametric models determining scattering properties from the gross ground and object parameters, in accordance with what was just stated. The treatment starts from electromagnetic first principles (i.e., Maxwell's equations), avoiding the pitfall in applying possible nonrelevant microwave models to the low-frequency case. Visibility as to the basic assumptions and their mathematical consequences will take precedence over exact results; there are many possible extensions of the present treatment, which better and more completely will cover various situations, but where the reader is provided only with the reference for further reading. Still, the present account will certainly shed light on why low-frequency radar has particular advantages compared with radar at higher frequencies.

An attempt is made to make the development of electromagnetic theory in the book self-contained, thus starting from the elements of electromagnetism. This is partly to avoid leaving the reader stranded in the search for the appropriate references, but more importantly because a logical and coherent development of the subject requires its fundaments and conventions to be adapted to what is to follow. The account of the elements will be brief, and it is assumed that the reader is already acquainted with these, though some refreshment may be in want. In composing the chapter, for general aspects of electromagnetic theory, the author is used to [1] and for homogeneous media [2]. The brief account in [3] has also proved valuable.

## 2.2 Basic Laws of Electromagnetism

In cosmic scale, gravitation is believed to play a dominant role and also to be the overwhelmingly strongest force when cosmologic amounts of mass are considered. Certainly, gravitation controls the Earth and thus terrestrial life. Still, it does so only in creating the kind of stationary conditions required for life to develop (e.g., holding the Earth in place with respect to the Sun and holding the Earth together by gravity). At extreme temperatures, nuclear and subnuclear forces play a major role and are believed to be decisive in the early formation of the universe. Just as gravitation, they are necessary for life on earth, but again in a background role of providing the right amount of warmth and energy.

In contrast, electromagnetism is at the root of an extreme diversity of phenomena, which are always occurring, always developing, and thus constantly affecting us. For instance, electromagnetism is the fundamental to the formation of matter, both the atoms, that is, the elements, and the molecules, that is, chemistry and thus life. It even controls how we perceive the world with our eyes through light and color. Understanding electromagnetism has changed the everyday life of humans and plays a major role in the modern world. Doubts may be raised if gravitation or nuclear theory will ever obtain a practical importance, even remotely approaching that of electromagnetism.

Celestial phenomena may radiate at radio frequencies and thus be observable by radio telescopes. However, electromagnetic effects on the Earth characteristically occur in a small/microscopic scale (e.g., in the formation of atoms and molecules). Correspondingly, Earthbound sources of electromagnetic radiation are mostly in the infrared region or at even higher frequencies. Radio waves stand out as a human artifact requiring artificial devices for their creation, while lacking natural correspondents. In the optical regime, a camera and the eye are very close counterparts, but no known living creature uses radio waves for sensing.

Just like gravitation, the electromagnetic field is a field of force existing within and between any material bodies. The two-ended (by necessity) nature of the coupling between a field of force and matter is noted. For one, there will be a quantity stating the degree to which the field of force *affects* some particular piece of matter considered. For gravitation, this quantity is the body mass. For the electromagnetic field, this quantity is the property of the body called its electrical charge. Secondly, there is another independent quantity pertaining to the field, stating the degree to which the field is *created* by a material body. For the gravitational field, this quantity is the mass of the body times the gravitational constant. Presently, it will be the body charge divided by the vacuum dielectric constant $\varepsilon_0$, as will be further analyzed.

Electric charge will be represented as a real-valued scalar *density* function $q(x)$ across three-dimensional space, which just as the mass density is additive, that is, two densities $q(x)$, $q'(x)$ superimposed give the new density $q(x) + q'(x)$. One reason for gravitation and electromagnetism to have significantly different nature is, however, that in contrast to mass density, $q(x)$ is bipolar (can be both negative and positive). Considering matter in large collections, the charge distribution typically tends to average out, whereas a mass distribution just accumulates, eventually causing gravitation to dominate other forms of interaction.

Charges in motion due to a velocity vector field $\boldsymbol{v}(\boldsymbol{x})$ give rise to a vector current $\boldsymbol{j}(\boldsymbol{x}) = q(x)\boldsymbol{v}(\boldsymbol{x})$, satisfying the continuity equation[1]

$$\nabla \bullet j = -\frac{\partial q}{\partial t} \tag{2.1}$$

This type of equation applies to any scalar field quantity, which is conserved in the sense that its volume integral over some given region will change in the course of time only if there is a flow of the quantity through the boundary of the region. Thus, continuity equations apply to mass and energy density. Charge is conserved in the same way. Later, it will be seen that electromagnetic field energy also satisfies a continuity equation.

Being a field of force, the electromagnetic field must be vector valued. Actually, the electromagnetic field of force can be separated into two vector components affecting matter in different ways. One part will just affect charges in motion (i.e., affecting charges by the vector current $\boldsymbol{j}$). This is defined as the magnetic force field $\boldsymbol{H}$. The complementary field, affecting charges independently of their movement, is defined as the electric field $\boldsymbol{E}$. The force *density* created by the combined action of the applied electric and magnetic fields is

$$\boldsymbol{F} = q\boldsymbol{E} + \boldsymbol{j} \times \boldsymbol{H} \tag{2.2}$$

Also, when it comes to how electromagnetic fields are created from charges, the distinction between electric and magnetic field components is possible. In fact, there is a basic reciprocity according to which the electric field component $\boldsymbol{E}$ field arises from charge, whereas the magnetic component $\boldsymbol{H}$ arises from $\boldsymbol{j}$.

A fundamental characteristic of electromagnetism is that due to the bipolar nature of charge, matter will possess electromagnetic affinity even when it does not carry any net electrical charge. Matter is held together by the mutual attraction between positive and negative charges, bound in atoms or molecules. For electrically neutral matter, in the absence of an exterior field, the positive and negative charges each constitute a charge field, which are each other's opposite and add to a zero net charge distribution. Applying an exterior electromagnetic force field causes a separation between the positive

[1] Notational conventions will be to use the symbol "∇" for the gradient operator ($\partial/\partial x$, $\partial/\partial y$, $\partial/\partial z$). The "•" sign refers to dot or scalar product and "×" to cross product. Hence, "∇•" refers to the divergence operator and "∇×" to the "curl" or rotation operator. The effect of the divergence and rotation operators on a vector field is often explained through Gauss's and Stokes's integral theorems. Sufficient intuitive understanding may be attained by just considering these two operators as derivatives longitudinal and transversal to the field.

and negative fields. The resulting displacement creates an electromagnetic field, commonly referred to as polarization (not to be confused with polarization state of electromagnetic waves). The polarization field combines with the original exterior field. The combined field is commonly referred to as the field due to electromagnetic induction.

The magnetic field of force $\boldsymbol{H}$ gives rise to the magnetic induction field $\boldsymbol{B}$ and the electric field of force $\boldsymbol{E}$ gives rise to the electric induction field $\boldsymbol{D}$. Very commonly, one experiences homogeneous materials (henceforward media), which have scalar and linear relations

$$\boldsymbol{D} = \varepsilon * \boldsymbol{E};\ \boldsymbol{B} = \mu * \boldsymbol{H} \tag{2.3}$$

The relations are valid for any point of time $t$ with "*" signifying convolution with $\varepsilon = \varepsilon(\Delta t)$, $\mu = \mu(\Delta t)$ defined for a time interval $\Delta t = t - t'$ of times $t'$ prior to $t$. The convolution kernel is interpretable as a "material memory" which is a characteristic of a particular medium. The quantity $\varepsilon$ is referred to as electric permittivity and $\mu$ as magnetic permeability. Collectively, they are referred to as medium susceptibility. In the Fourier domain, the convolution will be represented as susceptibility being a frequency-dependent multiplicative factor. As will be evident in the sequel, the frequency dependence has the effect that electromagnetic fields propagate at speeds that depend on their frequency, a phenomenon known as dispersion. Depending on the problem at hand, dispersion can sometimes be neglected, sometimes not.

Whenever electric permittivity is considered a constant, it is commonly referred to as the dielectric constant of the medium. For most materials and most applications, unity magnetic permeability $\mu = 1$ can be assumed.

Because there is no such thing as absolute rest, there can be no absolute distinction between a pure electric and a pure magnetic field. Consider a closed room where only the electric and magnetic fields are present, whereas the actual charges creating these are outside. From the fundaments of physics, there must be no difference in fields inside the room if the room moves uniformly with the charges at rest or the charges moves uniformly. Thus, the time variation of the electromagnetic fields and, in particular, the existence of a magnetic field due to the linear motion of the charge must be the same as the time variation of the electromagnetic field when the room moves, caused by spatial inhomogeneities of the field surrounding a charge at rest. Thus, for instance, the time variation of the electric field component must by itself create a magnetic field. These considerations are fully taken care of by Maxwell's equations determining the electromagnetic field from a given a charge density $\rho$ and a current density $\boldsymbol{j}$

$$\begin{aligned}
&\frac{\partial \boldsymbol{D}}{\partial t} = c^2 \nabla \times \boldsymbol{H} - \frac{\boldsymbol{j}}{\varepsilon_0} \\
&\frac{\partial \boldsymbol{B}}{\partial t} = -\nabla \times \boldsymbol{E} \\
&\nabla \bullet \boldsymbol{D} = \frac{q}{\varepsilon_0} \\
&\nabla \bullet \boldsymbol{B} = 0
\end{aligned} \tag{2.4}$$

Here $c$ is the vacuum speed of light—in the SI system defined to have the exact value $c$ = 299,792,458 m/s. The first two equations predicts the temporal development (left-hand side of the above equation) of the fields from their instantaneous (right-hand side of the above equation) distribution as well as any currents present. The latter two equations are relations the instantaneous fields have to satisfy; in particular, they determine the induction from any charge distribution present. It is only by (2.2) that (2.4) is turned into a set of equations, enabling the actual determination of the electromagnetic field.

As will be seen in the next section, temporal variations in charge current (i.e., charge accelerations) imply electromagnetic radiation, which dissipates power from the charge. The power dissipation enables the radiated power to be measured, without reference to the mutual force between charges. However, whenever the electromagnetic field manifests itself as measurable entities like force and power, these are invariably determined from quadratic combinations of electric charge divided by the dielectric constant $\varepsilon_0$. Therefore, the field created by a charge cannot be measured independently of the force on a charge created by the field. For gravitation, the situation is different, that is, the gravitational force created by a body mass and the body mass itself can be independently established (the latter by collision experiments only based on inertia of the bodies). The definitions of electric fields of force and induced fields are thus conventions within the arbitrariness given by what can be physically measured. The definitions made here have been chosen from the viewpoint of simplifying the notations required in the sequel. Rationalizing notation could, however, go even further like selecting units so that $\varepsilon_0 = c = 1$, but that route is not followed since results then obtained will not readily translate into other conventions.

In SI units, the left-hand side of (2.2) has the physical dimension N/m$^3$ according to which $\boldsymbol{E}$ field has dimension N/[$Q$], where [$Q$] is the physical dimension for electric charge (not just charge density). From (2.4), it follows that the physical dimension of $\varepsilon_0$ is [$Q$]/([$E$] m$^2$). The two-dimensional relations

$[Q]^2/([\varepsilon_0]\ m^2) = \mathrm{N}$ and $[E]^2[\varepsilon_0] = \mathrm{Nm/m^3}$ are implied: the first refers to the force between charges at a distance (i.e., Coulomb's law) and the second to the energy density of the electric field. The magnetic fields differ from the electric by the dimension of speed, but otherwise obey similar relations.

The electric and magnetic susceptibility are considered just scale factors and thus dimensionless constants. Accordingly, $\boldsymbol{E}$ and $\boldsymbol{D}$ have the same physical dimension and so have $\boldsymbol{H}$ and $\boldsymbol{B}$. A common alternative convention is to define $\boldsymbol{D}$ and $\boldsymbol{B}$ as the fields created, either by charge when there is just a surrounding vacuum or by a combination of charge and induction, in the interior of media. In this case, rather than using $\boldsymbol{D}$ and $\boldsymbol{H}$ as defined, one adopts the fields $\boldsymbol{D}' = \varepsilon_0 \boldsymbol{D}$ and $\boldsymbol{H}' = c^2\varepsilon_0 \boldsymbol{H}$. It is also common practice to introduce the vacuum magnetic permeability constant as $\mu_0 = 1/c^2\varepsilon_0$. Clearly, this enables two of Maxwell's equations to be restated

$$\begin{aligned} \frac{\partial \boldsymbol{D}'}{\partial t} &= \nabla \times \boldsymbol{H}' - j \\ \nabla \bullet \boldsymbol{D}' &= q \end{aligned} \tag{2.5}$$

There is no change to the other two. This alternative convention would have exact correspondence for the gravitational field in splitting its representation into consisting of an induced gravitational field, caused by mass without reference to the gravitational constant, and a gravitational force field, determined by the induced field from this constant. In both cases, there will be a difference in physical dimensions for the induced fields and the fields of force. The practical advantage of using the current convention instead is that in vacuum and approximately in air, $\boldsymbol{D} = \boldsymbol{E}$ and $\boldsymbol{H} = \boldsymbol{B}$. This notational convenience is further augmented since (it is recalled) $\mu = 1$ and thus $\boldsymbol{H} = \boldsymbol{B}$ for most materials, and not just vacuum.

It is noted that what type of motion the electromagnetic forces (2.2) cause will depend on the mobility of the charge distribution. In a dielectric, the charge distribution experiences only reversible displacements from an equilibrium distribution assumed whenever no electromagnetic fields are present. In a conductor, arbitrary changes from the original distribution can occur, but the rate of change and the associated currents are in proportion to the strength of the imposed field. The constant of proportion, in reality a coefficient of friction, is the resistivity $\rho$ entering into Ohm's law $\boldsymbol{j} = \rho \boldsymbol{E}$. In this case, the density of the power dissipated from the electromagnetic field due to the movement of charges is

$$\boldsymbol{F} \bullet v = q\boldsymbol{E} \bullet v = \boldsymbol{E} \bullet j \tag{2.6}$$

It is observed that no power dissipation can occur due to a purely magnetic field since the velocity vector of the charges is at right angles to the forces creating them (because of the cross product in (2.4)). When combined with Ohm's law, the power dissipation is always positive, that is, it draws power from the electromagnetic field and transfers it into heat.

For a homogeneous medium, it will be seen that power dissipation inside the medium can be incorporated in a way that is easy to handle, namely, by assuming susceptibility to assume complex values with the imaginary part accounting for the heat loss.

The general problem of determining the induced electromagnetic field in heterogeneous medium situations is very complex. In many situations, analysis can be simplified. For instance, an electromagnetic device is engineered to operate in a certain way. The engineering design, a transmitting antenna for instance, may have as its purpose to create a certain charge and/or current distribution and the backreaction of the created field on this distribution is negligible. Truly difficult cases arise in electromagnetic scattering from some heterogeneous, complex structure. A simplification (commonly adopted, but with limited realism) is that the induction, caused by the arising movements of charges within this object, is small compared to the original field. Hence, the movements can be computed approximately from the first field and by again approximately determining the secondary field. This is known in electromagnetic scattering as the Born approximation.

## 2.3 Homogeneous Media

Note that $\boldsymbol{E}$ and $\boldsymbol{H}$, like any other vector fields obey the identity

$$\nabla \bullet (\boldsymbol{E} \times \boldsymbol{H}) = \boldsymbol{E} \bullet \nabla \times \boldsymbol{H} - \boldsymbol{H} \bullet \nabla \times \boldsymbol{E} \tag{2.7}$$

Assume homogenous space where $\varepsilon$ and $\mu$ are constant. Then Maxwell's equations imply that

$$c^2 \nabla \bullet (\boldsymbol{E} \times \boldsymbol{H}) = \frac{1}{\varepsilon_0} \boldsymbol{E} \bullet j + \frac{1}{2} \frac{\partial}{\partial t} \left( \varepsilon E^2 + c^2 \mu H^2 \right) \tag{2.8}$$

The first term on the right-hand side of the above equation was previously recognized as the power dissipation to charges in the medium. If there is no such dissipation (e.g., if resistivity is zero), then (2.8) assumes the form of a continuity equation

$$\nabla \bullet \boldsymbol{S} + \frac{\partial U}{\partial t} = 0 \tag{2.9}$$

for the two quantities

$$\begin{cases} U = \dfrac{\varepsilon_0}{2}\left(\varepsilon E^2 + c^2 \mu H^2\right) \\ \boldsymbol{S} = c^2 \varepsilon_0 \boldsymbol{E} \times \boldsymbol{H} \end{cases} \tag{2.10}$$

Equation (2.9) states that the quantity $U$ is conserved in the sense that any change of the volume integral of this quantity over a given volume can be associated by a flow described by the quantity $\boldsymbol{S}$ through the surface bounding this volume. Since in (2.8) these relations are balanced by possible power dissipation to charges within the volume, it is clear that $U$ is the energy density of the electromagnetic field and $\boldsymbol{S}$ represents the power flux of electromagnetic energy. This flux is the well-known Poynting vector of the electromagnetic field.

The core of Maxwell's equations is that that changes in the field distribution caused by changes in the charge distribution propagate with exactly the speed of light $c$ in vacuum or the speed of light, diminished by $\sqrt{\varepsilon\mu}$ in a homogeneous medium with permeability $\varepsilon$ and $\mu$. This property of Maxwell's equations is readily demonstrated by the application of the identity

$$\nabla \times (\nabla \times \boldsymbol{v}) = \nabla(\nabla \bullet \boldsymbol{v}) - (\nabla \bullet \nabla)\boldsymbol{v} \tag{2.11}$$

valid for any two vector field $\boldsymbol{v}$. Alternatively to $\nabla \bullet \nabla$, the notation $\nabla^2$will be used. This operator, commonly known as the Laplace operator, is of fundamental importance. It now takes precedence when operating with $\nabla \times$ on the first two equations in (2.4), swapping this operator and $\partial/\partial t$, and making use of the last two equations in (2.4)

$$\begin{aligned} \nabla^2 \boldsymbol{E} - \frac{\mu\varepsilon}{c^2}\frac{\partial^2 \boldsymbol{E}}{\partial t^2} &= \frac{\mu}{c^2}\frac{\partial \boldsymbol{j}}{\partial t} + \frac{\nabla q}{\varepsilon} \\ \nabla^2 \boldsymbol{H} - \frac{\mu\varepsilon}{c^2}\frac{\partial^2 \boldsymbol{H}}{\partial t^2} &= -\frac{1}{c^2}\nabla \times \boldsymbol{j} \end{aligned} \tag{2.12}$$

The equations derived are wave equations, that is, they imply for the field variables the property of action at a distance through a delay determined by the propagation speed $c/\sqrt{\varepsilon\mu}$ . Indeed, consider the wave equation for any space- and time-dependent quantities $\chi$ and $\psi$

$$\nabla^2 \chi - \frac{1}{C^2}\frac{\partial^2 \chi}{\partial t^2} = \psi \tag{2.13}$$

Regard $\chi$ as a field to be determined by the wave equation, while $\psi$ is a given scalar field acting as a source to $\chi$. Moreover, $C$ is any constant value, determining the propagation speed of the field $\chi$ according to (2.13). Equation (2.13) has the well-known solution

$$\chi(\boldsymbol{x},t) = \frac{1}{4\pi}\int \frac{\psi\left(\boldsymbol{x}',t - |\boldsymbol{x} - \boldsymbol{x}'|/C\right)}{|\boldsymbol{x} - \boldsymbol{x}'|} dV' + \chi_0(\boldsymbol{x},t) \tag{2.14}$$

The volume integral is performed over any suitable region of space. Moreover, $\chi_0 = \chi$ is a solution to the homogenous equation

$$\nabla^2 \chi - \frac{1}{C^2}\frac{\partial^2 \chi}{\partial t^2} = 0 \tag{2.15}$$

representing the part of the field that has no sources within the space region considered. By means of (2.14), each component of $\boldsymbol{E}$ and $\boldsymbol{H}$ is determined, given the charge and current distributions. The time delay due spatial separation and propagation speed is also manifest. Note that because $\sqrt{\varepsilon\mu} \geq 1$, the propagation speed is always smaller than the vacuum speed of light $c$.

An important property of Maxwell's equations is the implied existence of single vector potential field from which the electric and magnetic fields are both obtained. This is already indicated by the fact that the scalar charge field and the vector current field determine the electromagnetic fields (apart from any additional source free field). Moreover the continuity equation states that only the current field is independent. Also formally from the fact that $\boldsymbol{B}$ is divergence free, it is the rotation of an underlying vector field $\boldsymbol{A}$ (i.e., $\boldsymbol{B} = \nabla \times \boldsymbol{A}$). In terms of this field, the last of Maxwell's equations is automatically fulfilled. Similarly, a vector field is rotation free implies that there is an underlying scalar field $\phi$ for which the vector field is a gradient. According to the first equation, there is a field $\phi$ such that

$$\boldsymbol{E} + \frac{\partial \boldsymbol{A}}{\partial t} = -\nabla\phi \tag{2.16}$$

Set $\boldsymbol{A}' = \boldsymbol{A} + \nabla\phi$. Then since $\boldsymbol{B} = \nabla \times (\boldsymbol{A} + \nabla\phi) = \nabla \times \boldsymbol{A}$

$$\begin{aligned} \boldsymbol{E} &= -\frac{\partial \boldsymbol{A}}{\partial t} \\ \boldsymbol{B} &= \nabla \times \boldsymbol{A} \end{aligned} \tag{2.17}$$

Thus, a single vector field $\boldsymbol{A}$ determines both $\boldsymbol{E}$ and $\boldsymbol{B}$. For a known dielectric, it also determines $\boldsymbol{D}$ and $\boldsymbol{H}$. One finds, for a homogeneous dielectric, through Maxwell's equations (2.4) and the continuity equation (2.1)

$$\begin{aligned} &\nabla^2 \boldsymbol{A} - \frac{\eta}{c^2}\frac{\partial^2 \boldsymbol{A}}{\partial t^2} = \nabla(\nabla \bullet \boldsymbol{A}) - \frac{\mu}{c^2} j \\ &\frac{\partial^2}{\partial t^2}\nabla \bullet \boldsymbol{A} = \frac{1}{\varepsilon}\nabla \bullet j \end{aligned} \tag{2.18}$$

Here $\eta = \sqrt{\mu\varepsilon}$ is known as the index of refraction. As will be seen in the next section, it controls the reflection and refraction of electromagnetic field passing a media discontinuity.

In the case of no charges or currents, one has $\nabla \bullet \boldsymbol{A} = 0$ and the electromagnetic field is entirely described by the homogeneous wave equation

$$\nabla^2 \boldsymbol{A} - \frac{\eta}{c^2}\frac{\partial^2 \boldsymbol{A}}{\partial t^2} = 0 \tag{2.19}$$

The solution can be represented as a plane wave expansion

$$\boldsymbol{A}(\boldsymbol{x},t) = \oint \boldsymbol{A}_{\boldsymbol{n}}\left(t - \eta\frac{\boldsymbol{n} \bullet \boldsymbol{x}}{c}\right) d\Omega \tag{2.20}$$

where $\boldsymbol{n}$ is a direction vector and integration occurs of is solid angle $\Omega$. Here $\boldsymbol{A}_{\boldsymbol{n}}$ are arbitrary functions of time subjected to the requirement

$$\nabla \bullet \boldsymbol{A}_{\boldsymbol{n}}\left(t - \eta\frac{\boldsymbol{n} \bullet \boldsymbol{x}}{c}\right) = 0 \tag{2.21}$$

For simplicity, consider the situation that $\boldsymbol{n}$ is directed along the $x$-axis and then extrapolate conclusions to arbitrary directions. In this case, constraint (2.15) assumes the form[2]

[2] Conventions will be to put component index of vectors as a superfix, saving the subfix location as a much needed space for other designations. Exceptions will be for reference symbols, such as direction **n**, having fixed meaning, and not attributed with further indices.

$$\frac{\partial A_n^x}{\partial x}\left(t-\eta\frac{x}{c}\right)=\frac{c}{\eta}\frac{\partial A_n^x}{\partial t}=0 \tag{2.22}$$

One has

$$\begin{aligned}
E_n^x &= -\frac{\partial A_n^x}{\partial t}=0\\
E_n^y &= -\frac{\partial A_n^y}{\partial t}\\
E_n^z &= -\frac{\partial A_n^z}{\partial t}\\
B_n^x &= \frac{\partial A_n^z}{\partial y}-\frac{\partial A_n^y}{\partial z}=0\\
B_n^z &= \frac{\partial A_n^y}{\partial x}-\frac{\partial A_n^x}{\partial y}=\frac{\eta}{c}\frac{\partial A_n^y}{\partial t}=-\frac{\eta}{c}E_n^y\\
B_n^y &= \frac{\partial A_n^x}{\partial z}-\frac{\partial A_n^z}{\partial x}=-\frac{\eta}{c}\frac{\partial A_n^z}{\partial t}=\frac{\eta}{c}E_n^z
\end{aligned} \tag{2.23}$$

The generalized conclusion is thus that in the source free case and for a homogenous medium, the electromagnetic fields can be realized as a superposition of plane waves traveling with the speed $c/\eta$. Furthermore for each such plane wave component the electric and magnetic fields are at right angles with each other and also at right angles with the direction of propagation. In vector notation, the conclusions can be summarized as

$$\begin{aligned}
\boldsymbol{H}_n &= \frac{1}{c}\frac{\eta}{\mu}\boldsymbol{E}_n\times\boldsymbol{n}\\
\boldsymbol{E}_n &= -c\frac{\mu}{\eta}\boldsymbol{H}_n\times\boldsymbol{n}\\
\boldsymbol{n} &= \frac{1}{E_nH_n}\boldsymbol{H}_n\times\boldsymbol{E}_n
\end{aligned} \tag{2.24}$$

A particular example is at distances that are large compared to the extension of the region containing the sources. The field emanating from these will then have single well-defined direction, that is, the direction pointing away from the source region and (2.24) will hold.

For a plane wave, the energy density and power flux, expressed by (2.10), are noted to be

$$\begin{cases} U = \varepsilon_0 \varepsilon E^2 \\ S = c\varepsilon_0 \dfrac{\eta}{\mu} E^2 \end{cases} \tag{2.25}$$

As expected energy density and power flux become reduced by the polarization occurring in a medium.

Define Fourier transforms as

$$\begin{aligned} F_\omega &= \int F(x) e^{-i\omega x}\, dx \\ F(x) &= \frac{1}{2\pi} \int F_\omega e^{i\omega x}\, d\omega \end{aligned} \tag{2.26}$$

Applying Fourier transforms to the homogenous version of (2.19), it is transformed into the frequency domain. The Fourier representation of (2.14) is in the case of dispersion

$$\nabla^2 \chi_\omega + \frac{\eta(\omega)}{c^2} \omega^2 \chi_\omega = 0;\ \eta(\omega) = \sqrt{\mu(\omega)\varepsilon(\omega)} \tag{2.27}$$

The possible frequency dependence $\mu(\omega)$ is not encountered in practice. In fact, $\mu = 1$ will be assumed in the later sections of this first chapter. Since $\chi$ can be expressed in plane wave expansion (2.22), it follows that the solution to (2.27) can be stated as

$$\chi_\omega(\boldsymbol{x}) = \int_{t=-\infty}^{\infty} e^{-i\omega t} \left[ \int_{\omega'=0}^{\infty} \oint \chi_{\omega,\boldsymbol{n}}(\omega') e^{i\omega'\left(t - \eta \frac{\boldsymbol{n} \bullet \boldsymbol{x}}{c}\right)} |\omega'|\, d\Omega_{\boldsymbol{n}}\, d\omega' \right] dt \tag{2.28}$$

with $d\Omega_{\boldsymbol{n}}$ denoting integration of $\boldsymbol{n}$ with respect to solid angle measure. From the definition of the delta function, the expression (2.26) reduces to

$$\chi_\omega(\boldsymbol{x}) = \oint \chi_{\omega,\boldsymbol{n}} e^{-i\omega\eta \frac{\boldsymbol{n} \bullet \boldsymbol{x}}{c}}\, d\Omega_{\boldsymbol{n}} \tag{2.29}$$

In the sequel, frequent use will be made of the so-called wave vector

$$\boldsymbol{k} = \frac{\eta}{c} \omega \boldsymbol{n} \tag{2.30}$$

With this notation, (2.29) is reexpressed

$$\chi_{\omega}(\boldsymbol{x}) = \oint \chi_{\omega,\boldsymbol{n}} e^{-i\boldsymbol{k}\bullet\boldsymbol{x}} d\Omega_{\boldsymbol{k}} \tag{2.31}$$

This is just the expansion of $\chi_{\omega}(\boldsymbol{x})$ into three-dimensional Fourier components $\boldsymbol{k}$ restricted to that $|\boldsymbol{k}| = \eta\,\omega/c$ is constant. An immediate application, which will be returned to at several instances in the sequel, are the electric and magnetic fields solving Maxwell's equations in the homogenous case, expanded in the fashion (2.29) or (2.31).

The common situation that media are nonelastic (in popular terms "lossy") can be incorporated into the wave solutions to Maxwell's equations. The nonelasticity means that because of the material properties of the medium, wave propagation inside the medium will be associated with power dissipation into heat. In (2.8), the power dissipation was accounted for by the term $\boldsymbol{E}\bullet\boldsymbol{j}$. The medium nonelasticity is for a very wide variety of media characterized by Ohm's law $\boldsymbol{E} = \rho\boldsymbol{j}$; $\rho \geq 0$, where is the $\rho$ resistivity of the medium (constant in the case of the medium being homogenous). As will now be demonstrated, a plane wave electric field propagating through a lossy homogeneous medium takes the same form as a lossless medium, namely, with Fourier components

$$e^{i\omega t}\boldsymbol{E}_{\omega}(\boldsymbol{x}) = \boldsymbol{E}_{\omega,\boldsymbol{n}} e^{i\omega\left(t-\eta\frac{\boldsymbol{n}\bullet\boldsymbol{x}}{c}\right)} \tag{2.32}$$

The one important difference is that $\eta$ will be complex with the imaginary part determined by $\rho$. Clearly by letting $\eta$ to be complex in (2.27), the form of the particular solution will not change. The imaginary part of $\eta$ causes the field (2.32) to decay exponentially. The real part of (2.32) is just the spatial part of a plane wave propagating in the direction $\boldsymbol{n}$. Assume for simplicity the propagation of a monochromatic plane wave along the $x$-axis. The field takes the form $\boldsymbol{E}(x,t) = \boldsymbol{E}_{\omega}\cos[\omega(t - \eta_{\text{re}}x)/c]e^{-\omega\eta_{\text{im}}x/c}$ for which it must be required that $\eta_{\text{im}} \geq 0$, that is, the wave attenuates in the direction of propagation. The magnetic field is obtained $H(x,t) = \eta_{\text{re}}E(x,t)/c\mu$ according to (2.24). Hence, for (2.8)

$$\begin{aligned}
&\boldsymbol{E}\times\boldsymbol{H} = -\frac{\eta_{\text{re}}}{c\mu}\boldsymbol{E}\times(\boldsymbol{n}\times\boldsymbol{E}) = -\frac{\eta_{\text{re}}}{c\mu}E^2\boldsymbol{e}_x \\
&\Rightarrow -\frac{c\eta_{\text{re}}}{\mu}\frac{\partial}{\partial x}E^2 = \frac{\rho}{\varepsilon_0}E^2 + \varepsilon\frac{\partial}{\partial t}E^2
\end{aligned} \tag{2.33}$$

where Ohm's law has been exploited. By time averaging over many periods with respect to the frequency $\omega$, the last term vanishes, whereas

$$-\frac{c\varepsilon_0\eta_{\rm re}}{\mu}\frac{\partial}{\partial x}e^{-2\omega\eta_{\rm im}x/c} = \rho e^{-2\omega\eta_{\rm im}x/c} \Rightarrow \eta_{\rm im}\eta_{\rm re} = \frac{\rho\mu}{2\omega\varepsilon_0} \quad (2.34)$$

The formula applies to a large number of dielectrics, for example, many types of soil as exemplified in Figure 2.1 (based on [4]), and wood for which the dielectric properties will be required in Section 2.7. For all these dielectrics, $\eta_{\rm im} << \eta_{\rm re}$. Another category of media is *metals* for which resistivity $\rho$ is practically zero, while field attenuation occurs very rapidly (i.e., $\eta_{\rm im} >> 0$). Also, this situation is allowable according to (2.34), which implies in this case $\eta_{\rm re} \approx 0$.

The fractional power attenuation per unit time, considered as a time average, is $\rho\langle E^2\rangle/\varepsilon_0\langle E^2\rangle = \rho/\varepsilon_0$. The loss per unit length is this value divided by the propagation speed in the medium, which is $\eta_{\rm re}c$. The power loss per unit length is therefore

$$C = \frac{\rho}{c\eta_{\rm re}} = k\frac{2\eta_{\rm im}}{\mu} \quad (2.35)$$

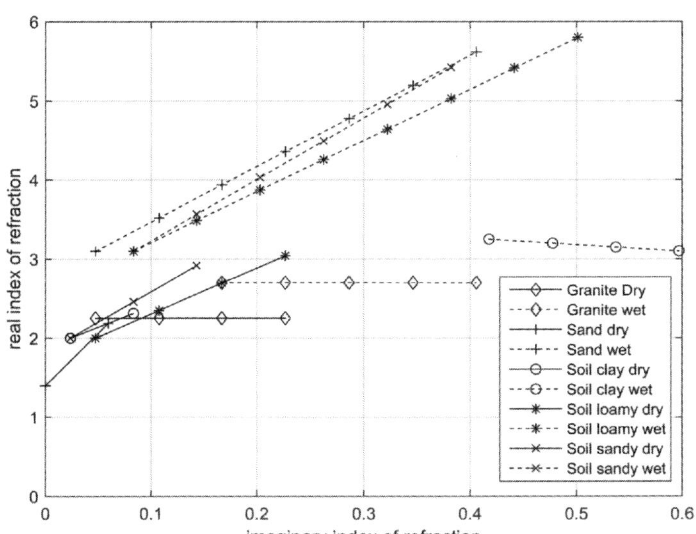


**Figure 2.1** Real and imaginary indices of refraction for common ground materials, established at 100 MHz. The graphs are obtained from values tabulated in [4].

In the literature, the treatment of signal attenuation is normally done in terms of a complex dielectric constant $\varepsilon$ rather than a complex index of refraction. The difference is trivial as long as $\eta_{\mathrm{im}} << \eta_{\mathrm{re}}$ in which case

$$\eta = \sqrt{\varepsilon_{\mathrm{re}} + i\varepsilon_{\mathrm{im}}} \approx \sqrt{\varepsilon_{\mathrm{re}}} + i\frac{\varepsilon_{\mathrm{im}}}{2\sqrt{\varepsilon_{\mathrm{re}}}} \Rightarrow \begin{cases} \varepsilon_{\mathrm{im}} \approx \dfrac{\rho\mu}{\omega\varepsilon_0} \\ C \approx k\dfrac{\varepsilon_{\mathrm{im}}}{\mu\sqrt{\varepsilon_{\mathrm{re}}}} \end{cases} \tag{2.36}$$

A final note is made on (2.27) in the case of dispersion. As was stated in Section 2.1, the dispersion arises due to a delayed polarization response in the medium, formally described by the induced field arising as a convolution, with the convolution kernel determined by the dielectric permeability (the magnetic permeability never shows dispersion). Since these kernels are real, the relations $\varepsilon(\omega)_{\mathrm{re}} = \varepsilon(-\omega)_{\mathrm{re}}$ and $\varepsilon(\omega)_{\mathrm{im}} = -\varepsilon(-\omega)_{\mathrm{im}}$ must be valid. While $\varepsilon$ is relatively frequency independent and can be assumed constant in most applications, the first correction to its real part will therefore be linear. For the imaginary part and thus for medium resistivity, the correction will be quadratic, also making it less likely that resistivity must be compensated for due to changing frequency.

## 2.4 Media Discontinuities: Media Divided by a Flat Boundary

The analysis of media discontinuities has it its roots in optics with the so-called Fresnel's equations being the main governing law. The equations were originally proposed in 1816 by the Augustin-Jean Fresnel by semitheoretical arguments combined with optical experiments. Any formal derivation was not possible before the advent of Maxwell's equations in 1860 and was first carried out by Gustav Kirchoff in 1882. Derivations of Fresnel's equations are found in [2] and [3], through below the treatment has been adapted for subsequent reference in the analysis of scattering from rough surfaces and subsurface targets.

Transforming Maxwell's equations into wave equations, the preceding section accounted for closed form solutions, obtained under the assumption of a homogeneous medium. In ground surveillance by radar, one commonly encounters homogenous media in the form soil and solid objects, making up the ground. The media boundaries imply dielectric discontinuities, either toward the atmosphere or between the objects and the surrounding soil. The dielectric properties of the atmosphere can in most cases be assumed those of vacuum. Thus, the closed-form solutions are applicable in each continuous region, but require the appropriate boundary conditions to be unique. The task ahead is to find such boundary conditions applicable for modeling the ground. Doing so, conclusions regarding ground surface properties can be drawn from electromagnetic field measurements, as will be achieved with the radar technologies described in this book.

Consider the stationary case, that is, when the only time dependencies are due to delay but not absolute time. It was just assumed that $\mu(\boldsymbol{x}),\varepsilon(\boldsymbol{x})$ remains constant in regions, but is discontinuous on the boundaries of these regions. For every point in space outside these discontinuity surfaces, (2.37) assumes the form

$$\begin{aligned} i\frac{\omega}{c^2}\varepsilon(\boldsymbol{x})\boldsymbol{E}_\omega &= \nabla\times\boldsymbol{H}_\omega; \quad \nabla\bullet\boldsymbol{E}_\omega = 0 \\ i\omega\mu(\boldsymbol{x})\boldsymbol{H}_\omega &= -\nabla\times\boldsymbol{E}_\omega; \quad \nabla\bullet\boldsymbol{H}_\omega = 0 \end{aligned} \tag{2.37}$$

For the wave equations (3.19)

$$\begin{aligned} \Delta\boldsymbol{E}_\omega + \frac{\mu(\boldsymbol{x})\varepsilon(\boldsymbol{x})}{c^2}\omega^2\boldsymbol{E}_\omega &= 0 \\ \Delta\boldsymbol{H}_\omega + \frac{\mu(\boldsymbol{x})\varepsilon(\boldsymbol{x})}{c^2}\omega^2\boldsymbol{H}_\omega &= 0 \end{aligned} \tag{2.38}$$

Maxwell's equations are the valid condition for the electromagnetic fields in all space, not just outside the dielectric discontinuity surfaces. However, if $\varepsilon(\boldsymbol{x})$ is discontinuous, while an electric or magnetic force component ($\boldsymbol{E}(\boldsymbol{x})$ or $\boldsymbol{H}(\boldsymbol{x})$) remains continuous across the discontinuity, the corresponding induction component ($\boldsymbol{D}(\boldsymbol{x})$ or $\boldsymbol{B}(\boldsymbol{x})$) must be discontinuous. The converse can also be true, if $\boldsymbol{D}(\boldsymbol{x})$ or $\boldsymbol{B}(\boldsymbol{x})$ is continuous $\boldsymbol{E}(\boldsymbol{x})$ or $\boldsymbol{H}(\boldsymbol{x})$must be discontinuous. It is possible to combine this situation with the validity of Maxwell's equations: since vector rotation determines the transversal variations of a vector field, its validity across a discontinuity surface requires that the vector field components parallel to the surface must remain continuous. Similarly, since vector divergence determines the longitudinal variation of a vector field, its validity across a discontinuity surface requires that the vector field component normal to the surface must remain continuous. Consequently, the tangential parts of the $\boldsymbol{E}$ and $\boldsymbol{H}$ fields must agree at either side but in the immediate vicinity of the boundary between two dielectrics for (2.37) to be definable across the discontinuity. Moreover, the normal parts of $\boldsymbol{D}$ and $\boldsymbol{B}$ must agree at either side but in the immediate vicinity of the boundary for (2.37) to be definable across the discontinuity. Imposing either the first set or equivalently the second set of conditions allows the field distribution on both sides of the discontinuity to be decided uniquely. The present work will categorically adhere to the first set of assumptions.

A first application is the fundamental case of the two homogenous half-spaces $z < 0$ and $z > 0$, where $z > 0$ has susceptibility constants $\hat{\mu}$ and $\hat{\varepsilon}$ and $z < 0$ has $\tilde{\mu}$ and $\tilde{\varepsilon}$. Assume the situation of an initial field in the form of a monochromatic plane wave impingent on the discontinuity plane $z = 0$ from $x, z > 0$, that is, propagating along negative $x$ and $z$ directions, at an oblique angle $0 \leq \theta < \pi/2$ with respect to the $z$-axis (the incidence angle). This is the same as saying that the field is incoming and moves in the direction of the negative $z$- and $x$-axes. The incident field can thus be described as

$$\boldsymbol{E}(\boldsymbol{x}) = \boldsymbol{E}_0 e^{-i\omega\sqrt{\mu\varepsilon}\frac{\boldsymbol{n}\cdot\boldsymbol{x}}{c}} = \boldsymbol{E}_0 e^{i\omega\sqrt{\mu\varepsilon}(x\sin\theta + z\cos\theta)/c} \tag{2.39}$$

Assume that charges and currents are only due to induction by the impinging field, and thus accountable for by complex susceptibility constants, as described in the end of the preceding section. The resulting field distribution will now be determined from the provided boundary conditions.

The overall field distribution must consist of one field outgoing along the negative $z$-axis (i.e., it has been refracted into the dielectric) $\tilde{\boldsymbol{E}}$ and another outgoing along the positive $z$-axis $\breve{\boldsymbol{E}}$ (i.e., it has been reflected by the discontinuity). The homogeneity at the surface $z = 0$ requires that the boundary conditions are invariant to translations. As a result, the reflected and refracted fields must be monochromatic plane waves. From the symmetry of the problem, the only directions in which these can propagate are in the plane spanned by the direction of incidence and the $z$-axis (i.e., in the $x,z$-plane), which we call the plane of incidence. We may thus write for the reflected and refracted fields

$$\begin{aligned}\tilde{\boldsymbol{E}}(\boldsymbol{x}) &= \tilde{\boldsymbol{E}}_0 e^{-i\omega\sqrt{\tilde{\mu}\tilde{\varepsilon}}\tilde{\boldsymbol{n}}\cdot\boldsymbol{x}/c} = \tilde{\boldsymbol{E}}_0 e^{i\omega\sqrt{\tilde{\mu}\tilde{\varepsilon}}(x\sin\tilde{\theta} + z\cos\tilde{\theta})/c} \\ \breve{\boldsymbol{E}}(\boldsymbol{x}) &= \breve{\boldsymbol{E}}_0 e^{i\omega\sqrt{\mu\varepsilon}\breve{\boldsymbol{n}}\cdot\boldsymbol{x}/c} = \breve{\boldsymbol{E}}_0 e^{i\omega\sqrt{\mu\varepsilon}(x\sin\breve{\theta} - z\cos\breve{\theta})/c}\end{aligned} \tag{2.40}$$

In (4.3), the $\boldsymbol{E}_0$-field can be arbitrarily directed in the plane orthogonal to $\boldsymbol{n}_0$. The outcome in the arbitrary case is obtained as a linear combination from two orthogonal cases, which are suitably selected to be that $\boldsymbol{E}_0$ either is in the $x,y$-plane—refer to this as *horizontal polarization*—or in the $y,z$-plane—refer to this as *vertical polarization* (in the general situation of a field impinging against plane that may not be horizontal, the terms *transversal electric* (TE) and *transversal magnetic* (TM) will be used). In either case the ***E*** and ***H*** field components are parallel to the plane z = 0, while for either polarization boundary conditions require the components $E_x$, $E_y$, $H_x$, and $H_y$ to be continuous. Due to (2.24)

$$
\begin{aligned}
H^x &= -\frac{1}{c}\sqrt{\frac{\varepsilon}{\mu}}n^z E^y \\
E^x &= \frac{1}{c}\sqrt{\frac{\mu}{\varepsilon}}n^z H^y
\end{aligned}
\tag{2.41}
$$

From the assumptions on the direction of propagation for the incident, refracted, and reflected waves, the relations of (2.41) imply

$$
\begin{aligned}
n_z &= -\cos\theta \Rightarrow \quad
\begin{aligned} H^x &= \frac{1}{c}\sqrt{\frac{\varepsilon}{\mu}}\, E^y \cos\theta \\ E^x &= -c\sqrt{\frac{\mu}{\varepsilon}}\, H^y \cos\theta \end{aligned} \\
\tilde{n}_z &= -\cos\tilde{\theta} \Rightarrow \quad
\begin{aligned} \tilde{H}^x &= \frac{1}{c}\sqrt{\frac{\tilde{\varepsilon}}{\tilde{\mu}}}\, \tilde{E}^y \cos\tilde{\theta} \\ \tilde{E}^x &= -c\sqrt{\frac{\tilde{\mu}}{\tilde{\varepsilon}}}\, \tilde{H}^y \cos\tilde{\theta} \end{aligned} \\
\breve{n}_z &= \cos\breve{\theta} \quad \Rightarrow \quad
\begin{aligned} \breve{H}^x &= -\frac{1}{c}\sqrt{\frac{\varepsilon}{\mu}}\, \breve{E}^y \cos\breve{\theta} \\ \breve{E}^x &= c\sqrt{\frac{\mu}{\varepsilon}}\, \breve{H}^y \cos\breve{\theta} \end{aligned}
\end{aligned}
\tag{2.42}
$$

The boundary condition for the ***E*** fields in the case of horizontal polarization can be stipulated as

$$
E^y + \breve{E}^y = \tilde{E}^y \Rightarrow E_0 e^{i\omega\sqrt{\mu\varepsilon}x\sin\theta/c} + \breve{E}_0 e^{i\omega\sqrt{\mu\varepsilon}x\sin\breve{\theta}/c} = \tilde{E}_0 e^{i\omega\sqrt{\tilde{\mu}\tilde{\varepsilon}}x\sin\tilde{\theta}/c} \tag{2.43}
$$

Since the expression must remain valid for all *x* values, there must be complete identity between all exponents, that is, (2.43) implies

$$
\breve{\theta} = \theta;\ \tilde{\eta}\sin\tilde{\theta} = \eta\sin\theta;\ \eta = \sqrt{\mu\varepsilon} \tag{2.44}
$$

The first expression states that the incidence is the same for the impinging and reflected fields. This is the law of reflection. The second expression is *Snell's law of refraction* determining the angle of propagation for the refracted signal. In this, the convention is followed in substituting $\mu\varepsilon$ for index of refraction $\eta$. Both quantities can be complex, according to (2.32).

According to Section 2.3, the index of refraction may be viewed as determining the rate of propagation speed and attenuation of electromagnetic waves in a certain medium. With this interpretation, Snell's law is quite general and applies for all phenomena where a propagating wave refracts into a medium with a different propagation speed. Also, the reflection law is generally valid. In particular, these two laws also hold for horizontal polarization, which follows by inserting the boundary condition for the ***H*** fields in (2.43).

Considering both polarizations and adopting (2.42) and (2.44), the full boundary conditions, involving both the electric and magnetic fields, result in the following relations:

$$\begin{aligned}
&\text{Horizontal (TE):}\\
&E^y + \breve{E}^y = \tilde{E}^y;\ H^x + \breve{H}^x = \tilde{H}^x \Rightarrow\\
&R_E = \frac{\breve{E}^y}{E^y} = \frac{\sqrt{\varepsilon\tilde{\mu}}\cos\theta - \sqrt{\tilde{\varepsilon}\mu}\cos\tilde{\theta}}{\sqrt{\varepsilon\tilde{\mu}}\cos\theta + \sqrt{\tilde{\varepsilon}\mu}\cos\tilde{\theta}}\\
&T_E = 1 + R_E = \frac{\tilde{E}^y}{E^y} = \frac{2\sqrt{\varepsilon\tilde{\mu}}\cos\theta}{\sqrt{\varepsilon\tilde{\mu}}\cos\theta + \sqrt{\tilde{\varepsilon}\mu}\cos\tilde{\theta}}\\
&\text{Vertical (TM):}\\
&H^y + \breve{H}^y = \tilde{H}^y;\ E^x + \breve{E}^x = \tilde{E}^x \Rightarrow\\
&R_H = \frac{\breve{H}^y}{H^y} = \frac{\sqrt{\mu\tilde{\varepsilon}}\cos\theta - \sqrt{\tilde{\mu}\varepsilon}\cos\tilde{\theta}}{\sqrt{\mu\tilde{\varepsilon}}\cos\theta + \sqrt{\tilde{\mu}\varepsilon}\cos\tilde{\theta}}\\
&T_H = 1 + R_H = \frac{\tilde{H}^y}{H^y} = \frac{2\sqrt{\mu\tilde{\varepsilon}}\cos\theta}{\sqrt{\mu\tilde{\varepsilon}}\cos\theta + \sqrt{\tilde{\mu}\varepsilon}\cos\tilde{\theta}}
\end{aligned} \tag{2.45}$$

As follows from these expressions, in the case of a perfectly reflecting surface (i.e., $\eta \to \infty$), the electric field component $\tilde{E}^y$ goes to zero. However, the magnetic field component $\tilde{H}^y$ does not do so (in fact it approaches twice the strength of the incident magnetic field). For a perfectly reflecting surface, the boundary condition is therefore only that the tangential components of the incident and reflected *electric* field should cancel. No further assumption on the magnetic field is required since there is no refracted field component and just half the number of parameters to determine in the problem.

The expressions of (2.45) are known as *Fresnel's equations* for the reflection and transmission coefficient. Note that they by means of (2.24) provide complete information on the refraction process for both the electric and the magnetic field, that is, phase relations as well as the power flux (or intensity). In the special case $\mu = 1$ (invariably assumed here), Fresnel's equations are seen to reduce to a relation between refraction indices $\eta = \sqrt{\varepsilon}$ and $\tilde{\eta} = \sqrt{\tilde{\varepsilon}}$. Considering the reversed scattering configuration of the impinging's signal traveling inside the medium toward the atmosphere $\eta$ and $\tilde{\eta}$ will substitute one another. To further analyze this situation, $\eta$ is kept in the formulas, whereas $\eta = 1$ when representing the atmosphere index of refraction.

The index of refraction inside the medium might be complex—Fresnel's equations and Snell's law are still valid. By Snell's law $\tilde{\eta}\sin\tilde{\theta}$ is real in any case, because the atmospheric quantities it equates are real. Therefore, $\tilde{\theta}$ is complex if $\tilde{\eta}$ is complex. To understand the implications, consider this situation for horizontal polarization

$$\tilde{\boldsymbol{E}}(\boldsymbol{x}) = \tilde{\boldsymbol{E}}_0 e^{i\omega\tilde{\eta}(x\sin\tilde{\theta} + z\cos\tilde{\theta})/c} \tag{2.46}$$

With both $\tilde{\theta}$ and $\tilde{\eta}$ complex, $\tilde{\eta}\cos\tilde{\theta}$ will also be complex. Its real part will determine the direction of propagation for the wave inside the medium, while its imaginary part determines the attenuation of the wave. The direction of propagation is the angle $\tilde{\theta}_0$ given by

$$\tilde{\theta}_0 = \tan^{-1}\left[\left(\frac{\tilde{\eta}\sin\tilde{\theta}}{\tilde{\eta}\cos\tilde{\theta}}\right)_{\text{re}}\right] = \tan^{-1}\left[\frac{\sin\theta}{\left(\tilde{\eta}\cos\tilde{\theta}\right)_{\text{re}}}\right] \tag{2.47}$$

For $\eta_{\text{im}} << \eta_{\text{re}}$, to the first order, $(\tilde{\eta}\cos\tilde{\theta})_{\text{re}} = \left(\sqrt{\tilde{\eta}^2 - \sin\theta^2}\right)_{\text{re}} \approx \sqrt{\tilde{\eta}_{\text{re}}^2 - \sin\theta^2}$, so

$$\tilde{\theta}_0 = \tan^{-1}\frac{\sin\theta}{\left(\sqrt{\tilde{\eta}^2 - \sin\theta^2}\right)_{\text{re}}} \approx \tan^{-1}\frac{\sin\theta}{\sqrt{\tilde{\eta}_{\text{re}}{}^2 - \sin\theta^2}} = \sin^{-1}\frac{\sin\theta}{\tilde{\eta}_{\text{re}}} \tag{2.48}$$

Evidently, as a first-order approximation, Fresnel's equations are applicable by just replacing the complex index of refraction with its real part, in determining the direction of propagation inside the medium. For the index of refraction values of Figure 2.1, the difference between the exact angle of propagation and its approximation has been checked to be less than 1%.

Fresnel's equations are of fundamental importance to many aspects of meter wavelength radar ground imaging. The underlying assumption of a dielectric discontinuity in the form of a plane remains appropriate whenever two dielectrics are separated by a transition zone that is thin and plane compared to the wavelength. Also, media must be homogenous compared to the wavelength. For radar wavelengths of a meter or more, the two assumptions are valid for open, generally flat and homogenous ground terrain, irrespective of whether the ground surface is grass, boulders, or water. The plane wave assumption adopted for (2.43) is appropriate since the radar is far away from the ground. Consequently, the radar signal illuminating the ground will be split up into two components, with one refracted into the ground according to Snell's law and another reflected by the law reflection. Their respective strengths are given by Fresnel's equations. The signal refracted into the ground will be attenuated by medium losses. As will be discussed in Section 2.7, the several attenuation stages that underground backscatter undergoes make the backscatter very weak and limit the possibility of imaging underground structures for all but very favorable ground compositions.

To the degree that the assumption of flatness is valid, no signal at all will be backscattered from the ground. In the next section, scattering from a gently undulating ground surface will be analyzed, deriving the levels of diffuse scattering, and thus backscattering, which will occur. Also, this surface backscatter will be very weak (i.e., most field power is scattered in the direction of specular reflection). The situation is totally different from the centimetre wave radar (say the $X$-band) case, where scattering from the ground is entirely diffuse and thus has a stronger backscattering component. For centimeter-wave radar, it is rather man-made structures, which can happen to be at least partly invisible because they contain surfaces that are flat also in centimeter scale and thus reflect away an impingent signal rather than causing backscatter.

The formulas in (2.43) have several important consequences, which must be analyzed further. First note that in the case $\theta = 0$, symmetry requires that there can be no difference between the two polarizations in (2.43) for the magnetic and electric fields for any polarization at normal incidence. It can easily be checked that this indeed in the case. For the situation $\theta \neq 0$, combining (2.45) with Snell's law

$$\begin{aligned} &\text{Horizontal: } R_E = -\frac{\sin\left(\theta - \tilde{\theta}\right)}{\sin\left(\theta + \tilde{\theta}\right)} \\ &\text{Vertical: } R_H = \frac{\tan\left(\theta - \tilde{\theta}\right)}{\tan\left(\theta + \tilde{\theta}\right)} \end{aligned} \tag{2.49}$$

Note that these formulas remain fully valid for a complex index of refraction with $\tilde{\theta}$ then being complex and obtained by Snell's law.

Consider the conservation of energy in the refraction process. Assume for the moment that the medium is lossless so that the index of refraction is real. It is noted that $E = cH/\tilde{\eta}$ by (2.24) and $S = c^2\varepsilon_0 EH$ by (2.25). The power flux is according to (2.10) a vector, with the interpretation that the component of this vector along any particular direction is the power flux through an area orthogonal to that direction. The power flux through $z = 0$ may thus be computed in two ways, that is, either as the difference between power flux projections on the $z$-axis of the incident and reflected field or as the power flux of the refracted field, projected on the z-axis. The two ways should agree. Using (2.45) and the boundary conditions, for either polarization, the two equivalent ways correspond to the expressions

$$\begin{aligned}
&\text{Horizontal:}\\
&\left(S - \breve{S}\right)\cos\theta = S\cos\theta\left[1 - \frac{\sin^2\left(\theta - \tilde{\theta}\right)}{\sin^2\left(\theta + \tilde{\theta}\right)}\right]\\
&\tilde{S}\cos\tilde{\theta} = \left(\sqrt{S} + \sqrt{\breve{S}}\right)^2 \frac{\tilde{\eta}}{\eta}\cos\tilde{\theta}\\
&\text{Vertical:}\\
&\left(S - \breve{S}\right)\cos\theta = S\cos\theta\left[1 - \frac{\tan^2\left(\theta - \tilde{\theta}\right)}{\tan^2\left(\theta + \tilde{\theta}\right)}\right]\\
&\tilde{S}\cos\tilde{\theta} = \left(\sqrt{S} + \sqrt{\breve{S}}\right)^2 \frac{n}{\tilde{\eta}}\cos\tilde{\theta}
\end{aligned} \tag{2.50}$$

The power transfer through $z = 0$ is obtained as

$$\begin{aligned}
&\text{Horizontal:}\\
&\tilde{S}\cos\tilde{\theta} = S\cos\theta\frac{\sin 2\theta \sin 2\tilde{\theta}}{\sin^2\left(\theta + \tilde{\theta}\right)} = S\cos\theta\left[1 - \frac{\sin^2\left(\theta - \tilde{\theta}\right)}{\sin^2\left(\theta + \tilde{\theta}\right)}\right]\\
&\text{Vertical:}\\
&\tilde{S}\cos\tilde{\theta} = S\cos\theta\frac{\sin 2\theta \sin 2\tilde{\theta}}{\sin^2\left(\theta + \tilde{\theta}\right)\cos^2\left(\theta - \tilde{\theta}\right)} = S\cos\theta\left[1 - \frac{\tan^2\left(\theta - \tilde{\theta}\right)}{\tan^2\left(\theta + \tilde{\theta}\right)}\right]
\end{aligned} \tag{2.51}$$

As noted in (2.51), the power transfer is invariably larger for vertical polarization than for horizontal. In fact, the power ratio is

$$\frac{\text{H-pol power transfer}}{\text{V-pol power transfer}} = \cos^2\left(\theta - \tilde{\theta}\right) \to \frac{\eta^2}{\tilde{\eta}^2};\ \theta \to \frac{\pi}{2} \tag{2.52}$$

According to (2.49), for vertical polarization there is no reflected field if $\theta + \tilde{\theta} = \pi/2$. Equivalently expressed, all impinged power is transferred through $z = 0$. This condition $\theta + \tilde{\theta} = \pi/2$ defines the so-called *Brewster angle* for $\theta$, which in terms of indices of refraction can be stated as

$$\theta_{\text{Brewster}} = \tan^{-1}\frac{\tilde{\eta}}{\eta} \tag{2.53}$$

Consider first $\theta > \tilde{\theta}$, which according to Snell's law means that the field propagates from a thinner to a denser medium (e.g., from air into the ground). According to (2.49), in this case, the reflected and the incident $\boldsymbol{E}$ fields are always in opposite phase for horizontal polarization. For vertical polarization, the reflected and impinging $\boldsymbol{H}$ fields have the same phase when $\theta < \theta_{\text{Brewster}}$ and opposite phase when $\theta > \theta_{\text{Brewster}}$. In the particular case $\theta \to 0$, the direction of propagation is close to opposite for the incident and reflected fields. The cross product relation between the electric and magnetic fields and the direction of propagation in (2.24) implies that if the magnetic or electric field components agree between the two fields, the other pair must have opposed signs. It thus follows from (2.49) that for normal incidence, the electric and magnetic components of the incident and reflected fields will coincide between vertical and horizontal polarizations. Of course, this is the expected result since the distinction between polarization states vanishes as incidence approaches the normal direction.

An imaginary index of refraction, even if small, can be influential on Fresnel's reflection and transmission coefficients for vertical polarization and near the Brewster angle. The horizontally polarized case is, however, insensitive to any small media losses. The approximation $\widehat{\eta}_{\text{re}} \sin\widehat{\theta}_0 \approx \sin\theta$ does not depend on polarization and seems valid also at and around the Brewster angle.

When the incidence approaches horizontal $\theta \to \pi/2$, the reflected and incident fields propagate in nearly the same direction. Again, because of (2.24), if (say) the magnetic field component agrees between the incident and reflected fields, the electric field component must agree, while if the magnetic field components have opposed signs, so must the electric field components. The latter is evidently the case for the horizontally polarized field. It is also the case for the vertically polarized field since $\theta \to \pi/2$ implies an incidence angle larger than the Brewster angle, that is, the denominator in the expression for magnetic field is negative. In fact, it is seen that $R_H \to R_E \to -1$ in this limit.

Also consider $\theta < \tilde{\theta}$, that is, when according to Snell's law the field propagates from denser to a thinner medium. Such a situation is the backscattered field from a buried object being refracted through the ground surface. In this case, there will be a limit incidence angle for there to be a refracted field at all, namely, when $\tilde{\theta} = \pi/2$. According to (2.51), one finds in this case that the power transfer through $z = 0$ is zero even though (2.49) implies a nonzero refracted field amplitude. This is the situation of *total reflection* which occurs when the impinging field attains an incidence angle

$$\theta_{\text{total reflection}} = \sin^{-1}\frac{\tilde{\eta}}{\eta} \tag{2.54}$$

It is noted that $\theta_{\text{Brewster}} > \theta_{\text{total reflection}}$. Since $\tilde{\theta}$ is bound to be less $\pi/2$ in (2.49), it follows that the Brewster angle is never attained for reflection at the surface of a medium thinner than that of the impinging field.

According to (2.49), field signs changes when substituting $\theta \leftrightarrow \tilde{\theta}$, that is, reversing causality between the impinging and the refracted fields. An example involving both the original and reversed case arises for the backscatter from an object $A$ buried in the ground and illuminated by a radiating antenna $B$ far off and above ground. The transmitted field must propagate at some incidence angle $\theta$ before hitting the ground and then propagate at an angle $\tilde{\theta}$ when hitting the object. The backscatter will move along $\tilde{\theta}$ below ground and then $\theta$ in order to hit the antenna. The two refraction processes follow from each other by the substitutions $\theta \leftrightarrow \tilde{\theta}$, though there is a difference in that the backscattered field reflects below ground and the transmitted field above. Notwithstanding that this is the case, the power transfer through the ground surface is seen to be identical for the two cases according to (2.51).

The phases of the reflected electric and magnetic fields are of fundamental importance and can readily be understood from first principles. If the field propagates from a thinner to a denser medium, the increased medium polarization in the denser medium creates a discontinuity in the induced electric field opposed in direction to that of the impinging field. This gradient is the cause of the reflected field. Hence, the electric field component of the reflected field will have opposed sign compared to the impinging field. Going into a thinner medium the polarization is relaxed and signs will be equal for the impinging and reflected electric fields, which act to enhance each other at the discontinuity.

Scattering from finite homogenous objects embedded in a homogenous medium of known dielectric properties will be considered in Section 2.7. A concern that will be settled now is the effect of the interaction between the object and the ground surface. A simple model would be to assume this to be split into the following: A) refraction through ground surface of the impinging field, B) backscattering from the subsurface object, and C) refraction through the ground surface of the scattered field. However, this model neglects the reflection component of the scattered field, again impinging on the object, which may significantly affect the scattered field (such an interaction between scattering objects will be referred to as *coupling*). The coupling effect can be investigated in the simple case of normal incidence toward the ground plane. The boundary value problem (taking coupling into account) in this case will consider the following.

1. an impinging plane wave field $\boldsymbol{E}$ vertically traveling down to the horizontal ground surface;
2. a subsurface field $\tilde{\boldsymbol{E}}$ traveling vertically downwards from the ground surface;
3. a scattered field $\breve{\boldsymbol{E}}_\eta$ from the buried object;
4. a field $\breve{\boldsymbol{E}}$ traveling upwards from the ground surface.

The field $\tilde{\boldsymbol{E}}$ has one component stemming from refraction of the impinging filed through the ground surface and another being the reflection at the ground surface of the scattered field. Similarly, $\breve{\boldsymbol{E}}$ has one component stemming from reflection of the impinging field at the ground surface and another being the refraction at the ground surface of the scattered field. The boundary conditions can be stipulated as

$$\boldsymbol{e}_z \times \left( \hat{\boldsymbol{E}} + \hat{\breve{\boldsymbol{E}}} \right) = \boldsymbol{e}_z \times \left( \hat{\tilde{\boldsymbol{E}}} + \hat{\breve{\boldsymbol{E}}}_{\text{obj}} \right); \boldsymbol{e}_z \times \left( \hat{\boldsymbol{H}} + \hat{\breve{\boldsymbol{H}}} \right) = \boldsymbol{e}_z \times \left( \hat{\tilde{\boldsymbol{H}}} + \hat{\breve{\boldsymbol{H}}}_{\text{obj}} \right) \tag{2.55}$$

Here, $\boldsymbol{e}_z$ is the vertical unit vector, while the use of a "$\wedge$" suffix understands field amplitudes at the ground surface. Though $\boldsymbol{E}$ is a plane wave fields, $\tilde{\boldsymbol{E}}$, $\breve{\boldsymbol{E}}_{\text{obj}}$, and $\breve{\boldsymbol{E}}$ are not, since $\breve{\boldsymbol{E}}_{\text{obj}}$ would radiate in all directions from the object. Hence, $\tilde{\boldsymbol{E}}$ and $\breve{\boldsymbol{E}}$ will also propagate in several directions. The fields $\breve{\boldsymbol{E}}_\eta$ and $\hat{\boldsymbol{E}}_\eta$ are related by the object scattering amplitude according to $\breve{\boldsymbol{E}}_{\text{obj}} = \gamma \tilde{\boldsymbol{E}}$. Expression for $\gamma$ will be derived in Section 2.7. If the buried object is dielectrically denser than the ground medium $\gamma$ will be negative.

A fact that will be repeatedly analyzed in the sequel is that while the fields $\boldsymbol{E}_\eta$ and $\breve{\boldsymbol{E}}_\eta$ spread in all directions, the field $\tilde{\boldsymbol{E}}$ is effectively determined by a ground surface patch immediately above the buried target. The phase of the field scattered from the target changes within this patch as $\phi = 2\pi\eta d/\lambda\cos\breve{\theta}_\eta$, where $\breve{\theta}_\eta$ is the scattering angle with respect to $\boldsymbol{e}_z$. The size of the patch will be restricted to phase variations less than $\pi/2$. The subsurface fields $\boldsymbol{E}_\eta$ and $\breve{\boldsymbol{E}}_\eta$ will be essentially parallel to the ground surface when restricted to this ground patch. The following scalar equations are therefore obtained for the scattered field at the ground surface, which also will be parallel to the surface itself:

$$
\begin{aligned}
\widehat{E} + \breve{E} &= \tilde{E} + \breve{E}_{\text{obj}} = \tilde{E}(1+\gamma) \\
\widehat{E} - \breve{E} &= \eta\widehat{E}_\eta - \eta\breve{E}_{\text{obj}} = \tilde{E}(1-\gamma)
\end{aligned}
\tag{2.56}
$$

Here in the second row, being the magnetic boundary conditions, fields traveling upward appear with negative sign and fields traveling downward with a positive. The following relation is implied:

$$
\breve{E} = \frac{\dfrac{1+\gamma}{1-\gamma} - \eta}{\dfrac{1+\gamma}{1-\gamma} + \eta}\widehat{E} \approx \frac{1-\eta}{1+\eta} + \frac{4\gamma\eta}{(1+\eta)^2}
\tag{2.57}
$$

The rightmost expression is the first-order approximation of the middle expression, assuming $\gamma << 1$. Its first term is the reflection from the ground surface itself according to Fresnel's equations. Its second term is the combination of the scattering coefficient with Fresnel's refraction coefficients for the passages into and out of the ground medium, that is, it states exactly the process A to C. For the weakly scattering subsurface objects, which are going to be considered in this book, this approximation will be adhered to.

## 2.5 Nonflat Media Discontinuities

The present section analyzes nonflat media boundaries by the so–called small perturbation method (SPM). The theory was developed in stages by Valenzuela [5], Fung [6], and Ulaby [7] in the fifties and sixties. However, the basic concepts were given by Rayleigh already in the late nineteenth century. As developed here, the theory exposes the polarimetric differences in scattering behavior, which are of importance for the later development.

According to the preceding section, a perfectly flat dielectrically homogeneous ground surface will not provide any backscatter. Since the degree of flatness is measured with respect to wavelength, meter-wave radar will in many cases encounter ground surfaces for which flatness is a fair representation. In these cases, a major part of the transmitted radar field obeys the laws of specular scattering of the foregoing section and is split into one part refracted into the ground (by Snell's law) and another part encountering reflection by the law of reflection. However, since the ground flatness is not perfect, a small part of the transmitted radar signal will undergo a nonspecular scattering process. It will scatter diffusely, with a component of the field backscattered into the radar. Generally, such backscatter is an unwanted and interfering part in the received signal, since the objective of using the radar in most cases is to find objects located on or below the ground surface. Subsurface objects in particular give generally very weak backscattering (as will analyzed in Section 2.7) and even a weak ground surface backscatter may overwhelm and hide such responses. The scattering properties of a nonflat (but dielectrically homogenous) ground surface will now be analyzed. It will be described how polarimetric properties of the diffusely scatted field of the ground surface differ from those of subsurface objects and could be used for discriminating between them.

For nonflat ground, the concepts of roughness and topography may be used to characterize chaotic and smooth shape variations. Natural ground is in most cases chaotic in the centimeter scale, whereas bare ground tends to be smooth and generally flat at scales of a meter and more. In this section, the boundary conditions of Maxwell's equations will be applied to a surface of known shape to determine backscatter magnitudes, whereas in the next section, a model for natural ground types is developed to which the results are applied.

The configuration of an impinging plane wave traveling from the positive $x$ and $z$ directions will be reused. The ground surface, even though it may possess chaotic properties, is assumed representable as a differentiable height function $z = h(x,y)$. It is assumed that $h(x,y)$ will be bound to deviations around $z = 0$, which are small in a sense to be further discussed, while $x$ and $y$ attain arbitrarily large values.

Recall the solutions of monochromatic wave equation (4.1), with wave vectors $\boldsymbol{k} = \omega\eta/c\boldsymbol{n}$ and $\eta$ index of refraction. The wave vectors are up to a sign uniquely determined by their $x$ and $y$ components $k_x$ and $k_y$, whereas $k_z = \pm\sqrt{k - k_x^2 - k_y^2}$ ; $k = \omega\eta/c$. Instead, with (2.31), the $x$ and $y$ components of $\boldsymbol{k}$ may be used to parametrize a distribution of plane waves, that is, these can be parametrized by horizontal wavenumbers rather than their direction in space. A difference will be, as the sign ambiguity indicates, that the parametrization restricts fields to propagate either in the positive or in the negative $z$ direction. For instance, in the situation of the positive, the monochromatic electric field can be decomposed as

$$\boldsymbol{E}(\boldsymbol{x},t) = e^{-i\omega t} \int_{-\infty}^{\infty} e^{i\left(k_x^2 x + k_y^2 y + \sqrt{k^2 - k_x^2 - k_y^2}\, z\right)} \hat{\boldsymbol{E}}_{k_x,k_y} dK \tag{2.58}$$

with a similar expression for the magnetic field. Here, the notation $dK = dk_{x\text{-}} dk_y/(2\pi)^2$ will be used to reduce repetitious and unnecessary national details. It is noted that the function $\hat{\boldsymbol{E}}$, that is, the inverse Fourier transform of $\hat{\boldsymbol{E}}_{k_x,k_y}$, is the electric field at $z = 0$ and $t = 0$.

Thus, (2.58) implies that the fields above or below the ground surface follows from a field distribution on the surface itself. The immediate objective is to find this distribution. Note that since the free charge distribution on the ground surface is zero, one of Maxwell's equations states $\nabla \bullet \boldsymbol{E}_\omega = 0$, which implies that

$$\begin{aligned}(\nabla \bullet \boldsymbol{E})_{z=0} &= \int_{-\infty}^{\infty} ik_x \hat{\boldsymbol{E}}^x_{k_x,k_y} e^{i(xk_x + yk_y)} dK + \int_{-\infty}^{\infty} ik_y \hat{\boldsymbol{E}}^y_{k_x,k_y} e^{i(xk_x + yk_y)} dK \\ &+ \int_{-\infty}^{\infty} i\sqrt{k^2 - k_x^2 - k_y^2}\, \hat{\boldsymbol{E}}^z_{k_x,k_y} e^{i(xk_x + yk_y)} dK = i \int_{-\infty}^{\infty} \boldsymbol{k} \bullet \hat{\boldsymbol{E}}_{k_x,k_y} e^{i(xk_x + yk_y)} dK = 0\end{aligned} \tag{2.59}$$

Hence, $\boldsymbol{k} \bullet \hat{\boldsymbol{E}}_{k_x,k_y} = 0$. Therefore, only two of the three components of $\hat{\boldsymbol{E}}_{k_x,k_y}$ are independent, while the third is an immediate consequence of the other two.

Notations will be lightly adjusted from the preceding section in that plain symbols refer to all field components above ground, while symbol crowned with "~" denotes field components below ground. The wave vectors $\boldsymbol{k} = \omega/c\boldsymbol{n}$ and $\tilde{\boldsymbol{k}} = \tilde{\eta}\omega/c\boldsymbol{n}$ represent the two cases. Hence, (2.58) can be used to set up two expressions: one for fields emerging from the ground surface and moving into the air and another for fields emerging from the ground surface and moving into the ground. Select signs for the wave vectors of the diffuse field to represent a radiation beam centered on the reflected and refracted specular components. The wave vectors above ground will thus have a general direction along the negative $x$-axis and the positive $z$-axis. Below ground, the direction will be negative for both $x$ and $z$. For the $y$ direction, any convention could be used. The following will be assumed:

$$
\begin{aligned}
\boldsymbol{k} \bullet \boldsymbol{x} &= -k_x x - k_y y + k_z z;\ k_z = \sqrt{\left(\frac{\omega}{c}\right)^2 - k_x^2 - k_y^2} \\
\tilde{\boldsymbol{k}} \bullet \boldsymbol{x} &= -k_x x - k_y y - \tilde{k}_z z;\ \tilde{k}_z = \sqrt{\left(\frac{\omega\tilde{\eta}}{c}\right)^2 - k_x^2 - k_y^2}
\end{aligned}
\tag{2.60}
$$

According to the assumption of small deviations from flatness, the electromagnetic fields can be determined from a low-order expansion of $h(x,y)$ with the zero-order terms determined by Fresnel's equations. Thus, considering horizontal polarization first, the field distribution above ground assumes the form

$$
\begin{aligned}
&E_\omega^y(\boldsymbol{x}) = \int_{-\infty}^{\infty} \hat{E}_{k_x,k_y}^y e^{i(k_x x + k_y y - k_z z)}\, dK + e^{ikx\sin\theta}\left(e^{ikz\cos\theta} + R_E e^{-ikz\cos\theta}\right) \\
&E_\omega^x(\boldsymbol{x}) = \int_{-\infty}^{\infty} \hat{E}_{k_x,k_y}^x e^{i(k_x x + k_y y - k_z z)}\, dK;\ E_\omega^z(\boldsymbol{x}) = \int_{-\infty}^{\infty} \hat{E}_{k_x,k_y}^z e^{i(k_x x + k_y y - k_z z)}\, dK
\end{aligned}
\tag{2.61}
$$

with the integrals providing small field contributions in comparison with the specular components. For the field having penetrated the ground surface, the following expressions apply:

$$
\begin{aligned}
&\tilde{E}_\omega^y(\boldsymbol{x}) = \int_{-\infty}^{\infty} \hat{\tilde{E}}_{k_x,k_y}^y e^{i(k_x x + k_y y + \tilde{k}_z z)}\, dK + T_E e^{i\tilde{k}x\sin\tilde{\theta} + i\tilde{k}z\cos\tilde{\theta}} \\
&\tilde{E}_\omega^x(\boldsymbol{x}) = \int_{-\infty}^{\infty} \hat{\tilde{E}}_{k_x,k_y}^x e^{i(k_x x + k_y y + \tilde{k}_z z)}\, dK;\ \tilde{E}_\omega^z(\boldsymbol{x}) = \int_{-\infty}^{\infty} \hat{\tilde{E}}_{k_x,k_y}^z e^{i(k_x x + k_y y + \tilde{k}_z z)}\, dK
\end{aligned}
\tag{2.62}
$$

The quantities $R_E$ and $T_E$ are Fresnel's reflection and transmission coefficients as stated in (2.45). In considering the air–ground interface, it will be assumed that $\mu = \varepsilon = 1$. For the ground, $\tilde{\mu} = 1$ is assumed. Then

$$
T_E = 1 + R_E = \frac{2k\cos\theta}{\tilde{k}\cos\tilde{\theta} + k\cos\theta}
\tag{2.63}
$$

Moreover

$$
\begin{aligned}
\boldsymbol{k} \bullet \hat{\boldsymbol{E}}_{k_x,k_y} &= k_x \hat{E}_{k_x,k_y}^x + k_y \hat{E}_{k_x,k_y}^y - k_z \hat{E}_{k_x,k_y}^z = 0 \\
\tilde{\boldsymbol{k}} \bullet \hat{\tilde{\boldsymbol{E}}}_{k_x,k_y} &= k_x \hat{\tilde{E}}_{k_x,k_y}^x + k_y \hat{\tilde{E}}_{k_x,k_y}^y + \tilde{k}_z \hat{\tilde{E}}_{k_x,k_y}^z = 0
\end{aligned}
\tag{2.64}
$$

Since diffuse scattering is expected, there will be a power loss in the specular directions. In (2.61) and (2.62), such a loss is allowed to arise since there may be components of the field due to the terms containing $\hat{\boldsymbol{E}}$ and $\hat{\tilde{\boldsymbol{E}}}$, which are in the specular directions but which will interfere destructively with the terms only depending on Fresnel's reflection and transmission coefficients.

In order to implement the boundary conditions and determine $\hat{\boldsymbol{E}}$ and $\hat{\tilde{\boldsymbol{E}}}$, the expressions (2.61) and (2.62) will now be restricted to the ground surface, that is, $z = h(x,y)$. The primary assumption on ground nonflatness is that its deviations from $z = 0$ are restricted by $kh(x,y) << 1$. It follows that the exponents in (2.61) and (2.62) can be expanded to first order in $z$. There will also be a condition that the $x$ and $y$ derivatives of $h(x,y)$ are small. This condition is less restrictive than may be thought at first. Actually, it is known that structures that extend in height as well as distance much less than the wavelength have little influence on the resulting scattering pattern. Under the assumption that height fluctuations are small, the large derivative fluctuations belong to this category.

From (5.4) and (5.5), it follows that:

$$
\begin{aligned}
\left(\boldsymbol{E}-\tilde{\boldsymbol{E}}\right)_\omega^y &= \int_{-\infty}^{\infty}\left(\hat{\boldsymbol{E}}-\hat{\tilde{\boldsymbol{E}}}\right)_{k_x,k_y}^y e^{i(k_x x+k_y y)}\,dK - iz\int_{-\infty}^{\infty}\left(k_z\hat{\boldsymbol{E}}+\tilde{k}_z\hat{\tilde{\boldsymbol{E}}}\right)_{k_x,k_y}^y e^{i(k_x x+k_y y)}\,dK + \mathrm{SPEC}\\
\left(\boldsymbol{E}-\tilde{\boldsymbol{E}}\right)_\omega^x &= \int_{-\infty}^{\infty}\left(\hat{\boldsymbol{E}}-\hat{\tilde{\boldsymbol{E}}}\right)_{k_x,k_y}^x e^{i(k_x x+k_y y)}\,dK - iz\int_{-\infty}^{\infty}\left(k_z\hat{\boldsymbol{E}}+\tilde{k}_z\hat{\tilde{\boldsymbol{E}}}\right)_{k_x,k_y}^x e^{i(k_x x+k_y y)}\,dK\\
\left(\boldsymbol{E}-\tilde{\boldsymbol{E}}\right)_\omega^z &= \int_{-\infty}^{\infty}\left(\hat{\boldsymbol{E}}-\hat{\tilde{\boldsymbol{E}}}\right)_{k_x,k_y}^z e^{i(k_x x+k_y y)}\,dK - iz\int_{-\infty}^{\infty}\left(k_z\hat{\boldsymbol{E}}+\tilde{k}_z\hat{\tilde{\boldsymbol{E}}}\right)_{k_x,k_y}^z e^{i(k_x x+k_y y)}\,dK
\end{aligned}
\tag{2.65}
$$

Here, *SPEC* is the term arising due to specular reflection, that is, according to the reflection law. Without any approximation

$$
e^{-ikx\sin\theta}\mathrm{SPEC} = \left(e^{ikz\cos\theta} + R_E e^{-ikz\cos\theta}\right) - T_E e^{i\tilde{k}z\cos\tilde{\theta}} \tag{2.66}
$$

where Snell's law $k\sin\theta = \tilde{k}\sin\tilde{\theta}$ has been used. Since Fresnel's equations obey the relation $T_E = 1 + R_E$, the expression can be stated as

$$
e^{-ikx\sin\theta}\mathrm{SPEC} = 2i\sin(kz\cos\theta) + T_E\left(e^{-ikz\cos\theta} - e^{i\tilde{k}z\cos\tilde{\theta}}\right) \tag{2.67}
$$

Consider the approximation of (2.65) by a Taylor expansion of $z = h(x,y)$. The diffuse scattering field components contained in the integrals in (2.65) arise only if $h(x,y)$ fluctuates. Hence, already when $h(x,y)$ is expanded to first order, the integrals will be second order in $h(x,y)$, so the second order is the lowest nontrivial order of expanding (2.65). In the expansion of (2.67) to second order in $h(x,y)$, the first-order term will vanish due to (2.63), so

$$\begin{aligned} e^{-ikx\sin\theta}\mathrm{SPEC} &= -\frac{1}{2}z^2k^2\left(\cos^2\theta - \tilde{\eta}^2\cos^2\tilde{\theta}\right)T_E = \\ &= -\frac{1}{2}z^2k^2\left(1-\sin^2\theta - \tilde{\eta}^2 + \tilde{\eta}^2\sin^2\tilde{\theta}\right)T_E \\ &= -\frac{1}{2}z^2k^2\left(1-\tilde{\eta}^2\right)T_E \end{aligned} \tag{2.68}$$

The rightmost expression is due to Snell's law. That SPEC does not contain first-order terms is expected, since the specular response is maximum for a perfectly flat surface, that is, when $z = 0$.

Boundary conditions will now be implemented. These are to be stated in terms of the ground surface normal direction

$$\boldsymbol{n}(x,y) = \nabla\left[z - h(x,y)\right] = \left(-\frac{\partial h}{\partial x}, -\frac{\partial h}{\partial y}, 1\right) \tag{2.69}$$

According to (4.1), $i\omega\mu\boldsymbol{H}_\omega = -\nabla \times \boldsymbol{E}_\omega$. Hence (with $\mu = 1$ and suppressing the index $\omega$ for notational simplicity)

$$\left[\boldsymbol{n}\times\left(\boldsymbol{E}-\tilde{\boldsymbol{E}}\right)\right]_{\boldsymbol{x}=[x,y,h(x,y)]} = 0;\ \left[\boldsymbol{n}\times\nabla\times\left(\boldsymbol{E}-\tilde{\boldsymbol{E}}\right)\right]_{\boldsymbol{x}=[x,y,h(x,y)]} = 0 \tag{2.70}$$

Each equation in (2.65) provides three scalar equations out of which two are independent (i.e., they provide relations between the fields above and below ground in the tangent plane to the ground surface but not in the normal direction). The independent equations are taken to be

$$\begin{aligned} &\frac{\partial h}{\partial y}\left(\boldsymbol{E}-\tilde{\boldsymbol{E}}\right)^z + \left(\boldsymbol{E}-\tilde{\boldsymbol{E}}\right)^y = 0 \\ &\frac{\partial h}{\partial x}\left(\boldsymbol{E}-\tilde{\boldsymbol{E}}\right)^z + \left(\boldsymbol{E}-\tilde{\boldsymbol{E}}\right)^x = 0 \\ &\frac{\partial h}{\partial y}\left(\frac{\partial\left(\boldsymbol{E}-\tilde{\boldsymbol{E}}\right)^y}{\partial x} - \frac{\partial\left(\boldsymbol{E}-\tilde{\boldsymbol{E}}\right)^x}{\partial y}\right) + \frac{\partial\left(\boldsymbol{E}-\tilde{\boldsymbol{E}}\right)^x}{\partial z} - \frac{\partial\left(\boldsymbol{E}-\tilde{\boldsymbol{E}}\right)^z}{\partial x} = 0 \\ &\frac{\partial h}{\partial x}\left(\frac{\partial\left(\boldsymbol{E}-\tilde{\boldsymbol{E}}\right)^y}{\partial x} - \frac{\partial\left(\boldsymbol{E}-\tilde{\boldsymbol{E}}\right)^x}{\partial y}\right) + \frac{\partial\left(\boldsymbol{E}-\tilde{\boldsymbol{E}}\right)^z}{\partial y} - \frac{\partial\left(\boldsymbol{E}-\tilde{\boldsymbol{E}}\right)^y}{\partial z} = 0 \end{aligned} \tag{2.71}$$

The expressions of (2.71) provide four equations for $\hat{\boldsymbol{E}}$ and $\hat{\tilde{\boldsymbol{E}}}$. Due to (2.64), two of these follow from the other four, which means that (2.71) determines the fields uniquely.

Since the spatial derivatives of $h(x,y)$ are small, the two first relations in (2.71) yield to the lowest, that is, the first order

$$E^x = \tilde{E}^x;\ E^y = \tilde{E}^y \Rightarrow \hat{E}^y_{k_x,k_y} = \hat{\tilde{E}}^y_{k_x,k_y};\ \hat{E}^x_{k_x,k_y} = \hat{\tilde{E}}^x_{k_x,k_y} \tag{2.72}$$

From (2.73) and (2.64)

$$k_z \hat{E}^z_{k_x,k_y} + \tilde{k}_z \hat{\tilde{E}}^z_{k_x,k_y} = 0 \Rightarrow \hat{\tilde{E}}^z_{k_x,k_y} = -\frac{k_z}{\tilde{k}_z}\hat{E}^z_{k_x,k_y} \tag{2.73}$$

From (2.72) and (2.73)

$$\begin{aligned}
\left(\boldsymbol{E} - \tilde{\boldsymbol{E}}\right)^y &= -iz\int_{-\infty}^{\infty}\left(k_z + \tilde{k}_z\right)\hat{E}^y_{k_x,k_y} e^{i(k_x x + k_y y)}\, dK - \frac{1}{2}z^2 k^2\left(1-\tilde{\eta}^2\right)T_E e^{ikx\sin\theta} \\
\left(\boldsymbol{E} - \tilde{\boldsymbol{E}}\right)^x &= -iz\int_{-\infty}^{\infty}\left(k_z + \tilde{k}_z\right)\hat{E}^x_{k_x,k_y} k_z e^{i(k_x x + k_y y)}\, dK \\
\left(\boldsymbol{E} - \tilde{\boldsymbol{E}}\right)^z &= \int_{-\infty}^{\infty}\left(1 + \frac{k_z}{\tilde{k}_z}\right)\hat{E}^z_{k_x,k_y} e^{i(k_x x + k_y y)}\, dK
\end{aligned} \tag{2.74}$$

The first terms in the two last expressions in (2.71) contain the $y$ component of the electric fields $\boldsymbol{E}$ and $\tilde{\boldsymbol{E}}$ and thus the specular parts that are not small. However, since these fields combine as $\boldsymbol{E} - \tilde{\boldsymbol{E}}$, it is seen in (2.68) that only the terms that are second order in $h(x,y)$ are left. Hence, the first terms are third order in $h(x,y)$. The second terms are second order in $h(x,y)$ and only the last two terms are first order. Keeping these in the next last relation in (5.14) yields to the first order in $h(x,y)$

$$\hat{E}^x_{k_x,k_y} = -\frac{k_x}{\tilde{k}_z}\hat{E}^z_{k_x,k_y} \tag{2.75}$$

For the last relation in (2.71), to the first order in $h(x,y)$

$$
\begin{aligned}
& i\int_{-\infty}^{\infty}\left(1+\frac{k_z}{\tilde{k}_z}\right)k_y \hat{E}^z_{k_x,k_y} e^{i(k_x x+k_y y)}\, dK \\
& \quad = -i\int_{-\infty}^{\infty}\left(1+\frac{\tilde{k}_z}{k_z}\right)k_z \hat{E}^y_{k_x,k_y} e^{i(k_x x+k_y y)}\, dK - zk^2\left(1-\tilde{\eta}^2\right)T_E e^{ikx\sin\theta}
\end{aligned}
\tag{2.76}
$$

Adopt the Fourier transform $h_{k_x,k_y}$ of $h(x,y)$

$$
\begin{aligned}
h(x,y) &= \int_{-\infty}^{\infty} h_{k_x,k_y} e^{i(k_x x+k_y y)}\, dK \\
&= \int_{-\infty}^{\infty} h_{k_x-k\sin\theta,k_y} e^{i[(k_x-k\sin\theta)x+k_y y]}\, dK
\end{aligned}
\tag{2.77}
$$

where it is noted that $k$ and $\theta$ are given constants in the problem. Insertion of them in (2.76) yields

$$
\hat{E}^y_{k_x,k_y} = -\frac{k_y}{\tilde{k}_z}\hat{E}^z_{k_x,k_y} + i\left(k_z-\tilde{k}_z\right)T_E h_{k_x-k\sin\theta,k_y} \tag{2.78}
$$

The $x$ and $y$ components of the $\hat{\boldsymbol{E}}$ field have been stated in terms of the $z$ component by (2.76) and (2.77). Thus, by adopting (2.64), it is possible to determine $\hat{E}^z_{k_x,k_y}$.

$$
\hat{E}^z_{k_x,k_y} = i\frac{k_y\tilde{k}_z\left(k_z-\tilde{k}_z\right)}{k^2-k_z\left(k_z-\tilde{k}_z\right)}T_E h_{k_x-k\sin\theta,k_y} \tag{2.79}
$$

The expressions for the components $\hat{\boldsymbol{E}}^x_{k_x,k_y}$ and $\hat{\boldsymbol{E}}^y_{k_x,k_y}$ follow directly by means of (2.75) and (2.78). For below ground, (2.72) and (2.73) determine the field. The solution reveals the perhaps remarkable fact that for a shallow ground topography, the electromagnetic field is determined by a linear operator on the ground height function, which is moreover diagonal in the Fourier domain. For an arbitrary nonshallow ground surface, the relation between height function and the scattered field is not linear. In fact, if the ground surface is considered as a superposition of elementary ground features (like the Fourier components of the ground surface), the scattered field from this superposition will not only depend on these features in combination but also on mutual interaction between their electromagnetic responses. A typical example for low-frequency radar is the scattering from a tree standing on the ground in which the ground plays a significant part.

Note that scattering in the plane of incidence (the $x,z$-plane) corresponds to $k_y = 0$. Then, according to (2.79) and (2.75), $E^z_\omega$ and $E^x_\omega$ will be zero all over space. Hence, depolarized backscatter is of lower magnitude order compared to copolarized backscatter.

The radar applications to be considered are only monostatic, that is, only the backscattered field component is considered. This situation is defined from the expressions above by restricting

$$k_x = -k\sin\theta,\ k_z = k\cos\theta,\ \tilde{k}_z = \tilde{k}\cos\tilde{\theta} = k\sqrt{\tilde{\eta}^2 - \sin^2\theta} \tag{2.80}$$

Thereby, from (2.45)

$$\hat{E}^y_{-k\sin\theta,k_y} = -ik\left(\tilde{\eta}\cos\tilde{\theta} - \cos\theta\right)T_E h_{k_x - k\sin\theta,k_y} = i2k\cos\theta R_E h_{-2k\sin\theta,k_y} \tag{2.81}$$

A corresponding derivation will now be carried out for vertical polarization. As for flat ground, the case of vertical polarization is derived by assuming the $\boldsymbol{H}$ component of the impinging field directed along the $y$-axis but geometry otherwise as was assumed for horizontal polarization. Hence, (2.61) and (2.62) remain valid with electric vector components substituted for magnetic components along the same directions and $T_E$ and $R_E$ substituted for $T_H$ and $R_H$. There is a strong duality between the case of horizontal and vertical polarizations, due to the similar appearance of the $\boldsymbol{E}$ and $\boldsymbol{H}$ fields in Maxwell's equations in the case of no charges or currents. However, for this symmetry to be valid, it must be assumed that both the magnetic susceptibility and the electric permeability are nonunity $\mu \neq 1 \neq \varepsilon$, whereas in practice $\mu = 1$. In this situation, the duality becomes broken. When $\mu = 1$, the scattering formulas become relatively straightforward for horizontal polarization, but less so for vertical polarizations. These somewhat more complex formulas for vertical polarization will now be derived.

Since $\nabla \bullet \boldsymbol{H}_\omega = 0$ is valid just as $\nabla \bullet \boldsymbol{E}_\omega = 0$, the expression (2.64) remains valid in this case with the $\boldsymbol{E}$ field components substituted for $\boldsymbol{H}$ field components, that is

$$\begin{aligned} \boldsymbol{k} \bullet \hat{\boldsymbol{H}}_{k_x,k_y} &= k_x \hat{H}^x_{k_x,k_y} + k_y \hat{H}^y_{k_x,k_y} - k_z \hat{H}^z_{k_x,k_y} = 0 \\ \tilde{\boldsymbol{k}} \bullet \hat{\tilde{\boldsymbol{H}}}_{k_x,k_y} &= k_x \hat{\tilde{H}}^x_{k_x,k_y} + k_y \hat{\tilde{H}}^y_{k_x,k_y} + \tilde{k}_z \hat{\tilde{H}}^z_{k_x,k_y} = 0 \end{aligned} \tag{2.82}$$

The reduced symmetry on vertical polarization is evident for the boundary conditions. The one of Maxwell's equations leading to (2.70) was $i\omega \boldsymbol{H}_\omega = -\nabla \times \boldsymbol{E}_\omega$, while for vertical polarization, $i\omega\varepsilon(\boldsymbol{x})\boldsymbol{E}_\omega = c^2\nabla \times \boldsymbol{H}_\omega$ should be used. Based on the latter, (2.70) is restated as

$$\left[\boldsymbol{n}\times\left(\boldsymbol{H}_\omega - \tilde{\boldsymbol{H}}_\omega\right)\right]_{\boldsymbol{x}=[x,y,h(x,y)]} = 0; \left[\boldsymbol{n}\times\nabla\times\left(\boldsymbol{H}_\omega - \frac{\tilde{\boldsymbol{H}}_\omega}{\tilde{\eta}^2}\right)\right]_{\boldsymbol{x}=[x,y,h(x,y)]} = 0 \tag{2.83}$$

Keeping in mind that only first-order terms in $h(x,y)$ need to be preserved

$$\begin{aligned}
&H_\omega^y - \tilde{H}_\omega^y = 0 \\
&H_\omega^x - \tilde{H}_\omega^x = 0 \\
&\frac{\partial h}{\partial y}\frac{\partial}{\partial x}\left(\boldsymbol{H} - \frac{\tilde{\boldsymbol{H}}}{\tilde{\eta}^2}\right)^y - \frac{\partial}{\partial x}\left(\boldsymbol{H} - \frac{\tilde{\boldsymbol{H}}}{\tilde{\eta}^2}\right)^z + \frac{\partial}{\partial z}\left(\boldsymbol{H} - \frac{\tilde{\boldsymbol{H}}}{\tilde{\eta}^2}\right)^x = 0 \\
&\frac{\partial h}{\partial x}\frac{\partial}{\partial x}\left(\boldsymbol{H} - \frac{\tilde{\boldsymbol{H}}}{\tilde{\eta}^2}\right)^y + \frac{\partial}{\partial y}\left(\boldsymbol{H} - \frac{\tilde{\boldsymbol{H}}}{\tilde{\eta}^2}\right)^z - \frac{\partial}{\partial z}\left(\boldsymbol{H} - \frac{\tilde{\boldsymbol{H}}}{\tilde{\eta}^2}\right)^y = 0
\end{aligned} \tag{2.84}$$

In contrast to (2.71), the $y$ components in these expression will contain residues of the specular component, which will be the first order in $h(x,y)$. For the magnetic boundary condition (the first two equations in (2.84)), the residue is calculated according to

$$\begin{aligned}
T_H &= \frac{2\tilde{k}\cos\theta}{k\cos\tilde{\theta} + \tilde{k}\cos\theta},\ T_H = 1 + R_H \Rightarrow e^{-ikx\sin\theta}\mathrm{SPEC}_H \\
&= e^{ikz\cos\theta} + \left(T_H - 1\right)e^{-ikz\cos\theta} - T_H e^{i\tilde{k}z\cos\tilde{\theta}} \\
&= 2i\sin(kz\cos\theta) + T_H\left(e^{-ikz\cos\theta} - e^{i\tilde{k}z\cos\tilde{\theta}}\right) \\
&\approx 2i\sin(kz\cos\theta) - izkT_H\left(\cos\theta + \tilde{\eta}\cos\tilde{\theta}\right) = -izk\frac{\tilde{\eta}^2 - 1}{\tilde{\eta}}\cos\tilde{\theta}\, T_H
\end{aligned} \tag{2.85}$$

For the electric boundary condition (the last two equations in (2.84)), the residue must be calculated to second order, since it will be reduced to first order by forming its derivatives. Using Snell's law, as well as $T_H = 1 + R_H$ and Fresnel's equation for $T_H$

$$
\begin{aligned}
e^{-ikx\sin\theta}\mathrm{SPEC}_E &= e^{ikz\cos\theta} + \left(T_H - 1\right)e^{-ikz\cos\theta} - \frac{1}{\tilde{\eta}^2}T_H e^{i\tilde{k}z\cos\tilde{\theta}} \\
&= 2i\sin(kz\cos\theta) + T_H\left(e^{-ikz\cos\theta} - \frac{1}{\tilde{\eta}^2}e^{i\tilde{k}z\cos\tilde{\theta}}\right) \\
&\approx T_H\left[1 - \frac{1}{\tilde{\eta}^2} - \frac{z^2k^2}{2}\left(\cos^2\theta - \cos^2\tilde{\theta}\right)\right] \\
&= \frac{\tilde{\eta}^2 - 1}{\tilde{\eta}^2}\left(1 + \frac{z^2k^2}{2}\sin^2\theta\right)T_H
\end{aligned}
\tag{2.86}
$$

Now pass to the Fourier domain in analogous to (2.65). By means of (2.85), for the first two equations of (2.84)

$$
\hat{\tilde{H}}^x_{k_x,k_y} = \hat{H}^x_{k_x,k_y}\,;\ \hat{\tilde{H}}^y_{k_x,k_y} = \hat{H}^y_{k_x,k_y} - ik\frac{\tilde{\eta}^2 - 1}{\tilde{\eta}}\cos\tilde{\theta}\,T_H\frac{h_{k_x - k\sin\theta,k_y}}{(2\pi)^2} = 0 \tag{2.87}
$$

Then from (2.82)

$$
\hat{\tilde{H}}^z_{k_x,k_y} = -\frac{k_z}{\tilde{k}_z}\hat{H}^z_{k_x,k_y} + \frac{k_y}{\tilde{k}_z}ik\frac{\tilde{\eta}^2 - 1}{\tilde{\eta}}\cos\tilde{\theta}\,T_H\frac{h_{k_x - k\sin\theta,k_y}}{(2\pi)^2} \tag{2.88}
$$

The last two equations in (2.84) can be given the following respective Fourier representation:

$$
\begin{aligned}
&\int_{-\infty}^{\infty}\left(\hat{X}^a_{k_x,k_y} - \hat{X}^b_{k_x,k_y} + \hat{X}^c_{k_x,k_y}\right)e^{i(k_x x + k_y y)}\,dK = 0 \\
&\int_{-\infty}^{\infty}\left(\hat{Y}^a_{k_x,k_y} + \hat{Y}^b_{k_x,k_y} - \hat{Y}^c_{k_x,k_y}\right)e^{i(k_x x + k_y y)}\,dK = 0
\end{aligned}
\tag{2.89}
$$

Their components are now computed individually to first order. Note that to the lowest order

$$
\frac{\partial}{\partial x}\left(\boldsymbol{H} - \frac{\tilde{\boldsymbol{H}}}{\tilde{\eta}^2}\right)^y \approx \frac{\partial}{\partial x}\mathrm{SPEC}_E = ik\sin\theta\,T_H\frac{\tilde{\eta}^2 - 1}{\tilde{\eta}^2}e^{ikx\sin\theta} \tag{2.90}
$$

Therefore

$$
\begin{aligned}
&\frac{\partial h}{\partial y}\frac{\partial}{\partial x}\left(H-\frac{\tilde{H}}{\tilde{\eta}^2}\right)^y \approx -\frac{k\sin\theta}{(2\pi)^2}T_H\frac{\tilde{\eta}^2-1}{\tilde{\eta}^2}\int_{-\infty}^{\infty} k_y h_{k_x-k\sin\theta,k_y} e^{i(k_x x+k_y y)}\,dK \\
&\frac{\partial h}{\partial x}\frac{\partial}{\partial x}\left(H-\frac{\tilde{H}}{\tilde{\eta}^2}\right)^y \approx -\frac{k\sin\theta}{(2\pi)^2}T_H\frac{\tilde{\eta}^2-1}{\tilde{\eta}^2}\int_{-\infty}^{\infty} \left(k_x-k\sin\theta\right) h_{k_x-k\sin\theta,k_y} e^{i(k_x x+k_y y)}\,dK
\end{aligned} \tag{2.91}
$$

It follows that:

$$
\begin{aligned}
\hat{X}^a_{k_x,k_y} &= -kk_y\sin\theta T_H\frac{\tilde{\eta}^2-1}{\tilde{\eta}^2}\frac{h_{k_x-k\sin\theta,k_y}}{(2\pi)^2} \\
\hat{Y}^a_{k_x,k_y} &= -k\left(k_x-k\sin\theta\right)\sin\theta T_H\frac{\tilde{\eta}^2-1}{\tilde{\eta}^2}\frac{h_{k_x-k\sin\theta,k_y}}{(2\pi)^2}
\end{aligned} \tag{2.92}
$$

Furthermore, by means of (2.88)

$$
\begin{aligned}
&\left(H-\frac{\tilde{H}}{\tilde{\eta}^2}\right)^z = \int_{-\infty}^{\infty}\left(\hat{H}^z_{k_x,k_y}-\frac{\hat{\tilde{H}}^z_{k_x,k_y}}{\tilde{\eta}^2}\right)e^{i(k_x x+k_y y)}\,dK \\
&= \int_{-\infty}^{\infty}\left[\frac{\tilde{\eta}^2\tilde{k}_z+k_z}{\tilde{\eta}^2\tilde{k}_z}\hat{H}^z_{k_x,k_y}-\frac{k_y}{\tilde{k}_z}ik\frac{\tilde{\eta}^2-1}{\tilde{\eta}^3}\cos\tilde{\theta}T_H\frac{h_{k_x-k\sin\theta,k_y}}{(2\pi)^2}\right]e^{i(k_x x+k_y y)}\,dK \\
&\Rightarrow \begin{cases}
\hat{X}^b_{k_x,k_y} = i\dfrac{k_x}{\tilde{\eta}^2\tilde{k}_z}\left[\left(\tilde{\eta}^2\tilde{k}_z+k_z\right)\hat{H}^z_{k_x,k_y}-ik_yk\dfrac{\tilde{\eta}^2-1}{\tilde{\eta}}\cos\tilde{\theta}T_H\dfrac{h_{k_x-k\sin\theta,k_y}}{(2\pi)^2}\right] \\
\hat{Y}^b_{k_x,k_y} = i\dfrac{k_y}{\tilde{\eta}^2\tilde{k}_z}\left[\left(\tilde{\eta}^2\tilde{k}_z+k_z\right)\hat{H}^z_{k_x,k_y}-ik_yk\dfrac{\tilde{\eta}^2-1}{\tilde{\eta}}\cos\tilde{\theta}T_H\dfrac{h_{k_x-k\sin\theta,k_y}}{(2\pi)^2}\right]
\end{cases}
\end{aligned} \tag{2.93}
$$

Finally

$$
\begin{aligned}
\frac{\partial}{\partial z}\left(H-\frac{\tilde{H}}{\tilde{\eta}^2}\right)^x &= \frac{\partial}{\partial z}\int_{-\infty}^{\infty}\left(e^{-ik_z z}-\frac{e^{i\tilde{k}_z z}}{\tilde{\eta}^2}\right)\hat{H}^x_{k_x,k_y}e^{i(k_x x+k_y y)}\,dK \\
&\approx -i\int_{-\infty}^{\infty}\frac{\tilde{\eta}^2k_z+\tilde{k}_z}{\tilde{\eta}^2}\hat{H}^x_{k_x,k_y}e^{i(k_x x+k_y y)}\,dk_x\,dk_y \Rightarrow \hat{X}^c_{k_x,k_y} \\
&= -i\frac{\tilde{\eta}^2k_z+\tilde{k}_z}{\tilde{\eta}^2}\hat{H}^x_{k_x,k_y}
\end{aligned} \tag{2.94}
$$

From (2.86) and (2.87)

$$\frac{\partial}{\partial z}\left(\boldsymbol{H}-\frac{\tilde{\boldsymbol{H}}}{\tilde{\eta}^2}\right)^y=\frac{\partial}{\partial z}\int_{-\infty}^{\infty}\left(e^{-ik_z z}\hat{H}^y_{k_x,k_y}-\frac{e^{i\tilde{k}_z z}}{\tilde{\eta}^2}\hat{H}^y_{k_x,k_y}+ik\frac{e^{i\tilde{k}_z z}}{\tilde{\eta}^2}\frac{\tilde{\eta}^2-1}{\tilde{\eta}}\cos\tilde{\theta}\,T_H\frac{h_{k_x-k\sin\theta,k_y}}{(2\pi)^2}\right)e^{i(k_x x+k_y y)}\,dK$$

$$+k^2\sin^2\theta\frac{\tilde{\eta}^2-1}{\tilde{\eta}^2}T_H e^{ikx\sin\theta}\int_{-\infty}^{\infty}\frac{h_{k_x,k_y}}{(2\pi)^2}e^{i(k_x x+k_y y)}\,dK$$

$$\Rightarrow \hat{Y}^c_{k_x,k_y}=-i\frac{\tilde{\eta}^2k_z+\tilde{k}_z}{\tilde{\eta}^2}\hat{H}^y_{k_x,k_y}+k\frac{\tilde{\eta}^2-1}{\tilde{\eta}^2}\left(k\sin^2\theta-\tilde{k}_z\frac{1}{\tilde{\eta}}\cos\tilde{\theta}\right)T_H h_{k-k\sin\theta_x,k_y} \tag{2.95}$$

The analysis henceforward is restricted to the plane of incidence (i.e., $K_y = 0$). The general case follows in the same way, just by assuming $k_y \neq 0$, but formulas will be complex and will not be required in the sequel.

Note that $k_x\hat{H}^x_{k_x,k_y} = k_z\hat{H}^z_{k_x,k_y}$ if $K_y = 0$ according to (2.82). Hence, the next to last equation in (2.84) yields in the plane of incidence

$$\hat{X}^a_{k_x,0}-\hat{X}^b_{k_x,0}+\hat{X}^c_{k_x,0}=0\Leftrightarrow \tilde{k}^2\left(\tilde{k}_z+k_z\right)\hat{H}^z_{k_x,0}=0\Leftrightarrow \hat{H}^z_{k_x,0},\hat{H}^x_{k_x,0}=0 \tag{2.96}$$

It follows that $H^x_\omega = H^z_\omega = 0$ all over space. Just as for horizontal polarization, keeping to the plane of incidence any depolarization will be of lower order in magnitude.

Consider the last equation in (2.84). From (2.95), (2.93), and (2.95)

$$\hat{Y}^a_{k_x,0}+\hat{Y}^b_{k_x,0}-\hat{Y}^c_{k_x,0}=0\Leftrightarrow \hat{H}^y_{k_x,0}=-ik\frac{\tilde{\eta}^2-1}{\tilde{\eta}}\frac{\tilde{\eta}k_x\sin\theta-\tilde{k}_z\cos\tilde{\theta}}{\tilde{\eta}^2k_z+\tilde{k}_z}T_H h_{k_x-k\sin\theta,0} \tag{2.97}$$

In the monostatic case (2.80), adopting the expression for $T_H$ in (2.45)

$$\hat{H}^y_{-k\sin\theta,0}=i2k\left(\tilde{\eta}^2-1\right)\frac{\sin^2\theta+\cos^2\tilde{\theta}}{\left(\tilde{\eta}\cos\theta+\cos\tilde{\theta}\right)^2}\cos\theta h_{-2k\sin\theta,0} \tag{2.98}$$

Re-expressing (2.98) in the independent variables $\theta$ and $\tilde{\eta}$

$$\hat{H}^y_{-k\sin\theta,0}=i2k\left(\tilde{\eta}^2-1\right)\frac{\tilde{\eta}^2\left(\sin^2\theta+1\right)-\sin^2\theta}{\left(\tilde{\eta}^2\cos\theta+\sqrt{\tilde{\eta}^2-\sin^2\theta}\right)^2}\cos\theta h_{-2k\sin\theta,0} \tag{2.99}$$

Going back and comparing (2.99) with (2.81), the formulas for the backscattered amplitude are seen to be significantly different for vertical and horizontal polarizations. In fact, for horizontal polarization, the amplitude, which is proportional to the horizontal Fresnel's reflection coefficient $R_E$, is intuitively satisfying. As seen, there is no analogy in the case of vertical polarization, and (2.99) bears little proximity to the vertical Fresnel's reflection coefficient $R_H$. This notwithstanding, by means of (2.49), the expression (2.98) can be manipulated to yield

$$\hat{H}^y_{-k\sin\theta,0} = -i2k\cos\theta \frac{\sin^2\theta + \cos^2\tilde{\theta}}{\cos^2\left(\theta - \tilde{\theta}\right)} R_E h_{-2k\sin\theta,0} \tag{2.100}$$

Here the expression (2.49) has been adopted for $R_E$. The expression (2.81) does exhibit some similarity in form with (2.100).

Even though $\hat{H}^y_{k_x,k_y}$ and $\hat{E}^y_{k_x,k_y}$ refer to electric and magnetic fields, they are the amplitudes of electromagnetic plane waves scattered off the ground surface, which contain both electric and a magnetic field components. These stand in the fixed proportion given by (2.24), so by knowing either the electric or magnetic component, the other component is also known. In the present treatment, it was assumed that the impinging field, either the magnetic or electric component, was of unit amplitude. Hence, (2.97) applies equally well to the electric field, stating in either case the proportion of the incident field, which is backscattered (be it its magnetic or electric component), but with the magnetic field direction parallel to the ground surface. Similarly, (2.100) is the backscattered proportion of the impinging field, equally the magnetic or electric component, but with the electric field direction parallel to the ground surface. The ratio between $\hat{H}^y_{-k\sin\theta,0}$ and $\hat{E}^y_{-k\sin\theta,0}$ is thus the ratio between the backscattered fractions of incident amplitudes, between the cases of vertical and horizontal polarizations. Combining (2.81) and (2.100) yields for this ratio

$$\frac{\text{Vertical}}{\text{Horizontal}} = -\frac{\sin^2\theta + \cos^2\tilde{\theta}}{\cos^2\left(\theta - \tilde{\theta}\right)} \tag{2.101}$$

Relation (2.101) is remarkable in neither depending on the wavelength nor on the ground topography.

Note that since $\tilde{\theta} \le \theta$, it follows that $\cos\tilde{\theta} \ge \cos\theta$ and thus that the ratio (2.100) is always larger than unity, that is, vertical polarization backscatters more strongly than horizontal. The difference increases with increased incidence angle. For $\theta \to \pi/2$

$$\frac{\text{Vertical}}{\text{Horizontal}} \rightarrow 1 - 2\tilde{\eta}^2 \tag{2.102}$$

The power ratio is the square of (2.101). It follows that already for a moderate index of refraction, the power ratio can be large, for example, if $\tilde{\eta} = 2$, the power ratio becomes 49, that is 17 dB.

The result (2.101) is of particular interest for imaging underground structures with low frequency radar. An issue in these applications is that the surface return tends to overwhelm the return from subsurface structures. In view of linear relation (2.101) for the surface return and given that exactly the same linear relation will not pertain to the returns from subsurface structures, it should be possible to discriminate between surface and subsurface returns. For subsurface objects, as was discussed in the preceding section, the refraction process in the ground surface has a well-determined polarimetric character. The relation between backscattering amplitudes for vertical and horizontally polarizations will also depend on intrinsic properties of the subsurface object. In principle, the relation between the horizontally and vertically polarized backscatters can be arbitrary. However, in the low-frequency radar regime, many underground objects have wavelength or near wavelength dimension, making them respond equal or nearly equal to different polarizations. In these cases, according to (2.52), the quotient of the power transfer ratio will be significantly larger for vertical than for horizontal polarization, just as it is for the surface return. The expression (2.52) is noted to be independent of whether the transfer goes from the dielectrically thinner to the denser material or in the other direction. The 2-way power transfer rate is thus affected by its square, and the following power ratio relation between surface and subsurface backscatter is obtained for polarization-neutral subsurface objects:

$$1 \leq \frac{\text{Backscatter V/H}}{\text{2-way power transfer V/H}} = \frac{\left[\tilde{\eta}^2 + \left(\tilde{\eta}^2 - 1\right)\sin^2\theta\right]^2}{\tilde{\eta}^4} \leq 4 \tag{2.103}$$

The maximum value is obtained when $\theta \rightarrow \pi/2$ and $\tilde{\eta} \rightarrow \infty$. But already at moderately shallow (and thus practical) incidence angles and for almost all practical ground dielectrics, the ratio (2.103) becomes not far from 4. For instance, if $\tilde{\eta} = 2$ and $\theta = 75°$, the ratio is 2.88. As will be described in Chapter 6, the ratio like this is sufficient for discrimination of the ground surface return.

The analytical formulas derived here are only an indication of the physical sensibility of this type of surface clutter discrimination process, in which the small perturbation model is not sufficiently accurate for such an application.

Also, the subsurface objects themselves are not accurately polarization neutral even in cases when they are nearly so. The method for discriminating the surface return is rather to look for the dominant linear relation between vertically and horizontally polarized data. This dominant relation almost certainly comes from the ever present surface return, compared to which subsurface responses are weak and scarce. Removing the dominant linear component, remaining data is thus from the subsurface structures, regardless of whether these are polarization-neutral or not; it is only required that they do not obey the same linear relation as the surface data. Chapter 6 develops this essentially statistical method further.

## 2.6 Ground Reflectivity

The scattering models for media discontinuities in the form of surfaces of infinite extension require one further step of development in order to be practically applicable. The preceding analysis assumed an impinging plane wave floodlighting the infinitely extended ground. Such a plane wave has, however, infinite power. Moreover, ground fluctuations cannot be expected to diminish at the infinity so the Fourier integrals providing the diffuse scattering will be divergent. Contrary to this, the preferable formulation would represent the scattering from the ground as arising by integration of a ground density function of the electromagnetic fields residing on the ground. For any point above the ground, the field to be considered will be that of the field from a particular finite ground region, rather than the field from the entire ground. This situation is in fact that of radar applications in which illumination with finite power is steered toward such a specific region and will be zero outside it. The required reformulation is certainly permissible, though only in the setting of statistical theory. To fit the preceding theory, the ground surface must be assumed to have a random character where any part of it which is illuminated by the radar is considered as a sample, sharing the same statistical properties as an infinite surface. It should be stressed that scattering from random surface is a wide discipline (see [8]). The development here is restricted to the demands of meter-wave ground imaging radar and to make good use if the SPM theory.

Wave equation solutions are epitomized by the so-called wave equation Green's function

$$G(\Delta\boldsymbol{x}) = \frac{1}{4\pi}\frac{e^{-ik|\Delta\boldsymbol{x}|}}{|\Delta\boldsymbol{x}|} \tag{2.104}$$

Green's theorem is a well-known integral expression allowing translation of the volume integral in (2.14) to a surface integral over the boundary of the region of integration by means of (2.104). As a result, the fields outside the surface follow (for a full derivation, see [9]) from field conditions on the boundary surface itself in the following way:

$$\begin{aligned} \boldsymbol{E}(\boldsymbol{x}+\Delta \boldsymbol{x}) &= \oint\left\{-i\omega G\, \boldsymbol{n}\times \boldsymbol{H}(\boldsymbol{x}) + \boldsymbol{n}\bullet \boldsymbol{E}(\boldsymbol{x})\nabla_{\boldsymbol{x}} G - \left(\nabla_{\boldsymbol{x}} G\right)\times\left[\boldsymbol{n}\times \boldsymbol{E}(\boldsymbol{x})\right]\right\} dA \\ \boldsymbol{H}(\boldsymbol{x}+\Delta \boldsymbol{x}) &= \oint\left\{ i\omega\frac{\eta^2}{c^2} G\, \boldsymbol{n}\times \boldsymbol{E}(\boldsymbol{x}) + \boldsymbol{n}\bullet \boldsymbol{H}(\boldsymbol{x})\nabla_{\boldsymbol{x}} G - \left(\nabla_{\boldsymbol{x}} G\right)\times\left[\boldsymbol{n}\times \boldsymbol{H}(\boldsymbol{x})\right]\right\} dA \end{aligned} \tag{2.105}$$

Here, $dA$ is the appropriate surface measure and $\boldsymbol{n}$ is the normal direction to the surface pointing outward. The integrals (2.105) are sometimes referred to as Kirchhoff's integral theorem and sometimes Huygens's principle, then understanding them in the context of wave motion in general rather than electromagnetic fields specifically. The reference to Huygens principle stems from (6.2) being a formal expression of the fact that a wave equation solution "propagates," that is, the wave equation implies that the field of a propagating wave at any point in space is the composition of contributions of elementary waves (represented by (2.104)) with amplitudes determined by the field distribution over any surface intersecting the propagating field.

Apply (6.2) to the present case, that is, the surface integrated over in (2.105) will be identified with the $x,y$-plane at $z = 0$, which becomes a closed surface by adding an infinite hemisphere for $z > 0$. Clearly, (2.105) allows the required interpretation of the scattered field at an arbitrary point $P$ above the ground surface as arising due to integration of a field density on the ground surface itself (rather than only as an effect of the global field across the surface). Consider the contribution of the field at $P$ from some area $X(Q) \subset S$ around a fixed point $Q$ on the ground surface. Let $\boldsymbol{x}$ denote any variable point in $X(Q)$ to be integrated over and let $\boldsymbol{\rho}$ represent a vector from $Q$ to this point over which the two-dimensional integration is performed. Thus, $\boldsymbol{x} = Q + \boldsymbol{\rho}$. If $P = \boldsymbol{x} + \boldsymbol{r}$, the relation $QP = \boldsymbol{r} + \boldsymbol{\rho}$ is implied. Assuming $|QP| >> |\boldsymbol{\rho}|$, the following first-order expansion is obtained:

$$k|\boldsymbol{r}| = k\sqrt{|QP|^2 + \rho^2 + 2QP\bullet\boldsymbol{\rho}} \approx k|QP| + \boldsymbol{k}\bullet\boldsymbol{\rho};\ \boldsymbol{k} = k\frac{QP}{|QP|} \tag{2.106}$$

Also assuming $|QP| >> \lambda$ and since $P = \boldsymbol{x} + \boldsymbol{r}$ with $P$ fixed

$$\nabla_x G(\boldsymbol{r}) \approx \frac{1}{4\pi|\boldsymbol{r}|}\nabla_x\left(e^{-i\boldsymbol{k}\bullet\boldsymbol{r}}\right) = i\boldsymbol{k}G(\boldsymbol{r}) \tag{2.107}$$

Now restrict (2.105) to the field contribution at $P$ coming from $X(Q)$, keeping $X(Q)$ sufficiently small so that only first-order terms in (6.3) are significant. Noting the cross-product identity $\nabla G \times (\boldsymbol{n} \times \boldsymbol{E}) = \nabla G \times (\boldsymbol{e}_z \times \boldsymbol{E}) = i(\boldsymbol{k} \bullet \hat{\boldsymbol{E}}\boldsymbol{e}_z - \boldsymbol{e}_z \bullet \hat{\boldsymbol{E}}\boldsymbol{k})$ and the corresponding expression for $\hat{\boldsymbol{H}}$

$$\begin{aligned} \boldsymbol{E}^{X(Q)}(P) &= \frac{ie^{-ik|QP|}}{4\pi|QP|}\int_{X(Q)}\left(-\omega\boldsymbol{e}_z \times \hat{\boldsymbol{H}} + \boldsymbol{k}\bullet\hat{\boldsymbol{E}}\boldsymbol{e}_z\right)e^{-i\boldsymbol{k}\bullet\boldsymbol{\rho}}d^2\boldsymbol{\rho} \\ \boldsymbol{H}^{X(Q)}(P) &= \frac{ie^{-ik|QP|}}{4\pi|QP|}\int_{X(Q)}\left(\omega\frac{\eta^2}{c^2}\boldsymbol{e}_z \times \hat{\boldsymbol{E}} + \boldsymbol{k}\bullet\hat{\boldsymbol{H}}\,\boldsymbol{e}_z\right)e^{-i\boldsymbol{k}\bullet\boldsymbol{\rho}}d^2\boldsymbol{\rho} \end{aligned} \tag{2.108}$$

The similarity between the integrals in (2.108) and Fourier transforms is noted. The one difference is that the integrals are restricted to a finite area, preventing that Fourier transforms for $\hat{\boldsymbol{E}}$ and $\hat{\boldsymbol{H}}$ can replace the integration in (2.108) in a general situation. The replacement is possible, however, if ground fluctuations are stationary, that is, the field distribution within $X(Q)$ is a good representation of the global field distribution, irrespective of the $X(Q)$ size. Stationarity is assumed in the subsequent analysis in this section, implying that the only change in (6.2) obtained when extending $X(Q)$ (by increasing the distance $|QP|$) is in the resulting net field amplitude. In particular, in the sense of expectation values, the following relations should be valid:

$$\boldsymbol{k}\bullet\int_{X(Q)}\hat{\boldsymbol{E}}e^{-i\boldsymbol{k}\bullet\boldsymbol{\rho}}d^2\boldsymbol{\rho} = \boldsymbol{k}\bullet\int_{X(Q)}\hat{\boldsymbol{H}}e^{-i\boldsymbol{k}\bullet\boldsymbol{\rho}}d^2\boldsymbol{\rho} = 0 \tag{2.109}$$

in accordance with (5.2). It follows from (2.109) that

$$\begin{aligned} \boldsymbol{H}^{X(Q)}(P) &= \frac{1}{\omega}\boldsymbol{k}\times\boldsymbol{E}^{X(Q)}(P) = \boldsymbol{k}\bullet\boldsymbol{e}_z\frac{ie^{-ik|QP|}}{4\pi|QP|}\int_{X(Q)}\hat{\boldsymbol{H}}e^{-i\boldsymbol{k}\bullet\boldsymbol{\rho}}d^2\boldsymbol{\rho} \\ \boldsymbol{E}^{X(Q)}(P) &= -\frac{c^2}{\eta^2\omega}\boldsymbol{k}\times\boldsymbol{H}^{X(Q)}(P) = \boldsymbol{k}\bullet\boldsymbol{e}_z\frac{ie^{-ik|QP|}}{4\pi|QP|}\int_{X(Q)}\hat{\boldsymbol{E}}e^{-i\boldsymbol{k}\bullet\boldsymbol{\rho}}d^2\boldsymbol{\rho} \end{aligned} \tag{2.110}$$

It will be convenient to reformulate (2.110) in polar coordinates. Let $\boldsymbol{k} = (k\sin\theta\cos\psi, k\sin\theta\sin\psi, k\cos\theta)$ and put $QP = \boldsymbol{r}_0$. It is noted that $\boldsymbol{r}_0 = r_0\boldsymbol{k}/k$, that is

$$QP = \boldsymbol{r}_0 = \left(r_0\sin\theta\cos\psi, r_0\sin\theta\sin\psi, r_0\cos\theta\right) \tag{2.111}$$

Thus

$$
\begin{aligned}
\boldsymbol{H}^{X(Q)}\left(\boldsymbol{r}_0\right) &= \frac{ik\cos\theta e^{-ikr_0}}{4\pi r_0} \int_{X(Q)} \hat{H} e^{-ik\sin\theta(x\cos\psi + y\sin\psi)} dxdy \\
\boldsymbol{E}^{X(Q)}\left(\boldsymbol{r}_0\right) &= \frac{ik\cos\theta e^{-ikr_0}}{4\pi r_0} \int_{X(Q)} \hat{E} e^{-ik\sin\theta(x\cos\psi + y\sin\psi)} dxdy
\end{aligned}
\tag{2.112}
$$

The formulas in (2.112) and their application to ground surface backscattering provide an excellent opportunity to define the concepts of radar cross section (RCS) and reflectivity. RCS is obviously the standard measure for assessing the received scattering from an object in radar applications and also thus the standard way of determining required radar transmit power. The concept of cross section assumes a situation of an illuminating plane wave hitting an object of finite dimensions and causing a reflected field from this object. At sufficient distance from the object, the scattered wave can be considered spherical, that is, attenuation will be inversely proportional to the distance from the measurement point to the object. By *differential* cross section is understood the ratio between the power flux measured in steradians, in a particular direction from the object, compared to the power flux per unit area of the illuminating plane wave. The physical dimension of differential cross section is thus area. By *total* scattering cross section is understood the differential cross section integrated over all angles. By *RCS* (the concept to be used henceforward) is understood the total cross section obtained by taking the measured differential cross-sectional value as representing isotropic scattering. Hence, RCS is just differential cross section multiplied with the factor $4\pi$.

If the two formulas in (2.112) are used for the cases of vertical and horizontal polarizations, respectively, they will represent the scattered field relative to the impinging field, since for the impinging field, the magnetic and electric components were assumed to have unit amplitude for each respective polarization case. The reduction in scattering flux measured at $P$ is therefore $|\boldsymbol{H}^{X(Q)}(\boldsymbol{r}_0)|^2$ and $|\boldsymbol{E}^{X(Q)}(\boldsymbol{r}_0)|^2$ for the vertical and horizontally polarized cases. The RCS is obtained by multiplying these figures with the area $4\pi r_0^2$ of the sphere of radius $r_0$, that is

$$
\begin{aligned}
\sigma_{X(Q)}^{VV}\left(\boldsymbol{r}_0\right) &= \frac{(k\cos\theta)^2}{4\pi} \left| \int_{X(Q)} \hat{H}^y e^{-ik\sin\theta(x\cos\psi + y\sin\psi)} dxdy \right|^2 \\
\sigma_{X(Q)}^{HH}\left(\boldsymbol{r}_0\right) &= \frac{(k\cos\theta)^2}{4\pi} \left| \int_{X(Q)} \hat{E}^y e^{-ik\sin\theta(x\cos\psi + y\sin\psi)} dxdy \right|^2
\end{aligned}
\tag{2.113}
$$

Along with the RCS, the concept of radar reflectivity will be used. Radar reflectivity is obtained by dividing the cross section by the illuminated area, that is, the area of $X(Q)$. As will be illustrated below, when the illuminated object extends over a large area, across which it exhibits stationary randomness in its properties, the expected growth of the integrals in (6.9) will be as the square root of the surface integrated over. As a consequence, reflectivity will be independent of the area of $X(Q)$. The limit value when this area grows smaller around $Q$ will thus provide the reflectivity at the ground point $Q$.

In (6.10), the results of the preceding section for horizontal and vertical polarizations may be inserted. Only the case of backscattering will be considered, though the methods provided can be applied to any bistatic angle at the cost of increased complexity in the expressions arrived at. The final relations of (2.81) and (2.100) between $\hat{E}^y$ and $\hat{H}^y$ and the ground height spectrum are noted. Denote for any function $f(x,y)$ on $S$ by

$$f_{\kappa_x,\kappa_y}^{X(Q)} = \int\limits_{X(Q)} f(x,y)e^{-i(\kappa_x x+\kappa_y y)}dxdy \tag{2.114}$$

Again, since ground fluctuations will be assumed stationary, any conclusions valid for the full Fourier integrals of the fields will also be applicable in the sense of expectation values, when integration is restricted to $X(Q)$. Therefore, (2.81) and (2.100) can safely be assumed valid when Fourier integrals are so restricted, so

$$\begin{aligned}\left\langle \sigma_{X(Q)}^{VV}\left(\boldsymbol{r}_0\right)\right\rangle &= \frac{k^4\cos^4\theta}{\pi}\left[\frac{\sin^2\theta+\cos^2\tilde{\theta}}{\cos\left(\theta-\tilde{\theta}\right)^2}\right]^2 R_E^2\left\langle\left| h_{2k\sin\theta,0}^{X(Q)}\right|^2\right\rangle \\ \left\langle \sigma_{X(Q)}^{HH}\left(\boldsymbol{r}_0\right)\right\rangle &= \frac{k^4\cos^4\theta}{\pi} R_E^2\left\langle\left| h_{2k\sin\theta,0}^{X(Q)}\right|^2\right\rangle\end{aligned} \tag{2.115}$$

Attention will now turn to setting up a parametric model characterizing different types of random ground surfaces.

A very characteristic feature of natural ground is that height variation is not simply a Gaussian process. The combined action of erosion and gravity will act to level out height differences and will thus put a penalty on slope steepness. Furthermore, gravity becomes a more powerful factor with increasing size, meaning that the leveling out effect becomes more pronounced in larger scales: chaotic structures are commonly found in a fine scale, gravel being an example, but become rarer and eventually nonexistent at scales of meters or tens of meters.

The statistical features of the ground surface will vary with the particular site considered. A grass field can be perfectly flat at large scales, but is chaotic considering a sufficiently small scale. Sand dunes can be smooth at both fine and large scales. A boulder field can be chaotic on the size of boulders but smooth at larger scales. The following general approach enables characterization of these different cases in a way in which (2.115) can be directly exploited for determining radar reflectivity.

The case of a one-dimensional ground height profile $h(x)$ will be considered as an introduction. This will be Fourier analyzed. First, it is observed that

$$\int_{-\lambda/4}^{\lambda/4} \cos(kx)\,dx = -\int_{-\lambda/4}^{\lambda/4} \sin(kx)\,dx = \frac{2}{k} \tag{2.116}$$

where $\lambda = 2\pi/k$. Note that in this and subsequent formulas concerning the ground height function, "wavelength" $\lambda$ has solely its meaning in relation to Fourier expansion of this function. To avoid any misunderstanding, such an expansion can be undertaken quite independently of considerations regarding the scattering properties of the ground when illuminated with an electromagnetic field, with a certain wavelength. As will be seen in the sequel, wavelength can be considered as a concept allowing subdivision of ground height fluctuation into components, relating to different scale lengths.

According to (2.13), the following two expressions

$$\begin{aligned} h_k^0\left(x_0\right) &= \frac{k}{2}\int_{-\lambda/4}^{\lambda/4} h(x)\cos\left[k\left(x+x_0\right)\right]dx \\ h_k^1\left(x_0\right) &= -\frac{k}{2}\int_{-\lambda/4}^{\lambda/4} h(x)\sin\left[k\left(x+x_0\right)\right]dx \end{aligned} \tag{2.117}$$

are weighted mean values, with $h_k^0(x_0)$ being the mean height within the cell $\ell_0 = \{x_0 - \lambda/4 \le x \le x_0 + \lambda/4\}$ and $h_k^1$ being the mean height difference between the end points of $\ell_0$. Note that when height variation occurs only at scales considerably larger than $\lambda$, neighboring values $h_k^{0/1}(n\lambda)$ and $h_k^{0/1}[(n+1)\lambda]$ will have nearly the same modulus and nearly always opposite signs. Thus, considering the sum

$$h_k^L\left(x_0\right) = \int_{-L/2}^{L/2} h(x)e^{-ik(x+x_0)}\,dx = \frac{2}{k}\sum_{n=-L/\lambda}^{L/\lambda}\left[h_k^0\left(n\lambda + x_0\right) + ih_k^1\left(n\lambda + x_0\right)\right] \tag{2.118}$$

the overall contribution will be negligible. Variations that occur at considerably finer scale than $\lambda$ smear out in the integrals (2.116), again resulting in that they will not largely affect the value of (2.118). The remaining variations, that is, those occurring at the scale order $\lambda$, will vary randomly and independently of each other with zero mean within a set of $x$ values, discretized by $\lambda$. Since the variance of a sum of independent terms with zero mean is the sum of their variances

$$\left\langle \left| \int_{-L/2}^{L/2} h(x) e^{-ik(x+x_0)} dx \right|^2 \right\rangle = \left\langle \left| \frac{2}{k} \sum_{n=-L/\lambda}^{L/\lambda} \left[ h_k^0(n\lambda + x_0) + i h_k^1(n\lambda + x_0) \right] \right|^2 \right\rangle$$
$$= \frac{2L}{\lambda} \left\langle 2 \left| \frac{2}{k} h_k^0(x) \right|^2 \right\rangle = \frac{8L}{\pi k} \left\langle \left| h_k^0(x) \right|^2 \right\rangle \tag{2.119}$$

Here, $\langle |h_k^0(x)|^2 \rangle = \langle |h_k^1(x)|^2 \rangle$ while $h_k^0(x)$ and $h_k^1(x)$ can be assumed independent. Moreover, the ground variation is assumed stationary, so $\langle |h_k^0(x)|^2 \rangle$ will be independent of $x$ (equivalently, it can be obtained as the variance of the weighted height (2.116) determined by averaging over many $x$ in the vicinity of $x_0$).

The result (2.119) is now extended to a two-dimensional height function $h(x,y)$. It will be assumed that $h(x,y)$ is statistically isotropic, that is, when generalized to two dimensions, the expectation values will only depend on $\boldsymbol{k} = (k_x, k_y)$ by its modulus $k = \sqrt{k_x^2 + k_y^2}$. The generalized expression may be obtained by considering the case $k_y = 0$ after which the case of arbitrary $\boldsymbol{k}$ is obtained by putting $k_x = k$. It is observed that

$$\int_{-\lambda/4}^{\lambda/4} \int_{-\lambda/4}^{\lambda/4} \cos(kx) dx dy = -\int_{-\lambda/4}^{\lambda/4} \int_{-\lambda/4}^{\lambda/4} \sin(kx) dx dy = \frac{2\pi}{k^2} \tag{2.120}$$

For two dimensions and thus subsequently the weighted height estimates, (2.116) thus become replaced by

$$h_k^0(\boldsymbol{x}_0) = \frac{k^2}{2\pi} \int_{-\lambda/4}^{\lambda/4} \int_{-\lambda/4}^{\lambda/4} h(x, y + y_0) \cos\left[ k(x + x_0) \right] dx dy$$
$$h_k^1(\boldsymbol{x}_0) = -\frac{k^2}{2\pi} \int_{-\lambda/4}^{\lambda/4} \int_{-\lambda/4}^{\lambda/4} h(x, y + y_0) \sin\left[ k(x + x_0) \right] dx dy \tag{2.121}$$

In two dimensions, consider a region $X(\boldsymbol{x}_0)$ (it may not by necessity be quadratic or even rectangular) centered on $\boldsymbol{x}_0$. Then for $\boldsymbol{k} = (k,0)$, (2.118) generalizes to

$$\begin{aligned} h_{k,0}^{X(\boldsymbol{x}_0)} &= \int_{X(\boldsymbol{x}_0)} h(x,y)e^{-i\boldsymbol{k}\bullet(x+x_0)}dxdy = \int_{X(\boldsymbol{x}_0)} h\left(x,y+y_0\right)e^{-ik(x+x_0)}dxdy \\ &= \frac{2\pi}{k^2}\sum_{n,m}\left\{h_k^0(\boldsymbol{x}) - ih_k^1(\boldsymbol{x});\ \boldsymbol{x} = \left(n\lambda + x_0\right)\boldsymbol{e}_x + m\lambda\boldsymbol{e}_y \in X\left(\boldsymbol{x}_0\right)\right\} \end{aligned} \tag{2.122}$$

Thereby

$$\left\langle\left|\int_{X(\boldsymbol{x}_0)} h(x,y)e^{-i\boldsymbol{k}\bullet(x+x_0)}dxdy\right|^2\right\rangle = \frac{4A_{X(\boldsymbol{x}_0)}}{\lambda^2}\left\langle 2\left|\frac{2\pi}{k^2}h_k^0(\boldsymbol{x})\right|^2\right\rangle = \frac{8A_{X(\boldsymbol{x}_0)}}{k^2}\left\langle\left|h_k^0(\boldsymbol{x})\right|^2\right\rangle \tag{2.123}$$

The notation $A_{X(\boldsymbol{x}_0)}$ is used for the area of $X(\boldsymbol{x}_0)$. Note that when $\langle|h_k^0(\boldsymbol{x})|^2\rangle$ is computed by averaging over many values $\boldsymbol{x}$ and $h(x,y)$ is statistically isotropic, the particular selection of $x$ and $y$ directions in (2.121) is irrelevant; any rigid rotation of $h(x,y)$ before performing the computation in (2.121) and the subsequent averaging would yield the same result.

It is observed that (2.123) is exactly the expression for the ground topography required in (2.115). Thus, the radar reflectivity, obtained by dividing the expressions in (2.115) by the area $A_{X(\boldsymbol{x}_0)}$, is readily obtained and seen to be independent of $A_{X(\boldsymbol{x}_0)}$. It is thus possible to let $X(\boldsymbol{x}_0)$ shrink around $\boldsymbol{x}_0$ so that the reflectivity will be a property specific to a ground point. In doing so, it must be clear that reflectivity is understood as an expectation value. Actually, establishing it requires averaging over multiple points $x$ in the vicinity of $\boldsymbol{x}_0$.

For the root mean square (RMS) ground height variation $\Delta h_k = \sqrt{\left\langle\left|h_k^0(\boldsymbol{x})\right|^2\right\rangle}$ to vary sharply with the frequency $k$, the ground surface must have a periodic structure, caused by some process (natural or man-made). As it happens, such processes do exist, water waves being an evident example. For dry ground, sand dune ripple is a counterpart. However, in most cases, with man-made ground formations included, ground surface shape will only depend on $k$ smoothly and slowly. To obtain parametric expressions for ground reflectivity for some general cases of terrain, the following simple model of the ground surface will be adhered to.

Understand the changes over distances of many or even tens of meters by large-scale ground height variations. At these scales, the parameter characterizing the ground surface shape is assumed to be slope steepness, that is, $\Delta h_k$ is assumed to be proportional to scale length $\lambda$, with the constant of proportionality valid for a certain type of terrain. By "fine scale," height fluctuations will be understood as height variations $\Delta h_k$ at the scale of centimeters or tens of centimeters. For these scale lengths, the ground texture will be chaotic in the sense that the scale length at which these fluctuations occur is $\lambda_F = 4\Delta h_k$. Different ground surface textures are characterized by different selections of $\lambda_F$. The two conditions can be combined into the single formula

$$\Delta h_k = \frac{1}{4}\left(C_L\lambda + \lambda_F\right) = \frac{\pi}{2}\left(\frac{C_L}{k} + \frac{1}{k_F}\right) \Leftarrow 2\pi/k = \lambda \geq \lambda_F \qquad (2.124)$$

Here, $C_L$ is the large-scale ground steepness, according to $\Delta h_k = C_L\lambda/4$. It is assumed that $C_L << 1$. The chaotic texture will take precedence when $\lambda$ approaches $\lambda_F$. The model cannot predict the spectrum at wavelengths shorter than $\lambda_F$, which, however, will be outside the wavelength regime where the SPM model is applicable.

For the purposes of this book, terrain is to be considered flat if mean slope steepness is of the order 1%. This means $C_L = 0.01$. For instance, $\Delta h_{0.16} \approx 0.5$m at the scale $\lambda = 40$m. If mean slope steepness is 5% (i.e., $C_L = 0.05$), the terrain is to be considered "hilly." The term "smooth ground" will refer to texture roughness of the ground surface at the level of gravel and correspond to a selection $\lambda_F = 2.5$ cm (i.e., $k_F = 250$ m$^{-1}$). "Rough" will correspond to textures like boulder fields represented by $\lambda_F = 10$ cm (i.e., $k_F = 62$ m$^{-1}$). Thus, four generic types of ground are defined: smooth-flat, smooth-hilly, rough-flat, and rough-hilly. A MATLAB script simulating surfaces and thus verifying the model generates surfaces of the type illustrated in Figure 2.2.

The SPM theory of the preceding section was derived with respect to a given frequency $\omega$. Now set $k = c/\omega$, that is, wavenumbers will relate to an impinging electromagnetic field rather than to ground surface scale lengths. In the derivation of the SPM scattering formulas, an arbitrary ground surface was allowed. However, ground height fluctuations were required to be small compared to the wavelength of the impinging field and the spatial derivatives small. It was argued that the result could be extended beyond the latter condition, since rapidly fluctuating structures are known to contribute negligibly, compared to those in Bragg resonance with the impinging field. From the arguments of the present section regarding random surfaces, the scattering solution can be made local and thus independent of ground height variation at scales much larger than the wavelength of the impinging field; in this case,

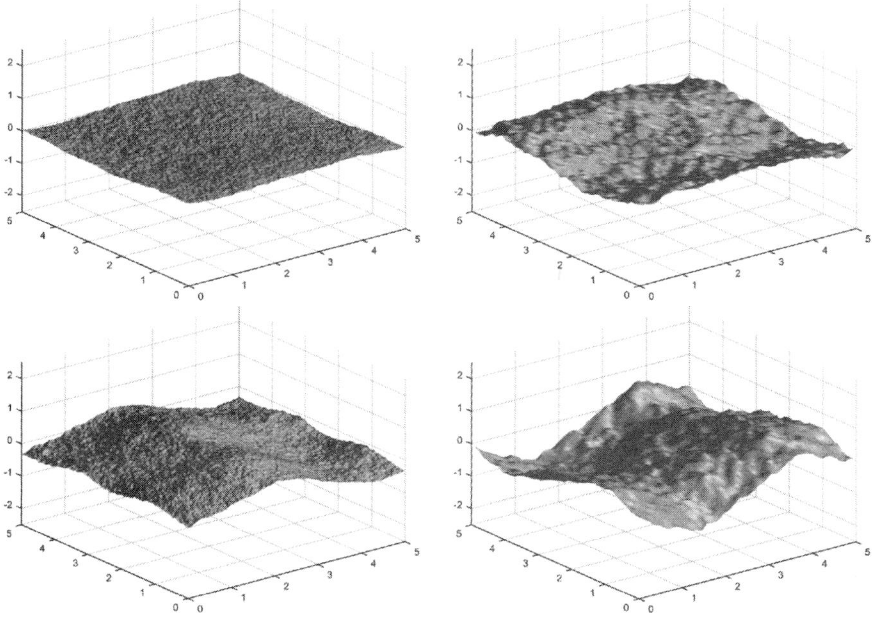

**Figure 2.2** Upper left to lower right: 5m × 5m sections of the four ground surface types smooth-flat, rough-flat, smooth hilly, and rough hilly. Height scale in meters. Surfaces are computed by making an initial assumption of a normal distributed ground surface spectrum with unity RMS spectral fluctuations. The spectrum is reweighted according to (2.123) and (2.124) with the parameters for each ground type, after which an inverse Fourier transform establishes the surface as a height function.

the solution will, however, reside on some slope of such a larger structure and may require adjustment of incidence and scattering angles with respect to the slope. In all, the conditions for the SPM theory to provide valid backscattering results must rest on the height fluctuations $h^0_{2k\sin\theta}(\boldsymbol{x})$ with spatial derivatives in the $x$ and $y$ directions of order $4h^0_{2k\sin\theta}(\boldsymbol{x})/\lambda$, where $\lambda = \pi/k\sin\theta$. It follows from this and the height fluctuation requirements imposed by the series expansion in (2.68) and (2.85):

$$\begin{aligned} A_1 &= k\cos\theta\Delta h_{2k\sin\theta} << 1 \\ A_2 &= \frac{4}{\pi/k\sin\theta}\Delta h_{2k\sin\theta} = \frac{4k\sin\theta}{\pi}\Delta h_{2k\sin\theta} << 1 \end{aligned} \tag{2.125}$$

Combined with (2.124), this leads to

$$A_1 = \frac{\pi}{2}\left(\frac{C_L}{2\tan\theta} + \frac{k\cos\theta}{k_F}\right) << 1$$
$$A_2 = C_L + \frac{2k\sin\theta}{k_F} << 1 \tag{2.126}$$

Evidently, the validity of the SPM model becomes more critical as the frequency of the iluminating field, i.e. value of $k$, grews higher. The second equation in (2.126) will determine the upper frequency limit for the SPM model to be applicable. Assuming, for instance, $A_1 = A_2 \lessapprox 0.25$, the SPM model should provide reasonably accurate results. In this case and in the case of allowing large incidence angles $\theta$

$$f_{\text{Limit}} = \frac{c}{8\lambda_F} \Rightarrow \begin{cases} 1.5 \text{ GHz for smooth ground} \\ 360 \text{ MHz for rough ground} \end{cases} \tag{2.127}$$

The first equation in (2.126) sets limit to the steepness of incidence of the impinging field. In fact, it is required that the first term of this equation is smaller than $\approx 0.25$ from which it follows:

$$\theta_{\text{Limit}} = a\tan\left(\frac{\pi}{4}C_L\right) \Rightarrow \begin{cases} 0.4^\circ \text{ for flat ground} \\ 2^\circ \text{ for hilly ground} \end{cases} \tag{2.128}$$

Evidently, there are no practical restrictions in incidence angle, whereas the wavelength restrictions exclude strict applicability to radar operating in the higher ultrahigh frequency band.

Within the limits of applicability of SPM, ground reflectivity parametric expressions are obtained by combining (2.115), (2.123), and (2.125). The result is

$$\left\langle \sigma_0^{VV} \right\rangle = \frac{\pi k^2}{2}\frac{\cos^4\theta}{\sin^2\theta}\left[\frac{\sin^2\theta + \cos^2\tilde{\theta}}{\cos\left(\theta - \tilde{\theta}\right)^2}\right]^2 R_E^2\left(\frac{C_L}{2k\sin\theta} + \frac{1}{k_F}\right)^2$$
$$\left\langle \sigma_0^{HH} \right\rangle = \frac{\pi k^2}{2}\frac{\cos^4\theta}{\sin^2\theta}R_E^2\left(\frac{C_L}{2k\sin\theta} + \frac{1}{k_F}\right)^2 \tag{2.129}$$

The reflectivity in the case of horizontal polarization agrees well with what would be intuitively expected: first of all, Fresnel's reflectivity coefficient $R_E$ is anticipated as a basic factor controlling the degree of scattering. The numerator cosine dependence to the fourth power is the effect of the field projections on the ground at oblique incidence of the illuminating and backscattered fields, as appearing in (2.100), (2.115), and (2.126). The denominator sine dependence is seen to correspond to the increasing coherency of the backscatter field when the incidence angle diminishes, the effect formally appearing in (2.123). If only large-scale ground features are considered, only the first term for the ground height model (2.125) matters and the reflectivity will be frequency independent, which is according to the expectations, that is, the mean scale length independent ground steepness $C_L$ is what controls reflectivity. For vertical polarization, the additional coefficient is of course not as obvious but is the well-known consequence of the boundary conditions in the vertically polarized case. The results are plotted in Figure 2.3 for a typical light soil index of refraction $\eta \approx 2.2$.

A final note is that since the SPM model does not take into account all scattering terms involved, it is expected to give conservative estimates in the case of terrain being rougher and more undulating than allowed by the model. The graphs are thus of interest also in these cases, in stating the lower reflectivity bounds when (2.127) and (2.128) no longer are applicable.

## 2.7 Generic Objects

Electromagnetic models suitable for low-frequency radar will be sought for the scattering from targets and vegetation elements. If wavelength is significantly larger than object size, the so-called Rayleigh scattering limit applies, according to which the return scattering field amplitude is given by target volume. If object size is much larger than wavelength, physical optics methods become reliable. Presently, the radar bands exploited imply radar wavelengths of the order of one or several meters. Hence, the structures to be modeled will have dimensions that can appear either larger or smaller, or in fact agreeing with the wavelength. Electrometric modeling in this transition zone is a significant challenge.

There are very few basic shapes for which there exist closed expressions for their scattering properties, valid through the wavelength/object size transition zone (the resonance region). One case is the dielectric sphere for which the Mie series constitute a complete exact solution to the scattering problem. Also for an infinite cylinder, exact solutions exist. The exact solutions emerged remarkably soon after the advent of Maxwell's equations. Gustav Mie published already in 1908 the first exact formulation of electromagnetic excitations of

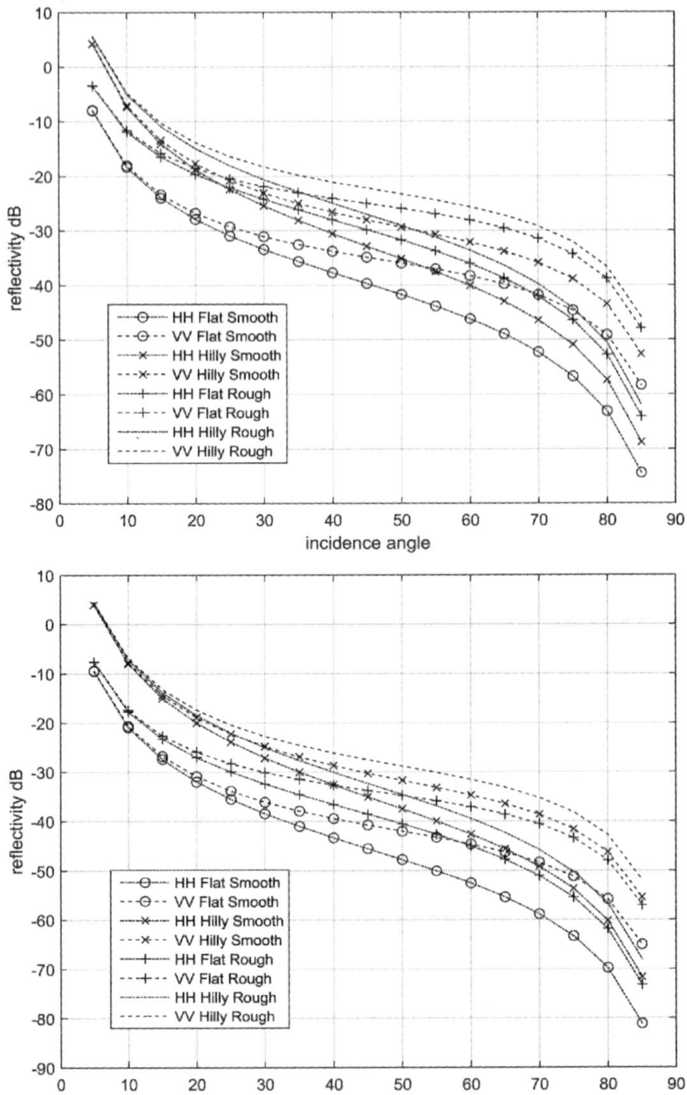


**Figure 2.3** Upper and lower: ground reflectivity for 1.5m (i.e., 200 MHz) and 5m, (i.e., 60 MHz). The index of refraction is assumed to be $\eta = 2.2$ (light dry soils per Figure 2.1). The invariably higher reflectivity for vertical polarization is noted, and also that the difference between polarizations becomes particularly strong at large incidence angles. Reflectivity is a few decibels lower at the longer wavelength, but ground texture has a much larger influence on reflectivity than wavelength. When the incidence angle tends to zero, there is no distinction between vertical and horizontal polarizations and the reflectivity tends to infinity in both cases.

the sphere. The primary application area seems to have been optics with radio waves and even more scattering of radio waves, still unexplored.

Since structures that are small compared to the wavelength scatter negligibly compared to any wavelength-sized structure, the exact shape of an object is not crucial for determining its gross scattering characteristics in the resonance region. Thus, while closed expressions for structures truly representing targets and vegetation do not exist, targets and vegetation structures may quite suitably be represented by spheres and cylinders. Indeed, spheres form good generic structures representing typical small targets below the ground (for which their detection is a challenge), whereas cylinders well model tree stems, branches, and targets standing on the ground surface. This section adopts this approach and accounts for the sphere and cylinder exact scattering solutions, followed by an analysis of the scattering behavior of surface and subsurface targets, as well as vegetation.

Two cases of scattering from the cylinder will be given an exact treatment. One is incidence at right angles to the cylinder, in which case arbitrary dielectric properties are allowed. The other is incidence at oblique angles for perfectly reflecting cylinders. The case of arbitrary dielectric properties leads to complications, which are not worthwhile to pursue in the derivation of exact expressions. In fact, it will be seen that the case of oblique incidence is a very straightforward generalization of normal incidence, thus also indicating the obvious generalization to oblique incidence for arbitrary dielectrics. Possible errors in this generalization are expected negligible, in particular since tree dielectric constants are quite high. The situation that the radar signal impacts tree elements from oblique angles is very important; the corner reflector formed by the tree stem and the ground plays a crucial role in low frequency radar phenomenology. There are all important differences between the scattering response for cylinders with finite and infinite reflectivity, which will be incorporated into the modeling of tree stem responses.

The case of cylinders of finite length is crucial for establishing tree RCS and forest reflectivity. The infinite cylinder scattering expressions are duly modified to expressions for these cases. The infinite cylinder expressions also form the basis for establishing a generic model of attenuation of radar signals in forest. The case considered is wave propagation through a structure of randomly distributed cylinders. By selecting diameter, the structures may represent twigs, boughs, or tree stems.

For spheres, arbitrary dielectic properties and a lossy surrounding medium are assumed. The theory can thus be applied to questions like detectability of buried nonmetallic objects (e.g., plastic mines) and the strength of unwanted responses from boulders causing inhomogeneities in soil. As for buried objects, another important situation, which can be treated, is a buried electric cable, represented by a slender cylinder with high conductivity, surrounded by a lossy medium. The treatment requires details of radar measurement process and will be provided in connection with the development of these in Chapter 3.

Scattering analysis of the cylinders and spheres starts off with the wave equation set up in cylindrical and spherical coordinates. The particular solutions in the respective coordinate systems allow direct implementation of the boundary conditions for each type of object. The analysis is straightforward for scalar fields (the case of acoustics, for instance) but less so in electromagnetics, where the field amplitudes are vectors. Here, in obtaining vector field particular solutions, the methods of [9] are adopted, though the present analysis of cylinders is not provided in it. The similar problem of scattering from spheres is detailed there, and the account here will be brief. The treatment of Mie theory in [10] is good and accessible. Final spherical and cylindrical section formulas are also found in [11] and [12]. Not providing full derivations, it is of course debatable if they can be trusted. Moreover, both works stem from times when reduction and simplification of exact results was a necessity to make them numerically useful. Modern computational tools make most of these efforts no longer required. The much more modern work of Balanis [13] is well adapted to modern computational resources and also provides derivations. However, it does not address the presently crucial topic of scattering of objects embedded in lossy media (i.e., subsurface targets), neither does it address forward scattering (i.e., vegetation attenuation). The present treatment is quite general and will also cope with these cases, while its format allows straightforward numerical implementation.

First, the elements of the theory for scalar solutions to the wave equation in curvilinear coordinates are summoned up. The theory follows in analogous to the case of Cartesian coordinates. In this latter case, by the Fourier transform techniques of (2.27), the general homogeneous source-free solution is obtained from any arbitrary distribution $X(\boldsymbol{x},0)$ at time $t = 0$. Obtaining its three-dimensional Fourier transform $X_{\boldsymbol{k}}$ of any electromagnetic field component, one finds (where the propagation speed in the medium $c$ is constant) that

$$X(\boldsymbol{x},t) = \frac{1}{(2\pi)^3} \int_{\forall \boldsymbol{k}} X_{\boldsymbol{k}} e^{i(ckt - \boldsymbol{k} \cdot \boldsymbol{x})} d^3k \tag{2.130}$$

that is to say, the temporal development of the distribution is determined. In the time stationary, that is, monochromatic, case (note that (2.58) constitutes an analogous expression for the scattering from the ground surface)

$$X(\boldsymbol{x},t) = \frac{k^2 e^{ickt}}{(2\pi)^3} \oint X_{\boldsymbol{k}} e^{-i\boldsymbol{k}\cdot\boldsymbol{x}} \, d\Omega_{\boldsymbol{k}} \tag{2.131}$$

In other words, based on plane waves, the Fourier transform determines the temporal development of an arbitrary initial assumption on the field distribution. In the monochromatic case, the distribution is only for directions, since wave vector magnitudes are fixed. Presently, the coordinate system must be changed from Cartesian to curvilinear. Plane waves are then no longer particular solutions, and the Fourier transform will no longer be the valid expansion. Sets of particular solutions for the chosen type of coordinates, corresponding to plane waves and allowing similar expansion of an initial distribution, must be found.

Coming to the spherical and cylindrical coordinates, a remark on sign conventions is required. A phase increasing with time seems the natural choice and is indeed what has been used up to (2.130). Treaties on spherical and cylindrical coordinate partial equations focus on spatial distributions, with the temporal development following trivially. The spatial distributions derived adopt the reversed sign conventions with field phase changing positively with space coordinates $\boldsymbol{r}$ and negatively with time $t$. The reversal of conventions has no practical impact. A possible remedy is to consider the field amplitudes derived as relating to the complex conjugates of previous field expressions. For details on the special functions adopted here, the author uses [9], [14], and [15]. Much more recent, there is the website [16] where the U.S. National Institute of Standards and Technology provides open source code for special functions. The same material is also available as [17]. The author has not used these references, however. The code for special functions, as provided in the MATLAB library and used in the ways recommended in the sequel, has been found sufficient.

For the spherical case, homogeneous stationary wave equation (2.27) is restated in spherical coordinates $r$, $\theta$, and $\phi$ (the Fourier index $\omega$ be suppressed for brevity) as

$$\frac{1}{r^2}\frac{\partial}{\partial r}\left(r^2\frac{\partial}{\partial r}\right)\chi + \frac{1}{r^2\sin\theta}\frac{\partial}{\partial\theta}\left(\sin\theta\frac{\partial}{\partial\theta}\right)\chi + \frac{1}{r^2\sin^2\theta}\frac{\partial^2\chi}{\partial\phi^2} + k^2\chi = 0 \tag{2.132}$$

with $k = \omega/c$. As is a well-known procedure, particular solutions to (2.132) are obtained by separating variables $\chi(r,\theta,\phi) = R(kr)\Theta(\theta)\Phi(\phi)$ to obtain three ordinary differential equations. Reflecting that $\Theta(\theta)\Phi(\phi)$ is indeed a function on the sphere, it must be periodic in $\theta$ and $\phi$. Independent solutions exist for integer numbers of polar and azimuthal periods $n$ and $m$. The correspondingly labeled solutions are expressible in special functions as

$$\chi^{p}_{n,m}(r,\theta,\phi) = R_n(kr)P^m_n(\cos\theta)\cos\left(m\phi - p\frac{\pi}{2}\right) \tag{2.133}$$

Here $p = 0,1$. The value selected implies the parity, that is, even or odd $\phi$-dependence. The functions $P^m_n(\cos\theta)$ are associated Legendre polynomials, obtainable from ordinary Legendre polynomials $P_n(\cos\theta)$, according to

$$\begin{aligned} P_n(\cos\theta) &= \frac{1}{2^n n!}\left(\frac{d}{d\cos\theta}\right)^n \left(-\sin^2\theta\right)^n \\ P^m_n(\cos\theta) &= \sin^m\theta\left(\frac{d}{d\cos\theta}\right)^m P_n(\cos\theta) \end{aligned} \tag{2.134}$$

where $n,m \geq 0$. It is seen that $P^m_n = 0$ if $m > n$ and that $P^0_0 = P_0 = 1$. Of particular interest is $P^1_n(\cos\theta)$ for which

$$P^1_n(\tau) = \frac{\sqrt{1-\tau^2}}{(-2)^{n-1}(n-1)!}\left(\frac{d}{d\tau}\right)^n \left[\tau\left(1-\tau^2\right)^{n-1}\right] \tag{2.135}$$

The radial behavior of (2.132) is governed by the so-called Helmholtz equation

$$\frac{1}{r^2}\frac{d}{dr}\left(r^2\frac{d}{dr}\right)R_n - \frac{n(n+1)}{r^2}R_n + k^2R_n = 0 \tag{2.136}$$

Its solutions are given by spherical Bessel functions, which are obtainable from half-integral order Bessel functions. The two solutions presently required are given by

$$R_n(kr) = \begin{cases} j_n(kr) = \sqrt{\dfrac{\pi}{kr}}J_{n+1/2}(kr) \\ h^{(1)}_n(kr) = \sqrt{\dfrac{\pi}{kr}}H^{(1)}_{n+1/2}(kr) \end{cases} \tag{2.137}$$

The upper solution is real valued for a real-valued argument. When combined with its time development, it represents a superposition of combined ingoing and outgoing spherical waves, just as $e^{-i\omega t}\sin(kx)$ is the superposition of two traveling waves moving in opposite directions along the $x$-axis. The lower solution is an outgoing spherical wave, in which $H_{n+1/2}^{(1)}$ is a half-integer order Hankel function (of the first kind, Hankel functions of the second kind will not be considered and the notation $H_{n+1/2} = H_{n+1/2}^{(1)}$ will therefore be used). As such, it has to be infinite at the origin but will presently be applied only to the scattered field, defined from the sphere surface and outward.

Note that $z = kr$ is complex in the case of a lossy medium. Bessel functions are obtainable through

$$J_{n+1/2}(z) = (-z)^n \sqrt{\frac{2z}{\pi}} \left( \frac{1}{z} \frac{d}{dz} \right)^n \frac{\sin z}{z} \tag{2.138}$$

whereas

$$H_{n+1/2}(z) = -i^n \frac{e^{in\pi/2} J_{n+1/2}(z) + ie^{-in\pi/2} J_{-n-1/2}(z)}{\cos(n\pi)} \tag{2.139}$$

Just as a general homogenous stationary wave solution can be decomposed into plane waves as in (2.131) (or for that matters (2.58)), it can also be represented as

$$\chi(r,\theta,\phi,t) = e^{-i\omega t} \sum_{n=0}^{\infty} \sum_{m=1}^{n} \sum_{p=0}^{1} a_n^{m,p} R_n(kr) P_n^m(\cos\theta) \cos\left( m\phi - p\frac{\pi}{2} \right) \tag{2.140}$$

An alternative representation in terms of combinations $P_n^{|m|} e^{\mathrm{i}m\phi}$ and including negative $m$ is of course possible, whereas the present representation highlights $\phi$-parity. The particular selection of $R_n(kr)$ depends on the wave problem considered, that is, whether the problem contains incoming or outgoing or standing waves (e.g., in a spherical cavity).

The Legendre polynomials as well as the functions $\cos(m\phi - p\pi/2)$ are orthogonal for different $n$, $m$, and $p$, and the expansion coefficients $a_{n,p}^{m}$ are to be found according to

$$a_n^{m,p} = \frac{(1+\delta_{n0})(2n+1)(n-m)!}{4\pi(n+m)!} \oint_{r=\mathrm{const}} \chi(r,\theta,\varphi,0) P_n^m(\cos\theta) \cos\left( m\phi - p\frac{\pi}{2} \right) d\Omega \tag{2.141}$$

Here, $\delta_{00} = 1$ and $\delta_{10} = 0$.

Finding solutions for the cylindrical coordinate scalar wave equation proceeds similarly. The cylindrical coordinates $\rho$, $\theta$, and $z$ are adopted. Here,

$\rho = \sqrt{x^2 + y^2}$ is cylinder radius, whereas the scalar time stationary homogenous wave equation reads

$$\frac{1}{\rho}\frac{\partial}{\partial\rho}\left(\rho\frac{\partial}{\partial\rho}\right)\psi + \frac{1}{\rho^2}\frac{\partial^2}{\partial\phi^2}\psi + \frac{\partial^2}{\partial z^2}\chi + \left(k^2 - \kappa^2\right)\psi = 0 \qquad (2.142)$$

Particular solutions are obtained by separation of variables

$$\psi^p_{n,k\cos\theta}(\rho,\varphi,z) = e^{ikz\cos\theta} Z_n(k\rho\sin\theta)\cos\left(n\phi - p\frac{\pi}{2}\right) \qquad (2.143)$$

with $k = \omega\sqrt{\mu\varepsilon}/c$ and $\theta$ arbitrary within $0 \le \theta \le \pi$. At this instance, solutions $Z_n$ are given by integer order Bessel functions. These can be real or complex valued representing incoming and outgoing waves or a mix, just as in the spherical case. A plane wave is represented entirely by Bessel functions of the first kind $Z_n = J_n$. In contrast to half integer order Bessel functions, the Bessel functions of the first kind for negative $n$ are regular at the origin and in fact dependent on those of positive $n$ according to $J_{-n} = (-1)^n J_n$. Independent particular solutions are found as Bessel functions of the second kind $Z_n = N_n$, which are infinite at the origin. Outgoing cylindrical waves are represented by the integer order Hankel function $Z_n = H_n$, for which two types exist (i.e., the two linear combinations $H_n = J_n \pm iN_n$ of Bessel functions of the first and the second kind). The plus sign represents outgoing waves, which are required here. Note that $Z_n e^{in\phi}$with summation over negative $n$ can only replace (2.143) in the case $Z_n = J_n$. With the other selections of $Z_n$, the two signs of $n$ represent independent Bessel functions.

For cylindrical coordinates, the expansion corresponding to (2.131) or (2.140) reads

$$\psi(\rho,\phi,z,t) = e^{-i\omega t}\sum_{n=0}^{\infty}\sum_{p=0}^{1}\cos\left(n\phi - p\frac{\pi}{2}\right)\int_{\kappa=0}^{\infty}\psi^{n,p}_{\kappa} e^{i\left(\sqrt{k^2-\kappa^2}z\right)} Z_n(\kappa\rho)\kappa d\kappa \qquad (2.144)$$

There are convergence issues involved in assessing (2.144) and a direct evaluation of the expansion coefficients $\psi^n_{\kappa,p}$ similar to (2.141) is not always possible. However, in the important case of $Z_n = J_n$, the well-known Fourier–Bessel theorem yields

$$\psi^{n,p}(\rho,z,0)=\oint_{\phi}\psi(\rho,\phi,z,0)\cos\left(n\phi-p\frac{\pi}{2}\right)d\phi=\pi\int_{\kappa=0}^{\infty}\psi_{\kappa}^{n,p}e^{i\left(\sqrt{k^2-\kappa^2}z\right)}J_n(\kappa\rho)\kappa d\kappa$$

$$\Rightarrow\psi_{\kappa}^{n,p}=\frac{1}{\pi}\int_{\rho=0}^{\infty}\oint_{z=0}\psi(\rho,\phi,z,0)J_n(\kappa\rho)\cos\left(n\phi-p\frac{\pi}{2}\right)\rho d\rho d\phi \tag{2.145}$$

For the scalar wave equation particular solutions to be applicable to electromagnetic scattering, they must be generalized to vector fields. An efficient of making the necessary generalizations is as follows.

For any coordinate system $\alpha,\beta,\gamma$, it is observed that its basis vectors $\boldsymbol{e}_\alpha$, $\boldsymbol{e}_\beta$, and $\boldsymbol{e}_\gamma$ at some space point are tangents to the coordinate lines through that point. For Cartesian coordinates *x,y,z*, the basis vectors are constant throughout space, while for curvilinear coordinates, they are as a rule not. Therefore, if $X(\boldsymbol{r})$ is a scalar solution to the time separated wave equation (2.27), vector field solutions in Cartesian coordinates are simply constructed as vector fields $\boldsymbol{x}X(\boldsymbol{r})$, where $\boldsymbol{x}$ is any fixed linear combination of $\boldsymbol{e}_x$, $\boldsymbol{e}_y$, and $\boldsymbol{e}_z$. Clearly, this way of constructing vector fields cannot be transferred into curvilinear coordinates.

Turning to vector fields in cylindrical coordinates, the basis vectors are noted to be $\boldsymbol{e}_\rho$, $\boldsymbol{e}_\phi$, and $\boldsymbol{e}_z$, where only $\boldsymbol{e}_z$ is constant in space. Thus, if $\psi(\boldsymbol{r})$ is a particular solution in cylindrical coordinates, so is $\boldsymbol{e}_z\psi(\boldsymbol{r})$, while $\boldsymbol{e}_\rho\psi(\boldsymbol{r})$ and $\boldsymbol{e}_\phi\psi(\boldsymbol{r})$ are not. It is, however, straightforward to construct two further independent vector fields, which are indeed particular solutions, since any differential of a solution remains a solution. Thereby, the two fields

$$\boldsymbol{M}=\nabla\times\left(\boldsymbol{e}_z\psi\right);\ \boldsymbol{N}=\frac{1}{k}\nabla\times\boldsymbol{M} \tag{2.146}$$

constitute two particular solutions. Note that from well-known rules for vector and scalar fields

$$\begin{aligned}\boldsymbol{M}&=(\nabla\psi)\times\boldsymbol{e}_z\\ \boldsymbol{N}&=k\psi\boldsymbol{e}_z+\frac{1}{k}\frac{\partial}{\partial z}(\nabla\psi)\end{aligned} \tag{2.147}$$

The vector fields $\boldsymbol{e}_z\psi(\boldsymbol{r})$, $\boldsymbol{M}$, and $\boldsymbol{N}$ are all independent, that is, they cannot be obtained as a linear combination of each other. However, performing a further rotation operation on $\boldsymbol{N}$, no new independent vector field is obtained. Indeed, it follows that

$$\boldsymbol{M} = \frac{1}{k}\nabla \times \boldsymbol{N} \tag{2.148}$$

Note that $\boldsymbol{M}$ and $\boldsymbol{N}$ both are rotation fields and thus divergence free. Just as for Cartesian coordinates, there will be three independent vector field solutions pertaining to any scalar cylindrical coordinate particular solution. The vector fields $\boldsymbol{e}_z\psi(\boldsymbol{r})$, $\boldsymbol{M}$, and $\boldsymbol{N}$ constitute such a set. Imposing the condition that the solution should be divergence free, just as in the case of Cartesian coordinates, there will be two independent fields, and $\boldsymbol{M}$ and $\boldsymbol{N}$ constitute such a reduced set.

The electromagnetic scattering problem of the cylinder can now be tackled. Being all divergence free solutions to the wave equation, it is expected that the vector potential $\boldsymbol{A}$ as well as the implied electric and magnetic fields $\boldsymbol{E}$ and $\boldsymbol{H}$ can be found as linear combinations of $\boldsymbol{M}$ and $\boldsymbol{N}$. The linear combinations will be interconnected through the vector potential since (it is allowable to assume the scalar potential zero in the stationary case)

$$\begin{aligned} \boldsymbol{E} &= -\frac{\partial}{\partial t}\boldsymbol{A} = i\omega\boldsymbol{A} \\ \boldsymbol{H} &= \nabla \times \boldsymbol{A} \end{aligned} \tag{2.149}$$

Thus, from (2.149) and with $a$ and $b$, undetermined constants

$$\begin{aligned} \boldsymbol{A} &= \frac{i}{\omega}(a\boldsymbol{M} + b\boldsymbol{N}) \Rightarrow \\ \boldsymbol{E} &= -(a\boldsymbol{M} + b\boldsymbol{N}) \\ \boldsymbol{H} &= \frac{ik}{\omega}(a\boldsymbol{N} + b\boldsymbol{M}) \end{aligned} \tag{2.150}$$

Consider an infinitely long cylinder, which is coaxial with the $z$-axis. Let the impinging field be a plane wave field moving in the direction of the positive $x$-axis at an angle $\theta$ with respect to the positive $z$-axis. As generic polarization states, the $\boldsymbol{E}$ field is either parallel to or at right angles with the $y$-axis. In cases where no confusion may arise, the two states will be referred to as horizontal and vertical polarizations. However, the $z$-axis may just as well represent the longitudinal direction of a tree branch or tree stem and thus horizontal and vertical directions. In such situations, the polarization states will be denoted TE and TM; the terminology was introduced already in Section 2.4.

As discussed for cylindrical coordinate particular solutions, a scalar plane wave is represented by Bessel functions of the first kind. With respect to these, a scalar plane wave possesses the well-known expansion

$$\psi(\boldsymbol{x},t) = e^{i(\boldsymbol{k}\cdot\boldsymbol{x}-\omega t)} = e^{i(kz\cos\theta-\omega t)} \sum_{n=-\infty}^{\infty} i^n J_n(k\rho\sin\theta)\cos(n\phi) \quad (2.151)$$

Here, $\boldsymbol{k} = k(\sin\theta,0,\cos\theta)$ in Cartesian coordinates, whereas $\boldsymbol{r} = (\rho\cos\phi,\rho\sin\phi,z)$. Since $J_{-n} = (-1)^n J_n$, the expansion can be reexpressed in a summation over positive $n$ and thus be brought in the form of (2.144)

$$\begin{aligned} \psi(\boldsymbol{x},t) &= e^{i(\boldsymbol{k}\cdot\boldsymbol{x}-\omega t)} = e^{-i\omega t}\sum_{n=0}^{\infty} i^n C_n \psi^{n,0}_{k\sin\theta}(\boldsymbol{x}); \\ \psi^{n,0}_{k\sin\theta} &= J_n(k\rho\sin\theta)e^{ikz\cos\theta}\cos(n\phi) \end{aligned} \quad (2.152)$$

The expansion coefficients $C_n$ simply become $C_n = 1$ for $n = 0$ and $C_n = 2$ for $n \geq 1$. Note that the plane wave is an even function in $\psi$ (i.e., the parity is $p = 0$). Also, observe that the scattering configuration considered in (2.152) is different from those in Sections 2.4 and 2.5 in which a plane wave was impinging in the negative $x$ direction. Assuming the latter, (2.152) would change by the substitution $\theta \to \pi - \theta$, $\phi \to \phi + \pi$ making expressions unnecessarily more complex.

Use (2.147) to derive $\boldsymbol{M}$ and $\boldsymbol{N}$ for the scalar plane wave with the direction of propagation given by $\boldsymbol{k}$. Assuming $\boldsymbol{k}$ to be in the $x,z$-plane and $\boldsymbol{n_k} \bullet \boldsymbol{e_z} = \cos\theta$

$$\begin{aligned} \boldsymbol{M} &= -ik\psi\sin\theta\boldsymbol{e}_y \\ \boldsymbol{N} &= k\psi\left(\boldsymbol{e}_z - \cos\theta\boldsymbol{k}\right) = k\psi\sin\theta\left[\sin\theta\boldsymbol{e}_z - \cos\theta\boldsymbol{e}_x\right] \end{aligned} \quad (2.153)$$

From (2.150)

$$\begin{aligned} \boldsymbol{E} &= k\psi\sin\theta\left[ia\boldsymbol{e}_y + b\left(\cos\theta\boldsymbol{e}_x - \sin\theta\boldsymbol{e}_z\right)\right] \\ \boldsymbol{H} &= \frac{\eta}{c}k\psi\sin\theta\left[ia\left(-\cos\theta\boldsymbol{e}_x + \sin\theta\boldsymbol{e}_z\right) + b\boldsymbol{e}_y\right] \end{aligned} \quad (2.154)$$

where $\eta$ is the index of refraction. According to the polarization convention

$$\begin{aligned} &\text{vertical polarization:} && a = 0;\ b \neq 0 \\ &\text{horizontal polarization:} && a \neq 0;\ b = 0 \end{aligned} \quad (2.155)$$

Applying the linear operations in (2.146) to (2.152), the vector fields of the impinging plane wave become

$$\widehat{\boldsymbol{M}}(\boldsymbol{x}) = e^{-i\omega t}\sum_{n=0}^{\infty} i^n C_n \widehat{\boldsymbol{M}}^n_{\hat{k}\sin\theta}(\boldsymbol{x});\ \widehat{\boldsymbol{N}}(\boldsymbol{x}) = e^{-i\omega t}\sum_{n=0}^{\infty} i^n C_n \widehat{\boldsymbol{N}}^n_{\hat{k}\sin\theta}(\boldsymbol{x}) \qquad (2.156)$$

The explicit expressions for $\boldsymbol{M}^{n,p}_{k\sin\vartheta}$ and $\boldsymbol{N}^{n,p}_{k\sin\vartheta}$ are derived from (2.147)

$$\begin{aligned}
\boldsymbol{M}^{n,p}_{k\sin\vartheta} &= \left(\nabla\psi^{n,p}_{k\sin\vartheta}\right)\times\boldsymbol{e}_z \\
&= e^{izk\cos\vartheta}\begin{bmatrix} -\dfrac{n}{\rho}Z_n(\rho k\sin\vartheta)\sin\left(n\phi - p\dfrac{\pi}{2}\right)\boldsymbol{e}_\rho \\ -k\sin\vartheta\, Z_n'(\rho k\sin\vartheta)\cos\left(n\phi - p\dfrac{\pi}{2}\right)\boldsymbol{e}_\phi \end{bmatrix} \\
\boldsymbol{N}^{n,p}_{k\sin\vartheta} &= k\psi^{n,p}_{k\sin\vartheta}\boldsymbol{e}_z + \frac{1}{k}\frac{\partial}{\partial z}\left(\nabla\psi^{n,p}_{k\sin\vartheta}\right) \\
&= e^{izk\cos\vartheta}\begin{bmatrix} ik\sin\vartheta\cos\vartheta\, Z_n'(\rho k\sin\vartheta)\cos\left(n\phi - p\dfrac{\pi}{2}\right)\boldsymbol{e}_\rho \\ -i\dfrac{n}{\rho}\cos\vartheta\, Z_n(\rho k\sin\vartheta)\sin\left(n\phi - p\dfrac{\pi}{2}\right)\boldsymbol{e}_\phi \\ +k\sin^2\vartheta\, Z_n(\rho k\sin\vartheta)\cos\left(n\phi - p\dfrac{\pi}{2}\right)\boldsymbol{e}_z \end{bmatrix}
\end{aligned} \qquad (2.157)$$

Here, $\vartheta$ denotes a general angle not required to agree with the plane wave $\theta$ of the impinging field. In the case of $\boldsymbol{M}^{n,0}_{k\sin\vartheta} = \widehat{\boldsymbol{M}}^n_{\hat{k}\sin\theta}$ and $\boldsymbol{N}^{n,0}_{k\sin\vartheta} = \widehat{\boldsymbol{N}}^n_{\hat{k}\sin\theta}$, it is understood that $Z_n = J_n$, $\vartheta = \theta$ and that the wavenumber $k = \hat{k}$ is set by the frequency of the field and the refraction index $\hat{\eta}$ of the medium surrounding the cylinder. As for the fields $\breve{\boldsymbol{M}}^{n,p}_{k\sin\vartheta}$ and $\breve{\boldsymbol{N}}^{n,p}_{k\sin\vartheta}$ scattered off the cylinder, $Z_n = H_n$, with the Hankel functions representing outgoing cylinder waves. Any fields $\tilde{\boldsymbol{M}}^{n,p}_{k\sin\vartheta}$ and $\tilde{\boldsymbol{N}}^{n,p}_{k\sin\vartheta}$ refracted into the cylinder must be finite along the $z$-axis, thus also requiring these fields to be defined with respect to the Bessel functions of the first kind. However, $\tilde{\chi}^{n,p}_{k\sin\vartheta}$ differs from $\breve{\chi}^{n,p}_{k\sin\vartheta}$ having $k$ replaced by the wave number inside the cylinder $\tilde{k}$. These various considerations must be implemented from the following general representation of the fields, implied by (2.150):

$$\begin{aligned} \boldsymbol{E} &= \sum_{n=0}^{\infty}\sum_{p=0}^{1} -i^n\left(a_{n,p}\boldsymbol{M}^{n,p}_{k\sin\vartheta} + b_{n,p}\boldsymbol{N}^{n,p}_{k\sin\vartheta}\right) \\ \boldsymbol{H} &= \frac{\eta}{c}\sum_{n=0}^{\infty}\sum_{p=0}^{1} i^{n+1}\left(b_{n,p}\boldsymbol{M}^{n,p}_{k\sin\vartheta} + a_{n,p}\boldsymbol{N}^{n,p}_{k\sin\vartheta}\right) \end{aligned} \tag{2.158}$$

Since only the scattered amplitude in relation to that of the impinging field is of concern, any amplitude can be assumed for the latter. It will be convenient to choose

$$\begin{aligned} &\text{vertical:} \begin{cases} \widehat{\boldsymbol{E}} = -\widehat{k}\widehat{\boldsymbol{N}} = -\widehat{k}e^{-i\omega t}\sum_{n=0}^{\infty} i^n C_n \widehat{\boldsymbol{N}}^n_{\widehat{k}\sin\theta} \\ \widehat{\boldsymbol{H}} = i\frac{\widehat{\eta}}{c}\widehat{k}\widehat{\mathbf{M}} = \frac{\widehat{\eta}}{c}\widehat{k}e^{-i\omega t}\sum_{n=0}^{\infty} i^{n+1} C_n \widehat{\boldsymbol{M}}^n_{\widehat{k}\sin\theta} \end{cases} \\ &\text{horizontal:} \begin{cases} \widehat{\boldsymbol{E}} = -\widehat{k}\widehat{\boldsymbol{M}} = -\widehat{k}e^{-i\omega t}\sum_{n=0}^{\infty} i^n C_n \widehat{\boldsymbol{M}}^n_{\widehat{k}\sin\theta} \\ \widehat{\boldsymbol{H}} = i\frac{\widehat{\eta}}{c}\widehat{k}\widehat{\boldsymbol{N}} = \frac{\widehat{\eta}}{c}\widehat{k}e^{-i\omega t}\sum_{n=0}^{\infty} i^{n+1} C_n \widehat{\boldsymbol{N}}^n_{\widehat{k}\sin\theta} \end{cases} \end{aligned} \tag{2.159}$$

The convention means according to (2.150) that $b = \widehat{k}$ and $b = \widehat{k}$ in the respective cases of vertical and horizontal polarizations. The intensity $|\hat{\boldsymbol{E}}|^2$ of the incident field follows by comparing (2.150) and (2.153):

$$\begin{aligned} &\text{Vertical polarization:} && \left|\widehat{\boldsymbol{E}}\right|^2 = |b|^2|\boldsymbol{N}|^2 = \widehat{k}^4\sin^2\theta \\ &\text{Horizontal polarization:} && \left|\widehat{\boldsymbol{E}}\right|^2 = |a|^2|\boldsymbol{M}|^2 = \widehat{k}^4\sin^2\theta \end{aligned} \tag{2.160}$$

Consider first the case of normal incidence $\theta = \pi/2$ (along the $x$-axis) and finite dielectric properties. The boundary conditions are therefore that for all $\phi$ and $z$ values on the cylinder radius $\rho = a$

$$\boldsymbol{e}_\rho \times \left(\widehat{\boldsymbol{E}} + \breve{\boldsymbol{E}} - \tilde{\boldsymbol{E}}\right)_{\rho=a,\forall\phi,z} = \boldsymbol{e}_\rho \times \left(\widehat{\boldsymbol{H}} + \breve{\boldsymbol{H}} - \tilde{\boldsymbol{H}}\right)_{\rho=a,\forall\phi,z} = 0 \tag{2.161}$$

In the case of normal incidence, (2.157) implies that

$$\begin{aligned} \boldsymbol{e}_\rho \times \boldsymbol{M}_{k\sin\vartheta}^{n,p} &= -kZ_n'(\rho k)\cos\left(n\phi - p\frac{\pi}{2}\right)\boldsymbol{e}_z = -kZ_n'(\rho k)\boldsymbol{\Xi}_{n,p}(\phi) \\ \boldsymbol{e}_\rho \times \boldsymbol{N}_{k\sin\vartheta}^{n,p} &= -kZ_n(\rho k)\cos\left(n\phi - p\frac{\pi}{2}\right)\boldsymbol{e}_\phi = -kZ_n(\rho k)\boldsymbol{\Phi}_{n,p}(\phi) \end{aligned} \tag{2.162}$$

The vector fields $\boldsymbol{\Xi}_{n,p}$ and $\boldsymbol{\Phi}_{n,p}$ (varying with $\phi$ but constant with respect to $z$) are all independent for different $n$ and $p$. Consider vertical polarization first. For the $\boldsymbol{E}$ field part of (2.161) to be valid, the coefficients of these fields must vanish identically and so

$$\begin{aligned} &C_n\hat{k}^2 J_n\left(a\hat{k}\right) + \breve{b}_{n,0}\hat{k} H_n\left(a\hat{k}\right) - \tilde{b}_{n,0}\tilde{k} J_n\left(a\tilde{k}\right) = 0 \\ &\breve{a}_{n,0}\hat{k} H_n'(ak) - \tilde{a}_{n,0}\tilde{k} J_n'\left(a\tilde{k}\right) = 0 \end{aligned} \tag{2.163}$$

Similarly, for the $\boldsymbol{H}$ field part

$$\begin{aligned} &C_n\hat{k}^2\hat{\eta} J_n'\left(\rho\hat{k}\right) + \breve{b}_{n,0}\hat{k}\hat{\eta} H_n'\left(a\hat{k}\right) - \tilde{b}_{n,0}\tilde{k}\tilde{\eta} J_n'\left(a\tilde{k}\right) = 0 \\ &\breve{a}_{n,0}\hat{k}\hat{\eta} H_n\left(a\hat{k}\right) - \tilde{a}_{n,0}\tilde{k}\tilde{\eta} J_n\left(a\tilde{k}\right) = 0 \end{aligned} \tag{2.164}$$

Do the same in the case of horizontal polarization. The following four equations are obtained:

$$\begin{aligned} &C_n\hat{k}^2 J_n'\left(\rho\hat{k}\right) + \breve{a}_{n,0}\hat{k} H_n'\left(a\hat{k}\right) - \tilde{a}_{n,0}\tilde{k} J_n'\left(a\tilde{k}\right) = 0 \\ &\breve{b}_{n,0}\hat{k} H_n(ak) - \tilde{b}_{n,0}\tilde{k} J_n\left(a\tilde{k}\right) = 0 \\ &C_n\hat{k}^2\hat{\eta} J_n\left(a\hat{k}\right) + \breve{a}_{n,0}\hat{k}\hat{\eta} H_n\left(a\hat{k}\right) - \tilde{a}_{n,0}\tilde{k}\tilde{\eta} J_n\left(a\tilde{k}\right) = 0 \\ &\breve{b}_{n,0}\hat{k}\hat{\eta} H_n'\left(a\hat{k}\right) - \tilde{b}_{n,0}\tilde{k}\tilde{\eta} J_n'\left(a\tilde{k}\right) = 0 \end{aligned} \tag{2.165}$$

The solutions for the scattered and refracted fields are thereby determined. Only the former is of interest here. The following solution to the cylinder scattering problem arises:

$$\begin{gathered}\text{Vertical polarization:}\\ \breve{a}_{n,0}=0;\ \breve{b}_{n,0}=C_n\widehat{k}\frac{\tilde{\eta}J_n\left(a\widehat{k}\right)J_n'\left(a\tilde{k}\right)-\widehat{\eta}J_n'\left(a\widehat{k}\right)J_n\left(a\tilde{k}\right)}{\widehat{\eta}H_n'\left(a\widehat{k}\right)J_n\left(a\tilde{k}\right)-\tilde{\eta}H_n\left(a\widehat{k}\right)J_n'\left(a\tilde{k}\right)}\\ \text{Horizontal polarization:}\\ \breve{b}_{n,0}=0;\ \breve{a}_{n,0}=C_n\widehat{k}\frac{\tilde{\eta}J_n'\left(a\widehat{k}\right)J_n\left(a\tilde{k}\right)-\widehat{\eta}J_n\left(a\widehat{k}\right)J_n'\left(a\tilde{k}\right)}{\widehat{\eta}H_n\left(a\widehat{k}\right)J_n'\left(a\tilde{k}\right)-\tilde{\eta}H_n'\left(a\widehat{k}\right)J_n\left(a\tilde{k}\right)}\end{gathered}\tag{2.166}$$

For a strongly reflecting cylinder, i.e., $\tilde{\eta} >> 1$, it is noted that

$$\begin{gathered}\text{Vertical polarization:}\\ \breve{a}_{n,0}=0;\ \breve{b}_{n,0}\approx -C_n\widehat{k}\frac{J_n\left(a\widehat{k}\right)}{H_n\left(a\widehat{k}\right)}\\ \text{Horizontal polarization:}\\ \breve{b}_{n,0}=0;\ \breve{a}_{n,0}\approx -C_n\widehat{k}\frac{J_n'\left(a\widehat{k}\right)}{H_n'\left(a\widehat{k}\right)}\end{gathered}\tag{2.167}$$

The coefficients $\breve{a}_{n,1}$ and $\breve{b}_{n,1}$ are both zero since parity will not change from that of the impinging field. From (2.166), it follows that no depolarization occurs in the scattering process.

The generalized case of oblique incidence can be analyzed with little extra effort for a perfectly reflecting cylinder. As was discussed in connection with Fresnel's equations in (2.45), the boundary conditions in this case is just the cancellation between the tangential components of the impinging and reflect fields, that is, $\boldsymbol{e}_\rho \times (\boldsymbol{E} + \breve{\boldsymbol{E}}) = 0$; the tangential magnetic field components $\boldsymbol{e}_\rho \times \tilde{\boldsymbol{H}}$ will not vanish but obtain values implied by the electric field boundary conditions.

With $\vartheta \neq \pi/2$, the tangential projections of $\boldsymbol{M}^{n,p}_{k\sin\vartheta}$ and $\boldsymbol{N}^{n,p}_{k\sin\vartheta}$ are no longer orthogonal as is evident from (2.155). The boundary conditions may thus imply equations for linear combinations between the coefficients $a_{n,p}$ and $b_{n,p'}$. Mixed combinations will occur if parities $p$ and $p'$ are such that the $\boldsymbol{M}^{n,p}_{k\sin\vartheta}$ and $\boldsymbol{N}^{n,p}_{k\sin\vartheta}$ field components have a common factor $\cos(n\phi)$ or $\sin(n\phi)$.

Clearly, for vertical polarization, the parity of $\breve{\boldsymbol{N}}^{n,p}_{\breve{k}\sin\vartheta}$ for the scattered $\boldsymbol{E}$ field must agree with that of $\widehat{\boldsymbol{N}}^{n}_{\widehat{k}\sin\vartheta}$ just as the parity of $\breve{\boldsymbol{M}}^{n,p}_{\breve{k}\sin\vartheta}$ in the scattered $\boldsymbol{H}$ field must agree with that of $\widehat{\boldsymbol{M}}^{n}_{\hat{k}\sin\vartheta}$ (for TE polarization, the parity assumptions are evidently swapped). Hence, for TM polarization, the boundary conditions imply the equations

$$\begin{aligned}
k\boldsymbol{e}_\rho \times \widehat{\boldsymbol{N}}^{n}_{\hat{k}\sin\theta} + \breve{a}_{n,0}\boldsymbol{e}_\rho \times \breve{\boldsymbol{M}}^{n,0}_{\hat{k}\sin\vartheta} + \breve{a}_{n,1}\boldsymbol{e}_\rho \times \breve{\boldsymbol{M}}^{n,1}_{\hat{k}\sin\vartheta} + \breve{b}_{n,0}\boldsymbol{e}_\rho \times \breve{\boldsymbol{N}}^{n,0}_{\hat{k}\sin\vartheta} &= 0 \\
k\boldsymbol{e}_\rho \times \widehat{\boldsymbol{M}}^{n}_{\hat{k}\sin\theta} + \breve{b}_{n,0}\boldsymbol{e}_\rho \times \breve{\boldsymbol{M}}^{n,0}_{\hat{k}\sin\vartheta} + \breve{a}_{n,0}\boldsymbol{e}_\rho \times \breve{\boldsymbol{N}}^{n,0}_{\hat{k}\sin\vartheta} + \breve{a}_{n,1}\boldsymbol{e}_\rho \times \breve{\boldsymbol{N}}^{n,1}_{\hat{k}\sin\vartheta} &= 0
\end{aligned} \tag{2.168}$$

It is evident that for the boundary conditions to be satisfied, the $z$ dependence can only be in the form of the factor $e^{iz\hat{k}\cos\theta}$ set by the impinging field (i.e., $\theta = \vartheta$). From (2.158) is then obtained

$$\left.\begin{aligned}
i\frac{n}{a}C_n\hat{k}\cos\theta J_n\left(a\hat{k}\sin\theta\right) + \breve{a}_{n,1}\hat{k}\sin\theta H_n'\left(a\hat{k}\sin\theta\right) + i\breve{b}_{n,0}\frac{n}{a}\cos\theta H_n\left(a\hat{k}\sin\theta\right) &= 0 \\
\breve{a}_{n,0}\hat{k}\sin\theta H_n'\left(a\hat{k}\sin\theta\right) &= 0 \\
C_n\hat{k}J_n\left(a\hat{k}\sin\theta\right) + \breve{b}_{n,0}H_n\left(a\hat{k}\sin\theta\right) &= 0
\end{aligned}\right\}$$

$$\Rightarrow \breve{a}_{n,0} = \breve{a}_{n,1} = \breve{b}_{n,1} = 0;\ \breve{b}_{n,0} = -C_n\hat{k}\frac{J_n\left(a\hat{k}\sin\theta\right)}{H_n\left(a\hat{k}\sin\theta\right)} \tag{2.169}$$

There is no depolarization. In fact, the coefficients $\breve{a}_{n,p}$ and $\breve{b}_{n,p}$ agree exactly with those of normal incidence. When the procedure is repeated for horizontal polarization, it is found that the coefficients again agree with those in (2.167). These results will be extrapolated to the case of oblique incidence for cylinders with finite dielectric properties; the errors will be very marginal when the index of refraction is large, as will invariably be the case in this work.

The opportunities to numerically assess the formulas of (2.166) and (2.167) are excellent. For instance, Bessel and Hankel functions are all available as MATLAB library routines. For numerically robust computation of the derivatives in (2.166) and (2.167), use can be made of the well-known relations

$$Z_n' = \begin{cases} -Z_1 \Leftarrow n = 0 \\ \dfrac{Z_{n-1} - Z_{n+1}}{2} \Leftarrow n \geq 1 \end{cases} \tag{2.170}$$

valid for any integer order cylinder function.

Now turn to scattering from a homogeneous dielectric sphere embedded in a medium of different though homogenous dielectric properties.

For the sphere, a suitable approach exploits methods similar to the cylinder case in defining vector field particular solutions to the wave equation from the scalar particular solutions. The vector fields are obtained in a way slightly different to that of the cylinder case, whereby they will be properly aligned to the sphere geometry and allow direct implementation of the boundary conditions. In the present account, the method is merely outlined; the full details are found in [9].

The vector fields to be exploited are based on the observation that if $\chi$ is a solution to the homogenous wave equation, so is the vector field $\boldsymbol{m} = \nabla \times (\boldsymbol{x}\nabla\chi) = \nabla\chi \times \boldsymbol{x}$. This statement can be proved by direct verification in Cartesian coordinates or by the more sophisticated argument given in [9]. Anyway, it follows that with respect to the spherical coordinate particular solutions (2.133), the following two independent divergence free vector field particular solutions can be defined:

$$\begin{aligned} \boldsymbol{m}^{p}_{n,m}(\boldsymbol{x}) &= \nabla\chi^{p}_{n,m} \times \boldsymbol{x} \\ \boldsymbol{n}^{p}_{n,m}(\boldsymbol{x}) &= \frac{1}{k}\nabla \times \boldsymbol{m}^{p}_{n,m}(\boldsymbol{x}) \end{aligned} \tag{2.171}$$

The explicit expressions for the two vector fields are

$$\boldsymbol{m}^{p}_{n,m}(\boldsymbol{x}) = \begin{cases} -\dfrac{m}{\sin\vartheta}R_n(kr)P^{m}_{n}(\cos\vartheta)\sin\left(m\phi - p\dfrac{\pi}{2}\right)\boldsymbol{e}_\theta \\ -R_n(kr)\dfrac{dP^{m}_{n}(\cos\vartheta)}{d\vartheta}\cos\left(m\phi - p\dfrac{\pi}{2}\right)\boldsymbol{e}_\phi \end{cases}$$

$$\boldsymbol{n}^{p}_{n,m}(\boldsymbol{x}) = \begin{cases} \dfrac{n(n+1)}{kr}R_n(kr)P^{m}_{n}(\cos\vartheta)\cos\left(m\phi - p\dfrac{\pi}{2}\right)\boldsymbol{e}_r \\ +\dfrac{1}{kr}\dfrac{d}{dr}\left[rR_n(kr)\right]\dfrac{dP^{m}_{n}(\cos\vartheta)}{d\vartheta}\cos\left(m\phi - p\dfrac{\pi}{2}\right)\boldsymbol{e}_\theta \\ -\dfrac{m}{kr\sin\vartheta}\dfrac{d}{dr}\left[rR_n(kr)\right]P^{m}_{n}(\cos\vartheta)\sin\left(m\phi - p\dfrac{\pi}{2}\right)\boldsymbol{e}_\phi \end{cases} \tag{2.172}$$

Since $\boldsymbol{m}^{p}_{n,m}$ and $\boldsymbol{n}^{p}_{n,m}$ follow from rotation operations, any divergence free vector field general solution can be expanded in terms of $\boldsymbol{m}^{p}_{n,m}$ and $\boldsymbol{n}^{p}_{n,m}$. In particular, this holds for the electromagnetic $\boldsymbol{E}$, $\boldsymbol{H}$, and $\boldsymbol{A}$ fields. Due to the sphere symmetry, just one polarization and one incidence angle have to be considered, in contrast to the cylinder case. Assume that the impinging plane wave moves in the direction of the positive $z$-axis (i.e., $\widehat{\boldsymbol{k}} \bullet \boldsymbol{x} = \widehat{k} r \cos\vartheta$). The $\boldsymbol{E}$ field is selected to be directed along the $y$-axis with $\boldsymbol{H}$ field along the $x$-axis. Since the $\boldsymbol{E}$ and $\boldsymbol{H}$ fields are aligned with the vectors $\boldsymbol{e}_y$ and $\boldsymbol{e}_x$, their decomposition in spherical base vectors will have $\cos\vartheta$ and $\sin\vartheta$ in their expansion coefficients. Thus, while the expansion of scalar spherical wave solution (2.140) may start at $n = 0$, the expansion of the impinging electromagnetic plane wave contains no $n = 0$ component and has $n = 1$ as its lowest order term. Then with the electrical field component of unit magnitude

$$\begin{cases} \widehat{\boldsymbol{E}} = \boldsymbol{e}_x e^{i(\widehat{k} r\cos\vartheta - \omega t)} = e^{-i\omega t} \sum_{n=1}^{\infty} i^n c_n \left( \widehat{\boldsymbol{m}}_n - i \widehat{\boldsymbol{n}}_n \right) \\ \widehat{\boldsymbol{H}} = \dfrac{\widehat{k}}{\omega} \boldsymbol{e}_y e^{i(\widehat{k} r\cos\vartheta - \omega t)} = -\dfrac{1}{\omega} e^{-i\omega t} \sum_{n=1}^{\infty} i^n c_n \left( \widehat{k} \widehat{\boldsymbol{m}}_n + i \widehat{k} \widehat{\boldsymbol{n}}_n \right) \end{cases} \tag{2.173}$$

Here, $k = \hat{k}$, $\widehat{\boldsymbol{m}}_n = \boldsymbol{m}^0_{n,1}$, and , while $R_n = j_n$, in accordance with the discussion in connection to (2.134). The expansion coefficients $c_n$ in (2.173) are obtained using the explicit expressions for the fields $\boldsymbol{m}^p_{n,m}$ and $\boldsymbol{n}^p_{n,m}$ in spherical coordinates, as detailed in [9]. For the impinging field of unit amplitude, that is, $|\breve{\boldsymbol{E}}| = 1$

$$c_n = \frac{2n+1}{n(n+1)} \tag{2.174}$$

The refracted and scattered field amplitudes can now be determined. By total analogy with the cylinder case, these are sought in the form

$$\begin{cases} \breve{\boldsymbol{E}} = e^{-i\omega t} \sum_{n=1}^{\infty} i^n \left( \breve{a}_n \breve{\boldsymbol{m}}^0_{n,1} - i \breve{b}_n \breve{\boldsymbol{n}}^1_{n,1} \right) \\ \breve{\boldsymbol{H}} = -\dfrac{1}{\omega} e^{-i\omega t} \sum_{n=1}^{\infty} i^n \left( \breve{b}_n k \breve{\boldsymbol{m}}^0_{n,1} + i \breve{a}_n k \breve{\boldsymbol{n}}^1_{n,1} \right) \end{cases}$$
$$\begin{cases} \tilde{\boldsymbol{E}} = e^{-i\omega t} \sum_{n=1}^{\infty} i^n \left( \tilde{a}_n \tilde{\boldsymbol{m}}^0_{n,1} - i \tilde{b}_n \tilde{\boldsymbol{n}}^1_{n,1} \right) \\ \tilde{\boldsymbol{H}} = -\dfrac{1}{\omega} e^{-i\omega t} \sum_{n=1}^{\infty} i^n \left( \tilde{b}_n \tilde{k} \tilde{\boldsymbol{m}}^0_{n,1} + i \tilde{a}_n \tilde{k} \tilde{\boldsymbol{n}}^1_{n,1} \right) \end{cases} \tag{2.175}$$

Here, $R_n = j_n$ in $\widehat{\boldsymbol{m}}^p_{n,m}, \widehat{\boldsymbol{n}}^p_{n,m}$ and $\tilde{\boldsymbol{m}}^p_{n,m}, \tilde{\boldsymbol{n}}^p_{n,m}$, while $R_n = h_n^{(1)}$ in $\breve{\boldsymbol{m}}^p_{n,m}, \breve{\boldsymbol{n}}^p_{n,m}$. Obviously, $k$ is replaced with $\tilde{k}$ in $\tilde{\boldsymbol{m}}^p_{n,m}, \tilde{\boldsymbol{n}}^p_{n,m}$.

With $r = a$, the sphere radius, the boundary conditions are now

$$\boldsymbol{e}_r \times \left(\boldsymbol{E} + \breve{\boldsymbol{E}} - \tilde{\boldsymbol{E}}\right)_{r=a,\forall\theta,\psi} = \boldsymbol{e}_r \times \left(\boldsymbol{H} + \breve{\boldsymbol{H}} - \tilde{\boldsymbol{H}}\right)_{r=a,\forall\theta,\phi} = 0 \quad (2.176)$$

Equation (2.153) implies four scalar equations from which the coefficients $\breve{a}_n$, $\breve{b}_n$, $\tilde{a}_n$, and $\tilde{b}_n$ are obtained. Such a solution is indeed possible since (2.172) implies (with $\boldsymbol{F}(\theta,\phi)$ and $\boldsymbol{G}(\theta,\phi)$ independent fields tangential to the sphere) that

$$\begin{aligned}
\boldsymbol{e}_r \times \widehat{\boldsymbol{m}}^0_{n,1} &= j_n\left(\widehat{k}r\right)\boldsymbol{F}(\theta,\phi) \\
\boldsymbol{e}_r \times \tilde{\boldsymbol{m}}^0_{n,1} &= j_n\left(\tilde{k}r\right)\boldsymbol{F}(\theta,\phi) \\
\boldsymbol{e}_r \times \breve{\boldsymbol{m}}^0_{n,1} &= h_n^{(1)}\left(\widehat{k}r\right)\boldsymbol{F}(\theta,\phi) \\
\boldsymbol{e}_r \times \widehat{\boldsymbol{n}}^1_{n,1} &= \frac{1}{\widehat{k}}\frac{\partial}{\partial kr}\left[\widehat{k}rj_n\left(\widehat{k}r\right)\right]_{r=a} G(\theta,\phi) \\
\boldsymbol{e}_r \times \tilde{\boldsymbol{n}}^1_{n,1} &= \frac{1}{\tilde{k}}\frac{\partial}{\partial \tilde{k}r}\left[\tilde{k}rj_n\left(\tilde{k}r\right)\right]_{r=a} G(\theta,\phi) \\
\boldsymbol{e}_r \times \breve{\boldsymbol{n}}^1_{n,1} &= \frac{1}{\widehat{k}}\frac{\partial}{\partial kr}\left[\widehat{k}rh_n^{(1)}\left(\widehat{k}r\right)\right]_{r=a} G(\theta,\phi)
\end{aligned} \quad (2.177)$$

The following four equations for the four unknowns $\breve{a}_n, \breve{b}_n, \tilde{a}_n$, and $\tilde{b}_n$ are implied:

$$\begin{aligned}
&c_n j_n\left(\widehat{k}a\right) + \breve{a}_n h_n^{(1)}\left(\widehat{k}a\right) - \tilde{a}_n j_n\left(\tilde{k}a\right) = 0 \\
&c_n\left[\tau j_n(\tau)\right]'_{\tau=\widehat{k}a} + \breve{a}_n\left[\tau h_n(\tau)\right]'_{\tau=\widehat{k}a} - \tilde{a}_n\left[\tau j_n(\tau)\right]'_{\tau=\tilde{k}a} = 0 \\
&c_n \widehat{k} j_n\left(\widehat{k}a\right) + \breve{b}_n \widehat{k} h_n\left(\widehat{k}a\right) - \tilde{b}_n \tilde{k} j_n\left(\tilde{k}a\right) = 0 \\
&c_n \tilde{k}\left[\tau j_n(\tau)\right]'_{\tau=\widehat{k}a} + \breve{b}_n \tilde{k}\left[\tau h_n(\tau)\right]'_{\tau=\widehat{k}a} - \tilde{b}_n \widehat{k}\left[\tau j_n(\tau)\right]'_{\tau=\tilde{k}a} = 0
\end{aligned} \quad (2.178)$$

Only the scattered fields are of interest presently. They determine the scattering solution for the dielectric sphere. It follows that:

$$\begin{aligned}\breve{a}_n &= c_n \frac{j_n(\tilde{k}a)\left[\tau j_n(\tau)\right]'_{\tau=\hat{k}a} - j_n(\hat{k}a)\left[\tau j_n(\tau)\right]'_{\tau=\tilde{k}a}}{h_n(\hat{k}a)\left[\tau j_n(\tau)\right]'_{\tau=\tilde{k}a} - j_n(\tilde{k}a)\left[\tau h_n(\tau)\right]'_{\tau=\hat{k}a}} \\ \breve{b}_n &= c_n \frac{\tilde{\eta}^2 j_n(\tilde{k}a)\left[\tau j_n(\tau)\right]'_{\tau=\hat{k}a} - \hat{\eta}^2 j_n(\hat{k}a)\left[\tau j_n(\tau)\right]'_{\tau=\tilde{k}a}}{\hat{\eta}^2 h_n(\hat{k}a)\left[\tau j_n(\tau)\right]'_{\tau=\tilde{k}a} - \tilde{\eta}^2 j_n(\tilde{k}a)\left[\tau h_n(\tau)\right]'_{\tau=\hat{k}a}}\end{aligned} \tag{2.179}$$

For a sphere that is strongly reflecting in the sense $\tilde{\eta} >> 1/ka$, one notes that $j_n(\tilde{k}a) \approx 0$. Also requiring $\tilde{\eta} >> 1$

$$\begin{aligned}\breve{a}_n &\approx -c_n \frac{j_n(\hat{k}a)}{h_n(\hat{k}a)} \\ \breve{b}_n &\approx -c_n \frac{\left[\tau j_n(\tau)\right]'_{\tau=\hat{k}a}}{\left[\tau h_n(\tau)\right]'_{\tau=\hat{k}a}}\end{aligned} \tag{2.180}$$

Evaluating (2.179) and (2.180) numerically, expressions for spherical function derivatives are evidently required. The following rule is an implication of a standard spherical Bessel function recursion expression:

$$\begin{aligned}\tau^{n+1}R_{n-1}(\tau) = \left[\tau^{n+1}R_n(\tau)\right]' &= \tau^n\left[\tau R_n(\tau)\right]' + n\tau^{n-1}\left[\tau R_n(\tau)\right] \\ &= \tau^n\left\{\left[\tau R_n(\tau)\right]' + nR_n(\tau)\right\} \Rightarrow \left[\tau R_n(\tau)\right]' \\ &= \tau R_{n-1}(\tau) - nR_n(\tau);\ n \geq 1\end{aligned} \tag{2.181}$$

From (2.137), it is an easy task to implement (2.179) and (2.180) by means of MATLAB Bessel function library calls.

Based on the derived scattering solutions for the sphere and cylinder, expressions for spherical and cylindrical RCSs will now be derived. The analysis is straightforward for the sphere; the sphere scattered field is represented in spherical functions, which is the prerequisite for establishing RCS expressions. The cylinder solution requires one further step of development, recasting the present solution to that of spherical waves from a cylinder of finite length.

The RCS for the sphere will only be considered for the monostatic case. Since the impinging plane wave was assumed traveling along the positive $z$-axis, monostatic or backscattering corresponds to a scatted filed at the angle $\theta = \pi$. The differential area at a distance $r$ from the sphere center is $dA = r^2 d\Omega$ for solid angle $d\Omega$. The power flux through that area of the expanding spherical wave is $\breve{S} \propto r^2 c^2 \varepsilon_0 |\breve{\boldsymbol{E}}|^2 d\Omega$, so the differential cross section is $d\sigma = r^2 |\breve{\boldsymbol{E}}|^2 / |\hat{\boldsymbol{E}}|^2 d\Omega$. The RCS is then $\sigma = 4\pi r^2 |\breve{\boldsymbol{E}}|^2 / |\hat{\boldsymbol{E}}|^2$. For (2.173), $|\hat{\boldsymbol{E}}| = 1$ holds.

At sufficient distance from the sphere, the asymptotic expressions for the Hankel functions, representing the radial behavior of the solution, are

$$
\begin{aligned}
R_n(kr) = h_n^{(1)}(kr) \approx \frac{(-i)^{n+1}}{kr} e^{ikr} \Leftarrow kr >> 1 \\
\Rightarrow \frac{d}{dr}\left[ rR_n(kr) \right] \approx i(-i)^{n+1} e^{ikr} = (-i)^n e^{ikr}
\end{aligned}
\tag{2.182}
$$

As for angular behavior following from (2.172), restrict attention to the relevant case of $\widehat{\boldsymbol{m}}_n = \boldsymbol{m}_{n,1}^0$ and $\widehat{\boldsymbol{n}}_n = \boldsymbol{n}_{n,1}^1$. For sphere backscatter, the solution at $\theta = \pi$ will be considered. This must be formed as a limit value, since the polar coordinate system becomes singular at the sphere poles. In fact, from (2.134), $P_n^1(-1) = 0$, that is, the term $P_n^1(\cos\theta)/(\sin\theta)$ in (2.172) becomes a limit value of the type 0/0. According to (2.134), when $\theta \approx \pi$

$$
\frac{dP_n^1(\cos\theta)}{d\theta} = \frac{d}{d\theta}\left[ \sin\theta \frac{dP_n(\cos\theta)}{d\cos\theta} \right] \approx -\frac{dP_n(\cos\theta)}{d\cos\theta} = -\frac{P_n^1(\cos\theta)}{\sin\theta}
\tag{2.183}
$$

Also note the following well-known relation for Legendre polynomials and its implication for (2.172):

$$
\begin{aligned}
\frac{dP_n^m(\cos\theta)}{d\theta} = -\sin\theta \frac{dP_n^m(\cos\theta)}{d\cos\theta} = \frac{(n-m+1)(n+m)P_n^{m-1}(\cos\theta) - P_n^{m+1}(\cos\theta)}{2} \\
\Rightarrow -\frac{P_n^1(\cos\theta)}{\sin\theta} = \frac{dP_n^1(\cos\theta)}{d\theta} = \frac{1}{2}n(n+1)P_n^0(-1) = \frac{1}{2}n(n+1)(-1)^n
\end{aligned}
\tag{2.184}
$$

Hence, adopting (7.53) and considering $\theta = \pi$

$$
\begin{aligned}
\boldsymbol{m}_{n,1}^0(\boldsymbol{x}) = -\frac{1}{2}n(n+1)(-1)^n \frac{(-i)^{n+1}}{kr} \boldsymbol{e}_y \\
\boldsymbol{n}_{n,1}^1(\boldsymbol{x}) = -\frac{1}{2}n(n+1)(-1)^n \frac{(-i)^n}{kr} e^{ikr} \boldsymbol{e}_y
\end{aligned}
\tag{2.185}
$$

In the case $\theta \to 0$, the $\boldsymbol{e}_\theta$ and $\boldsymbol{e}_\phi$ vectors form a positively oriented coordinate system in the $x,y$-plane and the relations $\boldsymbol{e}_x = \cos\phi\boldsymbol{e}_\theta - \sin\phi\boldsymbol{e}_\phi$ and $\boldsymbol{e}_y = \sin\phi\boldsymbol{e}_\theta + \cos\phi\boldsymbol{e}_\phi$ hold. In the current case $\theta \to \pi$, the $\boldsymbol{e}_\theta$ and $\boldsymbol{e}_\phi$ vectors form a negatively oriented coordinate system in the $x,y$-plane with the relations $\boldsymbol{e}_x = -\cos\phi\boldsymbol{e}_\theta - \sin\phi\boldsymbol{e}_\phi$ and $\boldsymbol{e}_y = -\sin\phi\boldsymbol{e}_\theta + \cos\phi\boldsymbol{e}_\phi$. It follows from (2.175) that:

$$\breve{\boldsymbol{E}} = i\frac{e^{i(\hat{k}r-\omega t)}}{2\hat{k}r}\sum_{n=1}^{\infty}(-1)^n n(n+1)\boldsymbol{e}_y\left(\breve{a}_n - \breve{b}_n\right) \tag{2.186}$$

Thereby, the sphere backscattering RCS becomes

$$\sigma = 4\pi r^2\frac{\left|\breve{\boldsymbol{E}}\right|^2}{\left|\hat{\boldsymbol{E}}\right|^2} = \pi\left|\frac{1}{\hat{k}}\sum_{n=1}^{\infty}(-1)^n n(n+1)\left(\breve{a}_n - \breve{b}_n\right)\right|^2 \tag{2.187}$$

To attain some physical insight in (2.187), assume that $a\hat{k}$ is small and $\hat{k} = k$ real. In this case, (2.187) has a simple dependence on sphere radius since it is possible to restrict the spherical Bessel function to first-order approximations (referred to as the *Rayleigh regime*). The definitions of (2.138) and (2.139) imply that

$$\begin{aligned} j_n(\tau) &\approx \sqrt{\frac{\pi}{\tau}}\frac{n!}{(2n+1)!}(2\tau)^n \\ h_n(\tau) &\approx \sqrt{\frac{\pi}{\tau}}\frac{n!}{(2n+1)!}(2\tau)^n - i\sqrt{\frac{\pi}{\tau}}\frac{(2n)!}{n!}\frac{1}{\tau(2\tau)^n} \end{aligned} \tag{2.188}$$

To the first order (invoking (2.181)), for a perfectly reflecting sphere

$$\left.\begin{aligned} \tau j_0(\tau) &\approx \sqrt{\frac{\pi}{\tau}}\tau;\ \tau h_0(\tau) \approx \sqrt{\frac{\pi}{\tau}} \\ j_1(\tau) &\approx \sqrt{\frac{\pi}{\tau}}\frac{1}{3}\tau;\ h_1(\tau) \approx -i\sqrt{\frac{\pi}{\tau}}\frac{1}{\tau^2} \end{aligned}\right\}$$

$$\Rightarrow \begin{cases} \breve{a}_1 = -\dfrac{3}{2}\dfrac{j_1(ka)}{h_1(ka)} = -\dfrac{3}{2}\dfrac{\frac{1}{3}\tau}{-i\frac{1}{\tau^2}} = -i\dfrac{\tau^3}{2} \\ \\ \breve{b}_1 = -\dfrac{3}{2}\dfrac{\tau j_0(\tau) - j_1(\tau)}{\tau h_0(\tau) - h_1(\tau)} = -\dfrac{3}{2}\dfrac{\frac{2}{3}\tau}{i\frac{1}{\tau^2}} = i\tau^3 \end{cases} \tag{2.189}$$

Therefore (where $V$ is sphere volume)

$$\sigma = \frac{\pi}{k^2}\left|2\left(\breve{a}_1 - \breve{b}_1\right)\right|^2 = \frac{9\pi}{k^2}(ak)^6 = \frac{81}{16\pi}k^4V^2 \tag{2.190}$$

In the opposite, that is, high frequency, limit $ak \to \infty$ (the *optical regime*), a perfectly reflecting sphere has mirror finish. Its cross section can thus be calculated from ray optics and the law of reflection. An optical ray impinging on the sphere offset a distance $\delta$ with respect to sphere center axis will be reflected at an angle $\theta = 2\delta/a$ (with $a$ the sphere radius). The differential and RCSs can be set up as

$$\begin{aligned}
\breve{p} &= \frac{dP}{d\breve{A}} = \frac{dP}{d\left(\pi\theta^2\right)} = \frac{1}{2\pi\theta}\frac{dP}{d\theta} = \frac{a^2}{8\pi\delta}\frac{dP}{d\delta} \\
\widehat{p} &= \frac{dP}{d\widehat{A}} = \frac{dP}{d\left(\pi\delta^2\right)} = \frac{1}{2\pi\delta}\frac{dP}{d\delta} \\
\breve{p} &= \Delta\sigma\widehat{p} \Rightarrow \Delta\sigma = \frac{a^2}{4} \Rightarrow \sigma_\infty = 4\pi\Delta\sigma_\infty = \pi a^2
\end{aligned} \tag{2.191}$$

Note that according to (2.179), an enlargement of the sphere $a \to \alpha a$; $\alpha > 1$ has the same effect on the coefficients $\breve{a}_n$ and $\breve{b}_n$ as a modification $\widehat{k} \to \alpha\widehat{k}, \tilde{k} \to \alpha\tilde{k}$, that is, making the sphere dielectrically denser and embedding it in a denser medium. In both cases, $\breve{a}_n$ and $\breve{b}_n$ will on average become larger. However, the RCS expression (2.187) contains $\widehat{k}^2$ in the dominator, balancing this increase. Indeed, according to (2.191), in the optical limit, the RCS do not at all depend on the surrounding medium. The means that the RCS is not expected to be greatly affected by the embedding of the target in medium. However, according to (2.190) in the Rayleigh regime, the RCS will grow as $\eta^4$ for a given sphere size.

For the reflecting sphere in the optical limit, it is straightforward to extend (2.191) to show that the RCS agrees with total cross section. Since according to (2.191), the RCS is the geometrical cross section, the latter determines the power lost by reflection, which is in full agreement with everyday experience. As expressed by (2.189), for spheres that are small compared to wavelength, the cross section is proportional to volume squared, that is, they will be next to invisible by comparison. It is close at hand to assume that also object shape, in detail small compared to wavelength, has negligible influence on scattering behavior. A formal expression for this supposition is that given a certain wavelength, that is, wave number $k$ and target size (radius $a$), there is a number $n = N$, setting an upper limit to scattering angular frequencies, that is, details in the scattering pattern. From the viewpoint of numerical evaluation, $N$ sets the limit at which (2.187) can be truncated. To find $N$, it is observed that the Stirling formula sets a lower bound for $n!$ The following relations apply (where presently $\tau = ak$):

$$
\begin{aligned}
j_n(\tau) &\approx \sqrt{\frac{\pi}{\tau}}\frac{n!}{(2n+1)!}(2\tau)^n \approx \frac{\sqrt{\pi}\sqrt{2\pi n}\, e^{n(\ln n-1)}e^{n\ln(2\tau)-0.5\ln\tau}}{\sqrt{2\pi(2n+1)}e^{(2n+1)[\ln(2n+1)-1]}} \\
&\approx \sqrt{\frac{\pi}{2}}e^{n\ln(2\tau)-0.5\ln\tau-(n+1)\ln n-(2n+1)(\ln 2-1)} \\
&\Rightarrow \left\{\begin{aligned} & j_n(\tau)\approx 0 \Leftrightarrow n\ln(2\tau)-0.5\ln\tau-(n+1)\ln n \\ & -(2n+1)(\ln 2-1)<0 \Leftrightarrow n > e^2\frac{\tau}{2} \end{aligned}\right\}
\end{aligned}
\tag{2.192}
$$

One way of analyzing the insensitivity of small details is by considering an irregular object as made up of a basic regular object with smaller details added. If the basic object and details scatter incoherently, the expected overall cross section is the sum of the basic object and detail cross sections. As the volume ratio between the two would be very large, (2.192) indicates a strong insensitivity of the overall cross section to the details.

In view of the scattering insensitivity to details, the sphere is a good representative of small buried targets in general (and buried mines in particular), allowing computation of the RCS also in the case of lossy ground and targets with finite conductivity. In the case that a lossy media surrounds the sphere, the RCS formula must be adapted to take this fact into account. In particular, the RCS expression (2.187) refers to an impinging plane wave and the outgoing spherical wave for which amplitudes are defined with respect to the origin at the sphere center (the spherical wave is infinite at the center and amplitudes must be understood as amplitude values multiplied with the distance from the center). For the impinging and scattered fields attenuating with distance traveled, the practical RCS (to be called *surface* RCS) is with respect to the point on the sphere surface closest to the source of the impinging field. From (2.33), at this point of contact, the impinging field is stronger by the amount $e^{-ka\eta_{\mathrm{im}}}$, while the scattered field is weaker by the same amount. Since the RCS relates to power rather than field quantities, the surface RCS decreases compared to the center RCS (2.187) by the amount $e^{-2ka\eta_{\mathrm{im}}}$. In all, the attenuation is $e^{-4ka\eta_{\mathrm{im}}}$. From Figure 2.1, typical index of refraction values are $\eta = 2.2 + 0.07i$ for dry soil, dry sand, and dry clay, while $\eta = 4.4 + 0.24i$ can be considered typical for wet soil and wet sand. For wet clay, $\eta = 3.2 +$

$0.5i$ is assumed. To obtain comparative RCS values for these cases, consider a buried sphere with $\tilde{\eta} = 10^8$, a value here representing an ideal reflector. Using the MATLAB library for Bessel functions and (2.181) for computing derivatives, the result is shown in Figure 2.4.

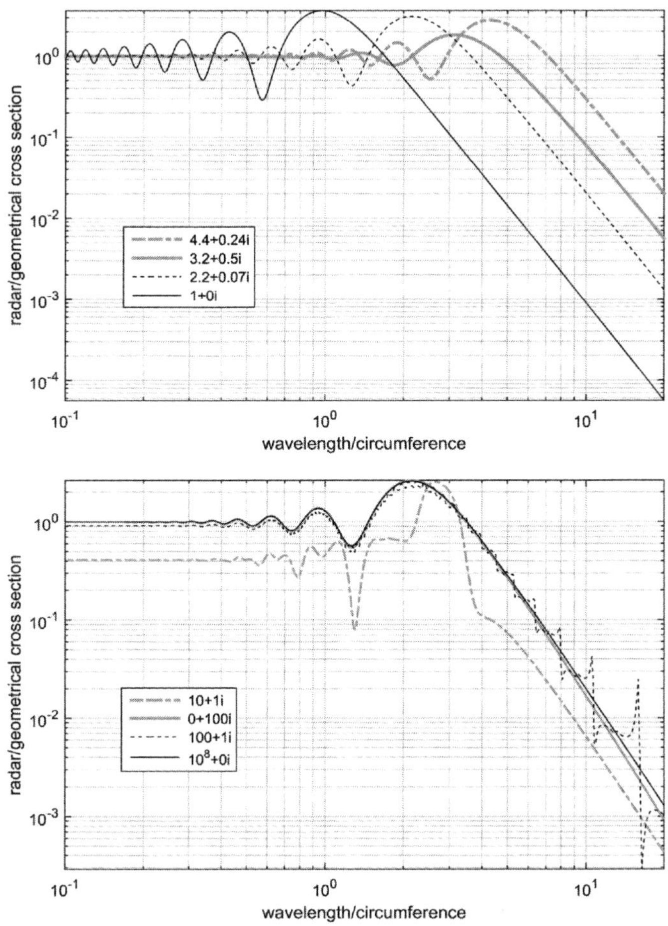


**Figure 2.4** Surface RCS for reflecting sphere, normalized by the sphere geometric cross section. Wavelength dependence normalized by sphere circumference. Upper: response for various lossy surrounding media (see Figure 2.1). Note that the onset of the weak Rayleigh scattering is pushed to longer wavelengths to a degree corresponding to the dialectic density of the medium. Lower: response for different assumptions on the sphere dielectric properties with $\eta = 2.2 + 0.07i$ being fixed for the surrounding medium. The narrow band resonance peaks are seen in the Rayleigh regime in the case $\eta = 100 + 1i$ but will also appear in the optical regime in the case of no imaginary part. They only go away with exceptionally large values for $\eta$.

Buried objects can be of metal (e.g., metal mines) or dielectric materials (e.g., underground boulders or plastic cans of buried explosives). Metals, as discussed in the end of Section 2.3, are characterized by a very large almost imaginary index of refraction, while an object becomes an ideal reflector also in the case of a purely real, very large index of refraction. When $ak$ is significantly larger than unity, some numerical stability problems are noted in computing the RCS when index of refraction is imaginary and exceptionally large ($\tilde{\eta} \geq 1000i$). For values of index of refraction that are large but not exceptionally so, the RCS exhibits occasional narrow band outlayers. The lower diagram of Figure 2.4 shows that independently of these detailed assumptions, the RCS response is essentially the same for high indices of refraction in the checked wavelength range $0.1 \leq ak \leq 10$. The diagram also includes a case of lower index of refraction $\tilde{\eta} = 10 + 1i$ representing plausible nonmetallic materials. It is noted that for this case, there is no reduction in the RCS peak value compared to an ideal reflector, but the peak has shifted to longer wavelength and become more narrow band. The mean response has dropped with about 4 dB.

The buried target case must also consider that the radar signal has to pass the air–ground interface. The ground surface affects the sphere scattering properties in various ways. For one, there are reflection losses in the ground surface, that is, part of the impinging and backscattered fields will undergo specular refection in the ground surface and be lost to the radar measurement. Secondly, the power densities of the impinging and scattered field are affected by the refraction process, thus also affecting RCS. Thirdly, reflections in the ground surface add to the field impinging the sphere, upsetting the sphere boundary conditions, making the formula (2.187) no longer exact. This third effect was previously estimated as negligible.

In (7.58), $r^2|\breve{\boldsymbol{E}}|^2$ can be substituted for radial power flux (flux per Steradian) and $|\hat{\boldsymbol{E}}|^2$ for transversal power flux (flux per surface area). Symbols crowned with "$\frown$" and "$\smile$" represent the impinging and scattered waves when below ground, while plain symbols represent propagation in air. According to the continuity equation, for the impinging plane wave, the power flux relation $\cos\widehat{\theta}_0\widehat{S} = \cos\theta\, S$ holds, with Snell's law $\widehat{\eta}_{\mathrm{re}}\sin\widehat{\theta}_0 = \sin\theta$ determining the relation between $\theta$ and $\widehat{\theta}_0$ (the angle $\widehat{\theta}_0$ of propagation for a subsurface wave in a weakly lossy medium was discussed in Section 2.4). For the radial power flux of the scattered field, the relation $\breve{S}d\breve{\Omega} = Sd\Omega$ is valid with $d\Omega = \sin\theta d\phi d\theta$ representing a solid angle segment around the angle of incidence. From Snell's law $d\breve{\theta} = \left(\cos\theta / \widehat{\eta}_{\mathrm{re}}{}^2\cos\widehat{\theta}_0\right)d\theta$, while $\phi = \breve{\phi}$ because refraction always occurs in a plane normal to the illuminated surface. It follows that $S = \left(\cos\theta / \widehat{\eta}_{\mathrm{re}}{}^2\cos\widehat{\theta}_0\right)\breve{S}$. Also incorporating the power lost by reflections in the ground (i.e., $1 - |R_{E/H}|^2$), the resulting (to be referred to as *ground*) RCS of a buried sphere follows from (2.49) that

Horizontal:

$$\sigma_g = e^{-4kD\eta_{im}/\cos\hat{\theta}_0}\left(\frac{\cos\theta}{\hat{\eta}_{re}\cos\hat{\theta}_0}\right)^2\left[1-\left|\frac{\sin(\theta-\hat{\theta})}{\sin(\theta+\hat{\theta})}\right|^2\right]\sigma$$

Vertical:

$$\sigma_g = e^{-4kD\eta_{im}/\cos\hat{\theta}_0}\left(\frac{\cos\theta}{\hat{\eta}_{re}\cos\hat{\theta}_0}\right)^2\left[1-\left|\frac{\tan(\theta-\hat{\theta})}{\tan(\theta+\hat{\theta})}\right|^2\right]\sigma \tag{2.193}$$

Here, to be certain on the validity of (2.193), $\tilde{\theta}$ is assumed complex and defined according to $\hat{\eta}\sin\hat{\theta} = \sin\theta$ with $\hat{\eta}$ complex. $D$ is the distance from the center of the buried sphere to the ground surface along the ground surface normal. The values of $\sigma$ follows from (2.187). As was discussed in Section 2.4, a difference arises in cross section between polarizations, with the vertical polarization yielding significantly more backscatter. In Figure 2.5, an example with a reflecting sphere with the center located at a depth equal to the sphere diameter is investigated for the three different soil types of Figure 2.4. As seen, the RCS shows a pronounced peak for wavelengths two to five times the sphere circumference. For a sphere of 0.2m diameter, wavelengths providing maximum response are thus between 1.5m and 3m. In the sequel in connection with statistical detection methods, the implication of the derived RCS values for the detectability of underground objects and particular mines will be discussed in detail. Just to pick one example, consider 60° incidence and vertical polarization, The upper diagram in Figure 2.5 states that in the case of the lightest type of soil, the RCS peak is $\sigma \approx 10^{-3}\text{m}^2 = -30$ dBm$^2$. Comparing to the upper diagram in Figure 2.3, valid for 1.5m, at 60° incidence and vertical polarizations, reflectivity is $\sigma_0 \approx -38$ dB for smooth flat ground. This means that with 1m resolution, the sphere response outweighs the ground surface response by 8 dB. However, if ground, the ground happens to be rougher or the soil less transparent, this positive margin will go away, as Figure 2.5 indicates. The unsafely small margin is in full accordance with practical experiences on subsurface imaging, as will be discussed in the sequel.

The discussed low-frequency scattering insensitivity to object details may be inferred to mean that scattering magnitudes are largely governed by object

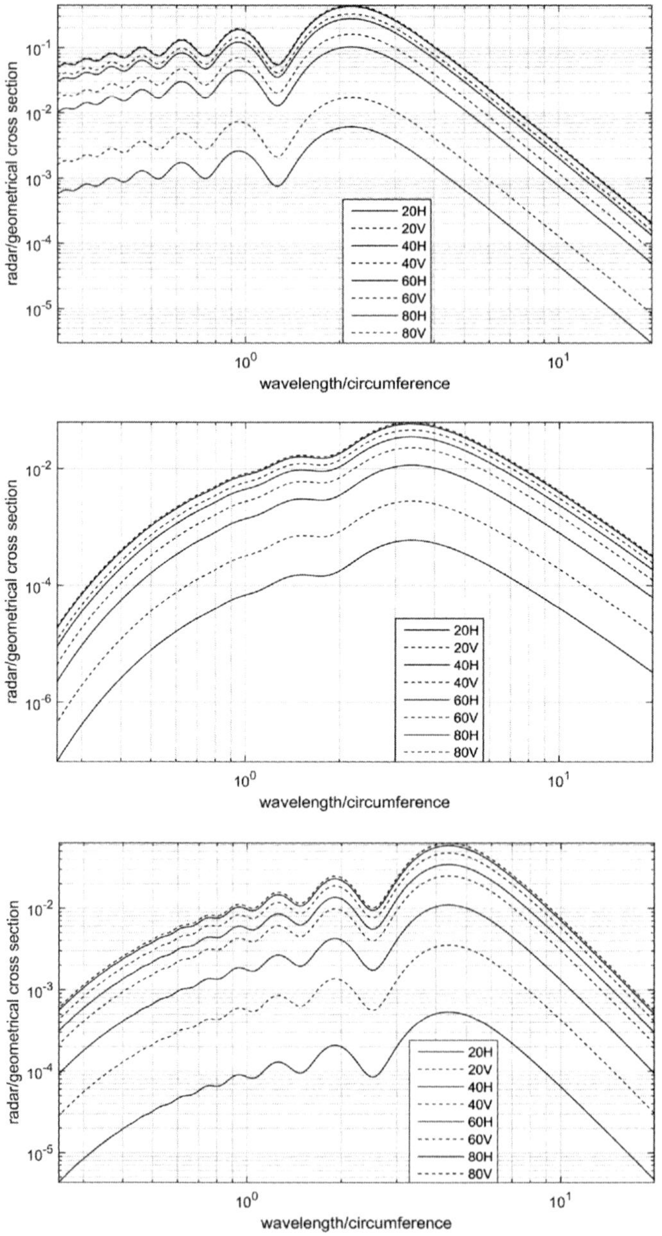


**Figure 2.5** Ground RCS for reflecting sphere, normalized by the sphere geometric cross section. Wavelength dependence normalized by sphere circumference. Response for various lossy surrounding media: $\eta = 2.2 + 0.07i$ (upmost), $\eta = 3.2 + 0.5i$ (middle), $\eta = 4.4 + 0.24i$ (bottom). The graphs refer to different angles of incidence and polarization as per the legends. They appear in order of incidence angle with the highest incidence angle the bottom graph for either polarization.

volume (as indeed is the case for the sphere) and only to a small degree by object shape. This might be true in many cases, but there are crucial exceptions. For one, for an object standing on the ground, the interaction between the object and the ground itself is decisive for target cross section. The interaction can be modeled in terms of the height of the object with respect to the ground, which largely affects the response, along with the object volume. Secondly, for cylinder–like objects, polarization plays a key role for cross section: depending on whether it is the electric or magnetic field, which is transversal to the extension of the cylinder (i.e., if polarization is TE or TM), the cylinder cross section can be widely different with TM polarization giving the larger values. In assessing the cross section of some structure composed of cylinders (trees, for instance), it may well be the parts of structure providing scattering with respect to TM polarization of the impinging field from which essentially all backscattering originates. To develop the analysis of these aspects on ground object scattering, the RCS for the cylinder will now be derived.

For a start, it is observed that the concept of (area) cross section requires a spherical wave expansion of the scattered field. Such expansions exist for any object of finite size and will be obtained by further development of the scattering solutions for the infinitely extended cylinder. In these solutions, the scattered fields are cylinder waves and fall off as $1/\sqrt{r}$, and the concept of *scattering width* applies instead of RCS. Radar scattering width is defined as $\Delta = 2\pi r|\breve{\boldsymbol{E}}|^2/|\hat{\boldsymbol{E}}|^2$ in analogous to the definition of RCS.

There is a rich and even bewildering literature on various special cases of finite cylinder scattering, with [18–20] being examples. Here, finite cylinder scattering is derived from the infinite case. The required passage from cylinder to spherical waves is achieved by averaging arguments, similar to what was done in Section 2.6, where expressions for an infinitely extend ground surface were recast into expressions for a finite ground area. Though derived different ways, the results obtained are in agreement with the cross-sectional expressions for finite cylinders in [13], also derived from the infinite case.

For the infinite cylinder, apart from the special case of normal incidence, there is no monostatic backscattering. At an oblique incidence angle $\theta$ (measured with respect to the $z$-axis), the only polar angle at which scattering occurs is $\pi - \theta$, what will be referred to as the *specular* reflection. This is a fair approximation also for finite cylinders. The basic mechanism for tree stem backscattering in low-frequency radar imaging of forest is thus that the backscattering contribution follows from the double bounce of the impinging radar signal being scattered off the tree stem at the angle $\pi - \theta$ (with $\theta$ the incidence angle for the radar imaging geometry) and then again reflected by ground specular reflection (obeying Fresnel's equations) at the incidence angle $\theta$

of the impinging radar signal. Also scattering from branches will be analyzed, based on the cylinder model, and with the ground bonce taken into account.

In order to obtain explicit expressions for cylinder scattering width and cross section, evaluate (2.157) for $p = 0$ and the asymptotic approximation of $Z_n(x) = H_n^{(1)}(x)$. When $x >> 0$

$$Z_n(x) = H_n^{(1)}(x) \approx \sqrt{\frac{2}{\pi x}} e^{i\left(x - \frac{2n+1}{4}\pi\right)} \Rightarrow Z_n'(x) \approx i\sqrt{\frac{2}{\pi x}} e^{i\left(x - \frac{2n+1}{4}\pi\right)} \approx iZ_n(x) \tag{2.194}$$

For the backscattering width and attenuation, the scattering azimuth angle will be $\phi = \pi$. From (2.155) and (2.194)

$$\begin{aligned} \boldsymbol{M}^n_{k\sin\theta} &= \frac{e^{i\pi/4}}{i^n}(-1)^{n+1}\sqrt{\frac{2k\sin\theta}{\pi\rho}} e^{ik(\rho\sin\theta + z\cos\theta)} \boldsymbol{e}_\phi \\ \mathbf{N}^n_{k\sin\vartheta} &= \frac{e^{-i\pi/4}}{i^n}(-1)^{n+1}\sqrt{\frac{2k\sin\theta}{\pi\rho}} e^{ik(\rho\sin\theta + z\cos\theta)}\left(\cos\theta \boldsymbol{e}_\rho + \sin\theta \boldsymbol{e}_z\right) \end{aligned} \tag{2.195}$$

Using (2.158), the scattered fields become

$$\begin{aligned} &\text{TM polarization:} \\ &\breve{\boldsymbol{E}} = e^{-i\pi/4}\sqrt{\frac{2k\sin\theta}{\pi\rho}} e^{ik(\rho\sin\theta + z\cos\theta)}\left(\cos\theta \boldsymbol{e}_\rho + \sin\theta \boldsymbol{e}_z\right)\sum_{n=0}^{\infty}(-1)^n \breve{b}_n \\ &\text{TE polarization:} \\ &\breve{\boldsymbol{E}} = e^{i\pi/4}\sqrt{\frac{2k\sin\theta}{\pi\rho}} e^{ik(\rho\sin\theta + z\cos\theta)} \boldsymbol{e}_\phi \sum_{n=0}^{\infty}(-1)^n \breve{a}_n \end{aligned} \tag{2.196}$$

The scattering width becomes, recalling (2.160) and setting $\rho = r\sin\theta$

$$\begin{aligned} &\text{TM polarization:} \quad \Delta = 2\pi r \frac{\left|\breve{\mathbf{E}}\right|^2}{\left|\widehat{\mathbf{E}}\right|^2} = \frac{4}{k^3\sin\theta}\left|\sum_{n=0}^{\infty}(-1)^n \breve{b}_n\right|^2 \\ &\text{TE polarization:} \quad \Delta = 2\pi r \frac{\left|\breve{\mathbf{E}}\right|^2}{\left|\widehat{\mathbf{E}}\right|^2} = \frac{4}{k^3\sin\theta}\left|\sum_{n=0}^{\infty}(-1)^n \breve{a}_n\right|^2 \end{aligned} \tag{2.197}$$

Note (see (6.35)) that the dimensions of $\breve{a}_n$ and $\breve{b}_n$ is $[\widehat{k}]$, that is, length$^{-1}$. It follows that the dimension of scattering width is length, as is required.

Presently, in (2.197), wavenumbers $k$ will refer to the atmosphere, whereas a single index of refraction $\eta$ is required in order to characterize propagation conditions inside the cylinder.

Referring to Section 2.6, the finite cylinder case can be modeled using (2.112), if substituting the $x,y$-plane in (2.112) with the $y,z$-plane at $x = a$ and (since the angle is measured with respect to the $z$-axis) by substituting $\theta \rightarrow \pi/2 - \theta$. Also recall the reversal of sign conventions for time and spatial exponents. In the present application, the flat surface $X(Q)$ is selected as a strip tangential to the cylinder and extending infinitely in the $y$ direction from $z = 0$ to $z = \Delta L$. The integrand in (2.112) represents the scattered electric field, restricted to this flat surface. So, being restricted, the field is denoted by $\breve{\boldsymbol{E}}_{x=a}$. The very definition of specular reflection is that $\breve{\boldsymbol{E}}_{x=a} = \hat{\boldsymbol{E}}e^{-ikz\cos\theta}$ is independent of $z$, making the integrand grow large in the direction of specular reflection. Taking these various considerations into account (2.112), at the specular angle $\vartheta = \pi - \theta$, takes the form

$$\breve{\boldsymbol{E}}_L(\boldsymbol{x}) = \frac{ik\sin\theta e^{ikr}}{4\pi r}\int_{y=-\infty}^{\infty}\int_{z=-L/2}^{L/2}\hat{\boldsymbol{E}}e^{-ikz\cos\theta}dydz = \frac{ikL\sin\theta e^{ikr}}{4\pi r}\boldsymbol{K} \tag{2.198}$$

By means of (2.196), the constant $\boldsymbol{K}$ can be estimated explicitly by obtaining the scattered field (through (2.157) and (2.158) and (2.166)) for the plane $x = a$ and performing integration with respect to $y$. A much more convenient route will be taken here.

Assume that the original infinite cylinder is subdivided into a set of sections, all of length $L$. Let $Q$ be a fixed space point with cylindrical coordinates $(\rho, 0, mL)$, for some integer $m$, while $P_nQ = [\rho, 0, (m-n)L]$ gives the vector to $Q$ from the $n$th cylinder. The overall scattered $\boldsymbol{E}$ field at $Q$ is

$$\breve{E}(Q) = \sum_{n=-\infty}^{\infty}\breve{E}_L\left(P_nQ\right) \tag{2.199}$$

If the fields $\breve{\boldsymbol{E}}_L(P_nQ)$ were axial waves, that is, attenuating as $1/\sqrt{r}$ and spreading only axially, $\breve{\boldsymbol{E}}(Q)$ would obtain its contribution only from the one term $m = n$. However, since $\breve{\boldsymbol{E}}_L(P_nQ)$ are spherical waves, attenuating as $1/r$, they spread in solid angle sectors. Therefore, also terms in the vicinity of $m = n$ will contribute to $\breve{\boldsymbol{E}}(Q)$. Regard, as before, scattering in the case of the plane wave impinging with angle $\theta$ in the $x,z$-plane, which implies that $\boldsymbol{k} \bullet \boldsymbol{x} = k\rho\sin\theta + kz\sin\theta$. Suitably selecting the origin along the $z$-axis, $\boldsymbol{x}_0 = P_0Q$ can be set to be coaligned with the propagation direction of the impinging field (i.e., $\boldsymbol{k}$).

Study scattering from cylinders neighboring $n = 0$, in which case the vectors $\boldsymbol{x}_n = P_nQ$ will deviate from $\boldsymbol{x}_0 = P_0Q$ by some angle $\Delta\theta$. To estimate (2.200), consider the difference (to second order)

$$kr_n - \left[k\rho_0 \sin\theta + knL\cos\theta\right] = kr_n - \boldsymbol{k} \bullet \boldsymbol{x}_n = kr_n(1 - \cos\Delta\theta) \approx -kr_0 \frac{\Delta\theta^2}{2} \tag{2.200}$$

The angle $\Delta\theta$ follows from the sine theorem. To the second order in $\Delta\theta$

$$\frac{\sin\Delta\theta}{nL} = \frac{\sin\theta}{r_n} \Rightarrow \Delta\theta \approx \frac{nL\sin\theta}{r_0} \tag{2.201}$$

Combining (2.198) and (2.199)

$$\breve{E}(Q) \approx \frac{ikL\sin\theta}{4\pi r_0} e^{ik[k\rho_0 \sin\theta + nL\cos\theta]} \boldsymbol{K} \sum_{n=-\infty}^{\infty} e^{i\frac{k}{2r_0}n^2L^2\sin^2\theta} \tag{2.202}$$

As is well known, assuming exponential terms with a quadratic increase in phase, only those with a phase value $\leq\pi/2$ contribute with any significance. For phase values larger than $\pi/2$, the exponential in the nominator oscillates so rapidly that the corresponding summation terms mutually cancel. Thus the summation in (2.203) can be terminated at values $n = \pm N$, where

$$\frac{k}{2r_0} N^2L^2\sin^2\theta = \frac{\pi}{2} \Rightarrow N = \pm\frac{1}{L\sin\theta}\sqrt{\frac{\pi r_0}{k}} \tag{2.203}$$

Within the remaining summation, the effective phase value will be approximately constant, in fact $\pi/4$ on average, while the modulus of the summation (in view of the unity module and $\pi$ radians total phase variation of the summation terms) will be close to $1/\sqrt{2}$ . Hence, as a fair approximation

$$\begin{aligned} \breve{E}(Q) &\approx \frac{ikL\sin\theta}{4\pi r_0} e^{ik[k\rho_0 \sin\theta + nL\cos\theta]} \boldsymbol{K} \frac{e^{i\pi/4}}{\sqrt{2}} \frac{2}{L\sin\theta}\sqrt{\frac{\pi r_0}{k}} \\ &= \frac{i}{2}\sqrt{\frac{k}{2\pi r_0}} e^{i\pi/4} e^{ik[k\rho_0 \sin\theta + nL\cos\theta]} \boldsymbol{K} \end{aligned} \tag{2.204}$$

In effect, the summation in (2.200) implies range attenuation according $1/\sqrt{r}$, that is, (2.200) is a cylindrical wave. Comparing with (2.196) for $\theta = \pi/2$, the constant $\boldsymbol{K}$ can be determined. As a result

$$\begin{aligned}
&\text{TM polarization:} \quad \boldsymbol{K} = -4\left(\cos\theta \boldsymbol{e}_\rho + \sin\theta \boldsymbol{e}_z\right)\sum_{n=0}^{\infty}(-1)^n \breve{b}_n \\
&\text{TE polarization:} \quad \boldsymbol{K} = -i4\boldsymbol{e}_\phi \sum_{n=0}^{\infty}(-1)^n \breve{a}_n
\end{aligned} \tag{2.205}$$

Both the spherical scattered field and the RCS of a finite cylinder thereby follows. In particular, the RCS becomes

$$\begin{aligned}
&\text{TM polarization:} \\
&\sigma = 4\pi r^2 \frac{\left|\breve{E}\right|^2}{\left|\widehat{E}\right|^2} = \frac{4L^2}{\pi k^2}\left|\sum_{n=0}^{\infty}(-1)^n \breve{b}_n\right|^2 = \frac{L^2 k \sin\theta}{\pi}\Delta \\
&\text{TE polarization:} \\
&\sigma = 4\pi r^2 \frac{\left|\breve{E}\right|^2}{\left|\widehat{E}\right|^2} = \frac{4L^2}{\pi k^2}\left|\sum_{n=0}^{\infty}(-1)^n \breve{a}_n\right|^2 = \frac{L^2 k \sin\theta}{\pi}\Delta
\end{aligned} \tag{2.206}$$

It is of interest to observe that the cross section is independent of the incidence angle $\theta$ of the impinging field (in contrast, the scattering width $\Delta$ becomes divergent as the impinging field becomes coaxial with the cylinder). The result is not as strange as it first might seem, keeping in mind that it is the specular reflection cross section and not the backscattering cross section, which has been determined. In fact, it is often assumed, as a simplification, that the bistatic cross section approximately equals monostatic cross section with the bisecting angle as incidence angle. The latter is noted to always be normal to the cylinder for specular scattering, so in the present case, this assumption has been demonstrated to be fully correct.

The appearance of the $L$-squared factor in (2.206) may be understood as double dependence on $L$; for one, the power of the impinging field scattered by the cylinder is proportional to $L$, but also the focusing of the scattered field around the specular reflection angle is proportional to $L$. The beam width of the scattered field follows from (2.198) just by replacing the specular angle $\vartheta = \pi - \theta$ with an arbitrary scattering angle $\vartheta$. Then

$$\breve{\boldsymbol{E}}_L(r,\vartheta) = \frac{ikL\sin\theta e^{ikr}}{4\pi r}\boldsymbol{K}\int_{z=-L/2}^{L/2} e^{ikz(\cos\theta+\cos\vartheta)}\,dz \tag{2.207}$$

Performing the Fourier transform, a sinc function is obtained. The result compared to specular scattering is

$$\breve{E}_L(r,\vartheta) = \Phi(\theta,\vartheta)\breve{E}_L(r,\pi-\theta);\ \Phi(\theta,\vartheta) = \frac{\sin\left(\dfrac{kL(\cos\theta+\cos\vartheta)}{2}\right)}{\dfrac{kL(\cos\theta+\cos\vartheta)}{2}} \tag{2.208}$$

As for the scattering beam width, define the beam angular limits $\vartheta = \pi - \theta \pm \Delta\theta_\pm$ as the first field zeros, at each side of the specular direction. The following exact result is obtained:

$$\begin{aligned} \Delta\theta_\pm &= \sin^{-1}\left[\cos\theta\sqrt{1-\left(\cos\theta-\frac{\lambda}{L}\right)^2} \pm \sin\theta\left(\cos\theta-\frac{\lambda}{L}\right)\right] \\ &\Rightarrow \left\{\theta = \frac{\pi}{2} \Rightarrow \Delta\theta_\pm = \sin^{-1}\left(\frac{\lambda}{L}\right)\right\} \end{aligned} \tag{2.209}$$

It is noted that unless wavelength $\lambda = 2\pi/k \le L$, no values $\Delta\theta_\pm$ can be determined from (2.209). On the other hand, since the scattered field is a vector field tangential to the sphere, it must as such have at least two zeros. The apparent contradiction seems to reside in that the finite cylinder solution derived, being based on (2.112), just considers scattering in the half space bounded by the plane tangential to the cylinder and containing either the $\boldsymbol{E}$ or $\boldsymbol{H}$ field, depending on polarization. It thus excludes the field conditions in the forward scattering half space. A related observation is that the current scattering solution assumes that the contribution to the scattered field from a finite cylinder mantle surface is the same as that from the corresponding finite part of the mantle of surface of an infinite cylinder. In particular, it excludes (an indeed make no assumption on) the scattering contributions from the cylinder ends and will therefore not be reliable for directions approaching the $z$-axis.

The finite cylinder solution is thus applicable to cylinders that are "slender" in some sense. The applicability conditions for cylinders of different lengths and widths, in cases of different wavelengths, will be returned to.

Just as was done for the sphere, further insight into (2.187) is achieved by assuming that $ak$ is small. For integer order Bessel functions presently adopted, the following series expansions are valid:

$$J_n(\tau) \approx \frac{1}{n!}\left(\frac{\tau}{2}\right)^n; \quad \frac{1}{H_n(\tau)} \approx \begin{cases} \dfrac{1}{1 - i\dfrac{2}{\pi}\ln(0.890535\,\tau)} & \Leftarrow n = 0 \\ i\dfrac{\pi}{(n-1)!}\left(\dfrac{\tau}{2}\right)^n & \Leftarrow n \geq 1 \end{cases} \tag{2.210}$$

Comparing (2.188) and (2.210), it is seen that arguments similar to those in (2.192) also pertain to integer order Bessel functions. Thus, the series in (2.196) could be truncated at essentially the same value $n = N$, which was found in (2.192). For small $ak$, to find the dominant scattering terms, the expansion in (2.196) up to second order must be considered. Also invoking (2.171)

$$\left.\begin{array}{r} J_0(\tau) \approx 1;\ H_0(\tau) \approx 1 - i\dfrac{2}{\pi}\ln(0.890535\,\tau) \\ J_1(ak) \approx \dfrac{\tau}{2};\ H_1(\tau) = \dfrac{2}{i\pi\tau} \\ J_2(\tau) \approx \dfrac{1}{2}\left(\dfrac{\tau}{2}\right)^2;\ H_2(\tau) \approx \dfrac{1}{i\pi}\left(\dfrac{2}{\tau}\right)^2 \end{array}\right\}$$

$$\Rightarrow \begin{cases} -\breve{a}_0 = k\dfrac{J_0'(ak)}{H_0'(a\widehat{k})} = k\dfrac{-J_1(ak)}{-H_1(a\widehat{k})} \approx i\pi k\left(\dfrac{ak}{2}\right)^2 \\ -\breve{a}_1 = 2k\dfrac{J_1'(ak)}{H_1'(a\widehat{k})} = 2k\dfrac{J_0(a\widehat{k})}{-H_2(a\widehat{k})} \approx -i2\pi k\left(\dfrac{ak}{2}\right)^2 \\ -\breve{b}_0 = k\dfrac{J_0(ak)}{H_0(a\widehat{k})} \approx \dfrac{i\pi k/2}{\ln(0.890535\,ak) + i\dfrac{\pi}{2}} \end{cases} \tag{2.211}$$

Insertion of these parameters into scattering width formulas (2.197) and (2.198) yield to lowest order

$$\text{TM polarization: } \Delta = \frac{4}{k^3\sin^2\theta}\left|\breve{b}_{0,0}\right|^2 = \frac{\pi^2}{k\sin^2\theta}\frac{1}{\ln^2(0.890535\,ak) + \dfrac{\pi^2}{4}}$$

$$\text{TE polarization: } \Delta = \frac{4}{k^3\sin^2\theta}\left|\breve{a}_{0,0} - \breve{a}_{1,0}\right|^2 = \frac{\pi^2}{k\sin^2\theta}\frac{9}{4}(ak)^4$$

(2.212)

For the RCS

$$\begin{aligned} &\text{TM polarization: } \sigma = \pi \frac{L^2}{\ln^2(0.890535\,ak) + \dfrac{\pi^2}{4}} \\ &\text{TE polarization: } \sigma = \frac{9}{4}\pi L^2 (ak)^4 = \frac{9}{4}\frac{k^4}{\pi} V^2 \end{aligned} \tag{2.213}$$

Evidently, the scattering width and cross section *increase* with wavelength $\lambda = 2\pi/k$ for TM polarization, while they *decrease* with $\lambda$ for TE polarization. However, the rate of change of the logarithm is much slower than for any power function, so in effect TM polarization cross section will be relatively insensitive to frequency (see the graph for a perfect reflector in Figure 2.8).

At the high frequency limit $ak \to \infty$, the assumption of a perfectly reflecting cylinder implies that it must have the true circular shape, up to and including a surface mirror finish. In analogous to the sphere, an optical ray impinging on the cylinder surface along the $x$-axis but offset a distance $\Delta y$ with respect to the $z$-axis (the cylinder axis) will be reflected at an angle $\Delta\theta = 2\Delta y/a$ according to the law of reflection. The power ratio between the impinging field within a segment of width $\Delta y$ and the corresponding scattered field (which has an angular extension of $\Delta\theta = 2\Delta y/a$) becomes $\Delta y/\Delta\theta = a/2$. The scattering width in the high frequency (optical) limit is therefore $\Delta_\infty = \pi a$.

Very commonly ground objects are characterized by vertical "walls," whereas surrounding ground is approximately horizontal. This situation prevails not only for tree stems, but also for buildings and vehicles and even for humans standing erect. In these cases, the combination of specular scatter from these walls and the ground produces backscatter enabling radar detection of the objects. Analyzing the scattering process for these situations, standing cylinders are an overall good generic model for objects residing on the ground surface with a circumference smaller or about equal to wavelength. In particular, when the target circumference is much smaller than the wavelength, the Rayleigh limit just considered provides accurate results. As the object grows thicker, more terms in the expansion (2.206) need to be included. The accuracy of the result will then, to an increasing degree, depend on how close the target shape resembles the cylinder. If only requesting how target detectability relates to target size and independently shape details, it seems that the cylinder is the "average" shape to use. To obtain the gross expectation values for a certain type of target, the scattering model is suitably a cylinder equaling the target as regards horizontal cross section area and height.

Since the radar response from objects is so insensitive to their shape, the detection of man-made objects (henceforward targets) located in a forest environment will depend on little else than how the cross-sectional responses compare between the targets and the forest. The detection possibilities will thus depend on:

1. Object RCS;
2. Tree cross section:
    a. Tree stem response;
    b. Branch response;
3. The objects will also be screened from the radar by the trees. To what extent depends on:
    a. Stem screening;
    b. Branch screening.

These different parameters all depend on incidence angle, frequency, and polarization, which should be selected so that 1) object RCS is as high as possible, 2) the difference between object and tree cross section is in favor, and 3) vegetation screening has minimum influence. These various aspects on detectability will now be discussed. There is, however, another very important aspect, namely, that object response should be as stable as possible. By "stable" it is meant that the same RCS values for a tree distribution will be obtained if the radar measurement is repeated at a later time. The measurement must thus not be sensitive to the imaging geometry of the radar or everyday changes to the vegetation itself (like the action wind and humidity). The reason, as will be plain in the sequel, is that the tree RCS values will be large and in no way can be reduced to levels sufficiently small for targets to be detected just by the strength of their response. A solution is to analyze pairs of images in which objects that have appeared or disappeared between the registrations can be detected as changes given the stability in imaging the ground features that have not changed. Targets are thus distinguished not only by their RCS but also in that they are temporary while the background is permanent.

Understanding the ground objects as cylinders of finite height standing erect on a flat ground surface, the first topic to be analyzed is RCS frequency, polarization, and incidence angle dependence. As cylinder height may be of wavelength order or smaller, not only specular scattering will affect the scattering characteristics. According to (2.208), there will be additional diffuse scattering components producing backscattering without the intermediate specular reflection. This basic model of meter-wave radar phenomenology has been gradually developed over the years of collecting such data, in Sweden with the CARABAS range of SAR systems, mentioned in Section 2.1. As shown in [21]–[24], the model exhibits very good fit for tree scattering for which (as will be analyzed) the tree stem is the major source of backscattering.

According to the model, in the overall scattering process for a cylinder ground target, the following four scattering components can be discerned.

1. The impinging field hits the cylinder and a field component back-scatters diffusely.
2. The impinging field hits the cylinder and a component bounces to the ground by specular scattering and a component of that field scatters specularly in the direction of the impinging field.
3. A part of the impinging field hitting the ground bounces by specular reflection hitting the cylinder, whereupon it specularly scatters back in the direction of the impinging field.
4. A part of the impinging field hitting the ground bounces by specular reflection to hit the cylinder, whereupon it diffusely scatters back to the ground, with a component scattered specularly in the direction of the impinging field.

The four components can be calculated by combining Fresnel's reflection equation (2.49) with (2.208). Note that the phase dependence of the four field components, determined by their different path lengths, plays a crucial role for the overall scattering behavior. From the geometry of Figure 2.6, it follows for each component that:

1. Path length: $\Delta r_1 = \dfrac{L}{\cos\theta}\sin\left(2\theta - \dfrac{\pi}{2}\right) = -\dfrac{L\cos(2\theta)}{\cos\theta}$.
2. Path length: $\Delta r_2 = \dfrac{L}{2\cos\theta} + \dfrac{L}{2\cos\theta}\sin\left(2\theta - \dfrac{\pi}{2}\right) = \dfrac{L}{2\cos\theta} - \dfrac{L\cos(2\theta)}{2\cos\theta}$.
3. Path length: $\Delta r_3 = \Delta r_2$.
4. Path length: $\Delta r_4 = \dfrac{L}{\cos\theta}$.

Here, $R$ is the reflection coefficient for the particular polarization considered. The overall backscattered field magnitude can thereby be estimated as

$$\left|\breve{\boldsymbol{E}}_{\text{tot}}(r,\theta)\right| = \left|\breve{\boldsymbol{E}}_{L}(r,\theta)\right| \left|\Phi(\theta,\theta)e^{-ik\frac{L}{2\cos\theta}} + 2R + \Phi(\theta,\theta)R^2 e^{ik\frac{L[1+\cos(2\theta)]}{2\cos\theta}}\right| \quad (2.214)$$

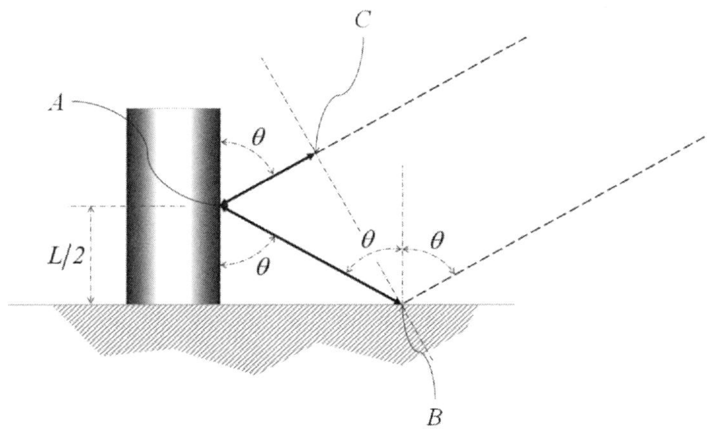


**Figure 2.6** Geometry of ground fading for vertical cylinder.

Note that if the ground is inclined at some angle $\delta\vartheta$, the reflection law implies that the factors $\Phi(\theta,\theta)$ are substituted by factors $\Phi(\theta,\theta + \delta\vartheta/2)$, whereas a factor $\Phi(\theta,\pi - \theta + \delta\vartheta/2)$ appears in the middle term. However, as long as $\delta\vartheta$ is small and cylinder height sufficiently short, the effect of an inclined ground plane remains insignificant. For $kL\cos\theta >> 1$, there is no diffuse backscattering, that is, $\Phi(\theta,\theta) \approx 0$ according to (2.208). The first and last terms are then insignificant and $|\breve{\boldsymbol{E}}_{\text{tot}}(r,\theta)| \approx 2|R||\breve{\boldsymbol{E}}_L(r,\theta)|$.

The dependence of $R$ on polarization and incidence angle will make (2.214) quite varied with respect to these parameters, as is illustrated in Figure 2.7. According to (2.49), at near normal incidence, the $\boldsymbol{E}$ field changes sign, while the $\boldsymbol{H}$ field does not, resulting in a Poynting vector determining an essentially reversed direction of propagation for the reflected field. At high incidence angles, that is, beyond the Brewster angle, both the $\boldsymbol{E}$ and $\boldsymbol{H}$ fields change sign according to (2.49). This is consistent with the directions of propagation of the impinging and reflected fields being essentially the same. Since the magnitude of $R$ also approaches unity as the incidence angle approaches 90°, (2.214) implies that as the direct and reflected backscattered field component from an object located on the ground eventually cancel.

As $R = 0$ at the Brewster angle for vertical polarization, the scattered field obtains a dip; the sharper the dip, the less diffuse the scattering becomes, that is, the higher the cylinder stands from ground. For horizontal polarization, the response depends on incidence angle smoothly, though at large incidence angles, the response from shallow objects will be weak, as per Figure 2.7.

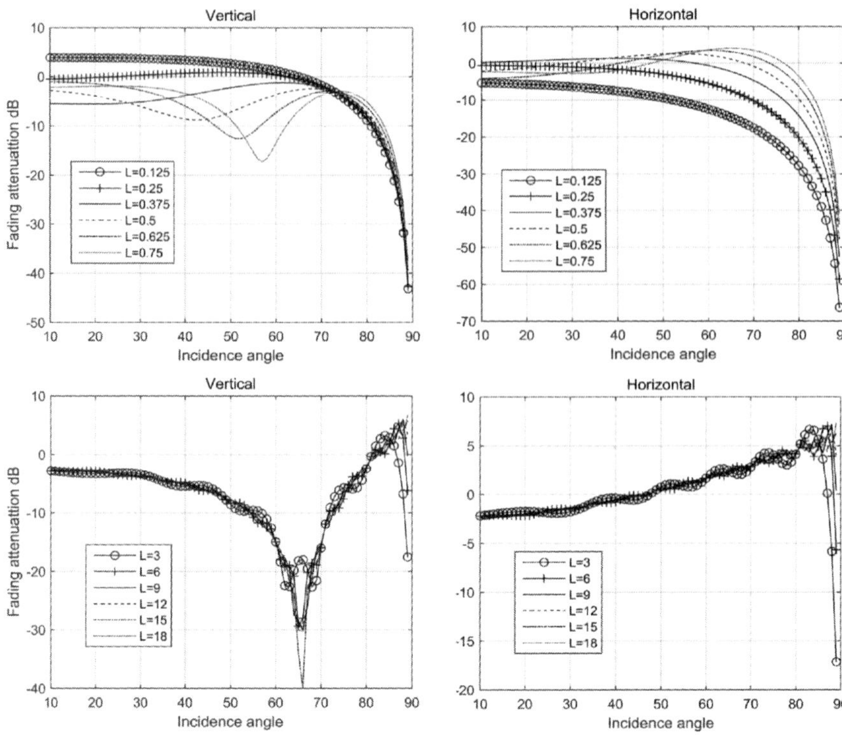


**Figure 2.7** Ground fading for shallow and high objects.

These different mechanisms are very evident and important in actual low-frequency SAR images. Horizontally polarized images will generally be very "clean." Shallow ground objects do not provide any radar response and thus do not show up in the image, with the consequence that those ground objects that satisfy the criterion of sufficient height stand out against a nearly zero clutter background. In contrast, the vertically polarized images are generally cluttered. For horizontal polarization, to avoid severe signal losses due to ground fading, Figure 2.7 indicates that incidence angle must be selected to $\theta \leq 60°$ and $\theta \leq 70°$ for ground object elevations of $L = 0.25\lambda$ and $L = 0.375\lambda$. The attenuation is then limited to at most a few decibels. For horizontal polarization, the requirement on incidence angle limits has a consequence that SAR surveillance range must be no larger than around twice the operating altitude of the radar aircraft. While the index of refraction in Figure 2.7 is assumed to be $\eta = 2.2$ (light soil as per Figure 2.1), this conclusion for surveillance range remains valid for all realistic values of $\eta$.

In the situation of a target surrounded by trees, the fact that typical targets are relatively shallow makes the ground surface interaction highly critical in the competition between the target and tree scattering responses. Selecting a definite example is probably the best way to illustrate how different parameters interact, together determining the possibilities for actually detecting the target. The target will be assumed perfectly reflecting. Dielectric constants for wood are fetched from [25]. They vary with frequency, with high values at low frequencies, and reduce when frequency increases. Above 1 MHz, they apparently stabilize with 50 MHz being the highest frequency measured in the report. Since wood is an anisotropic material, dielectric constants are given as tensorial quantities, that is, they depend on grain versus field directions. They also depend on temperature and will (just as is the case with water [4]) radically decrease at freezing temperatures. For soaked wood, here assumed to correspond to wood in living trees, dielectric $\varepsilon_{\mathrm{re}}$ constants at moderate temperatures, depending on wood type and direction, are given as varying in an interval 36 to 61 (indices of refraction approximately in the interval 6 to 8) with loss angle $\varepsilon_{\mathrm{im}}/\varepsilon_{\mathrm{re}}$ in the interval 0.28 to 0.56. Thus, numerical evaluations will here be made for $\eta_{\mathrm{im}}/\eta_{\mathrm{re}}$ in the interval 0.15 to 0.3 (see the discussion at the end of Section 2.3). Determining the object and stem scattering responses, the Bessel functions involved are available by MATLAB library calls, with their derivatives determined by (2.170).

The scattering widths (2.197), normalized by division with the optical scattering width $\Delta_\infty = \pi a$, are illustrated in Figure 2.8. The abscissa is wavelength $\lambda$ divided by radius $a$, and the trend toward the optical scattering width for short wavelengths is obvious. As seen (and which is expected), there is no difference between horizontal and vertical polarizations in the optical limit. It is less obvious that scattering width agrees well with the optical limit to around $\lambda/a \approx 15$. At larger wavelengths, the TM and TE scattering widths thereafter rapidly diverge. A suitable wavelength selection for reducing tree stem backscatter is consequently TE polarization with $\lambda > 15a$, where $a$ is the tree stem radius. For instance, for a stem radius of $\alpha \approx 0.2$m, the wavelength must be $\lambda > 3$m. Selecting the wavelength to be $\lambda \approx 6$m, the TE stem response has dropped off by 18 dB from optical backscattering and is more than 20 dB below the TM stem response. At this wavelength, targets with a radius $> \lambda/15 = 0.4$m have essentially their full optical scattering width, which favors their response in the radar image, compared to the stem responses.

The situation that scattering widths and thus RCS are the same for horizontal and vertical polarizations will be taken as indicative that the scattered field in (2.198) is made up of multiple local contributions, each emanating from a region of the mantle surface of linear extension not larger than cylinder

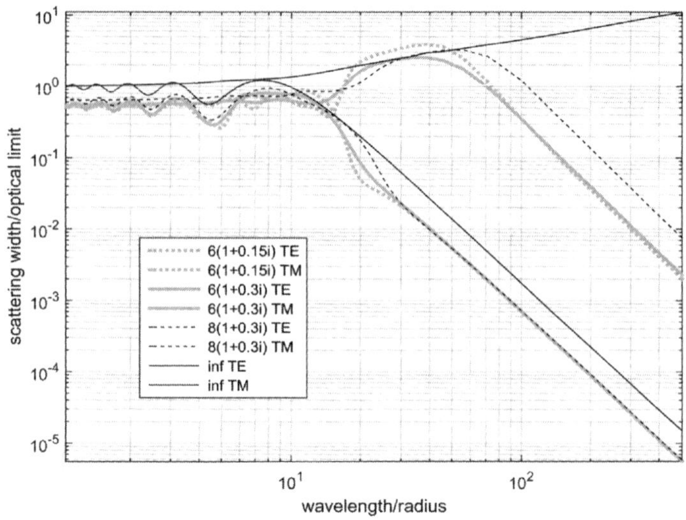


**Figure 2.8** Cylinder scattering widths as a function of $\lambda/a$, at normal incidence, normalized by division with the optical limit $\pi a$ for a perfectly reflecting cylinder ($\eta$ = 10,000 is required for graph to not deviate from the limit value), and homogenous cylinders with indices of refraction as labeled. Upper: TM. Lower: TE polarization.

width (hence, the contributions will not sense the difference between length and width of the cylinder). In other words, as long as wavelength is less than about 15 times cylinder radius, the finite cylinder approximation for RCS may be assumed valid down to cylinders with length equal to their width. For larger wavelengths, the RCS will be depending on polarization by (2.206). This difference in polarization response will then not be applicable to a cylinder of width and length equal, for which, as said, a (nearly) polarization neutral RCS is expected. Note, however, that the graphs in Figure 2.8 show a close similarity between the RCS at TM polarization and the RCS at TE polarization, with cylinder radius increased. Indications are thus that by increasing cylinder length by this amount, the TM response will no longer be truncated by a too short cylinder. In Figure 2.8, the cylinder radius increase required for TM response to mimic TE response for wood is about eight times. Thus, it seems natural to assume that cylinders made of wood must exhibit cylinder lengths four times the width to obey the finite cylinder RCS formula.

If objects to be detected are characterized in being wider than competing tree stems, the RCS difference is optimized simply by the use of horizontal polarization and the selection of appropriate frequency. Counteracting this effect is the ground fading of Figure 2.7, suppressing the response from shallow objects compared to that from objects raised. Assume a typical mid-size tree stem, represented by a vertical cylinder with height $L \approx$ 15m, radius $\alpha \approx$ 0.2m, and $\eta$ = 6(1 + 0.3$i$). Let a perfectly reflecting vertical cylinder, with the height $L \approx$ 15m and radius $\alpha \approx$ 0.6m, represent a target to be detected by the radar measurement. To investigate the options for detectability, evaluate numerically the RCS of the two objects using (2.206). Include the diffuse backscattering (2.208) and the consequential ground fading effect (2.214) (a light soil ground index of refraction $\eta$ = 2.2 is assumed). The object/stem RCS ratios are obtained in Table 2.1.

The columns are the cases of different incidence angles and rows are the cases of different wavelengths. For vertical polarization, the cross-sectional differences are obtained in Table 2.2.

In no case is the tree stem response sufficiently small to ensure target detection by strength alone (a target/stem RCS ratio of better than 10 dB would be required). Evidently, the change technique of detecting objects appearing or disappearing between successive radar registrations must be relied on.

For the two polarizations, there three situations of favorable cross-sectional ratio are evident: 1) horizontal polarization for a 6m wavelength and an incidence angle of less than 60°, 2) vertical polarization, for a 6m wavelength and an incidence angle of 60°, and 3) alternatively for a 1.5m wavelength and an incidence angle of 70°. The reasons behind the first and last two favorable situations are entirely different, however: for horizontal polarization, the explanation is the difference in diameter between the tree stem and the ground

**Table 2.1**
Object/Stem RCS Ratios

| | **50°** | **60°** | **70°** |
|---|---|---|---|
| 6m | 0.2/0.2 dBm$^2$ | −2.3/2.0 dBm$^2$ | −7.0/4.1 dBm$^2$ |
| 1.5m | 7.9/22.3 dBm$^2$ | 10.4/23.5 dBm$^2$ | 13.6/25.0 dBm$^2$ |
| 0.5m | 14.0/24.0 dBm$^2$ | 14.9/25.2 dBm$^2$ | 16.1/26.6 dBm$^2$ |

**Table 2.2**
Cross-Sectional Differences for Vertical Polarization

| | **50°** | **60°** | **70°** |
|---|---|---|---|
| 6m | 4.4/13.8 dBm$^2$ | 3.8/7.4 dBm$^2$ | 1.6/10.1 dBm$^2$ |
| 1.5m | 3.5/12.6 dBm$^2$ | −6.3/5.7 dBm$^2$ | 2.9/6.8 dBm$^2$ |
| 0.5m | 4.7/17.1 dBm$^2$ | −1.7/10.1 dBm$^2$ | −2.7/10.4 dBm$^2$ |

object (as illustrated in Figure 2.8), whereas for vertical polarization, it is the possibility of advantageous fading of tree stem responses, as seen in Figure 2.7. As is evident from Figure 2.7, such fading will be sensitive to incidence angle as well as tree height. Actual target detection will depend on the condition that these parameters happen to work in favor. Moreover, in view of the requirements for stability of the RCS values when applying change techniques, variable RCS outcome in repeating radar imaging is not acceptable. In contrast, the option of using horizontal polarization for target detection, based on different target and stem widths, is seen to depend only slowly on parameters such as incidence angle and tree height. It is therefore the better option for reliable target detection.

In the example above, target height was deliberately chosen to create a case at the limit of detectability (assuming change methods). Any more shallow and the target RCS rapidly disappears, whereas increasing the height just slightly, the target becomes comparable to or stronger than the stem responses.

It was noted in connection with (2.213) that at least for ideally reflecting cylinders, TM backscattering is very strong and slowly increasing with wavelength. Tree branches being near horizontal and thus prone to backscatter in the TM mode for an impinging horizontally polarized field may thus be the source of very strong backscatter. In particular, may this be the case when $\lambda$ is large, which was just concluded the best option for optimizing the target/stem RCS ratio. The contribution of tree branch scattering to tree RCS therefore needs to be carefully investigated.

Note that there is a crucial difference between the case of infinite and finite indices of refraction, well exhibited in Figure 2.8. As illustrated, when cylinder radius is $a \lesssim \lambda/30$, the TM backscatter will dominate entirely over TE backscatter. However, for a cylinder radius of $a \lesssim \lambda/300$, TM backscatter width (still dominating) is seen to have dropped 20 dB from optical backscatter levels.

The question to be analyzed is whether the branches on a tree collectively add to dominate the tree stem response. For a spruce tree, the branches form a fairly homogenous set as regards diameter length and also axil angle (the angle between the tree stem and the branch), typically a 15m spruce tree may carry 100 branches of mean radius 0.02m and mean length of 2m. A typical axil angle is 70°. The extent to which a tree carries branches significantly larger than this varies for different tree species. Spruce lacks capital branches entirely, whereas pine and deciduous trees carry large branches but in low numbers along with a massive distribution of small branches. The fact that TM scattering width is so insensitive to frequency compared to radius means that the large branches will not be that much stronger scatterers than the small ones. Being relatively few in number, they will not dominate the backscatter. The branch numbers and dimensions assumed will therefore be considered providing cross-sectional values, typical also for other tree species than spruce.

Different from tree stems, branches appear with variable orientation. The required adaption of (2.206) to branches will depend on two angles: 1) the angle between the direction of propagation of the impinging field and the direction of the branch and 2) the angle between the direction of the impinging $\boldsymbol{E}$ field and the plane defined by the branch and propagation directions. The former determines the amount of diffuse backscattering (as given by (2.208)) and the latter the projected fields for TM and TE backscatter. Denote by $\boldsymbol{e}_C$ the direction of the cylinder representing the branch, by $\boldsymbol{e}_P$ the direction of propagation (i.e., $\boldsymbol{e}_P = \boldsymbol{k}/k$), and by $\boldsymbol{e}_H$ and $\boldsymbol{e}_V$ the directions of the $\boldsymbol{E}$ field for horizontal and vertical polarizations, respectively. Figure 2.9 illustrates the situation. The backscattered signal will be depolarized; the backscattered field will contain both vertically and horizontally polarized components irrespective of the polarization of impinging field. The angles determining the backscatter are the inverse cosine of $\boldsymbol{e}_C \bullet \boldsymbol{e}_P$ and the inverse sine of $\boldsymbol{e}_H \bullet (\boldsymbol{e}_C \times \boldsymbol{e}_P)/|\boldsymbol{e}_C \times \boldsymbol{e}_P|$ or $\boldsymbol{e}_V \bullet (\boldsymbol{e}_C \times \boldsymbol{e}_P)/|\boldsymbol{e}_C \times \boldsymbol{e}_P|$, depending on the polarization of the impinging field. As a simplification, it is assumed that the dominant TM mode is the only source for backscattering. The copolarized branch backscattering RCS components (always stronger than the cross-polarized ones) become

$$\begin{aligned} \sigma_{\mathrm{CH}} &= \Psi_H \sigma_{\mathrm{TM}};\ \Psi_H = \left[\frac{\sin\left(kL\boldsymbol{e}_C \bullet \boldsymbol{e}_P\right)}{kL\boldsymbol{e}_C \bullet \boldsymbol{e}_P}\right]^2 \left\{1 - \left[\frac{\boldsymbol{e}_H \bullet \left(\boldsymbol{e}_C \times \boldsymbol{e}_P\right)}{\left|\boldsymbol{e}_C \times \boldsymbol{e}_P\right|}\right]^2\right\} \\ \sigma_{\mathrm{CV}} &= \Psi_V \sigma_{\mathrm{TM}};\ \Psi_V = \left[\frac{\sin\left(kL\boldsymbol{n}_C \bullet \boldsymbol{n}_P\right)}{kL\boldsymbol{n}_C \bullet \boldsymbol{n}_P}\right]^2 \left\{1 - \left[\frac{\boldsymbol{e}_V \bullet \left(\boldsymbol{e}_C \times \boldsymbol{e}_P\right)}{\left|\boldsymbol{e}_C \times \boldsymbol{e}_P\right|}\right]^2\right\} \end{aligned} \tag{2.215}$$

Here, $\sigma_{\mathrm{TM}}$ is the RCS in the TM mode obtained by (2.206), whereas $\sigma_{\mathrm{CH}}$ and $\sigma_{\mathrm{CV}}$ are the cross sections for horizontal and vertical polarizations, respectively. Denote by $\theta,\chi,\varphi$ the incidence angle of the impinging field, the polar angle of the branch, and the angle between the impinging field and the branch in the azimuth plane, as per Figure 2.9. According to the geometry in the figure

$$\begin{aligned} \boldsymbol{e}_C &= -\sin\chi \boldsymbol{e}_y + \cos\chi \boldsymbol{e}_z \\ \boldsymbol{e}_P &= \sin\theta\cos\varphi \boldsymbol{e}_x - \sin\theta\sin\varphi \boldsymbol{e}_y + \cos\theta \boldsymbol{e}_z \\ \boldsymbol{e}_V &= -\cos\theta\cos\varphi \boldsymbol{e}_x + \cos\theta\sin\varphi \boldsymbol{e}_y + \sin\theta \boldsymbol{e}_z \\ \boldsymbol{e}_H &= \sin\varphi \boldsymbol{e}_x + \cos\varphi \boldsymbol{e}_y \end{aligned} \tag{2.216}$$

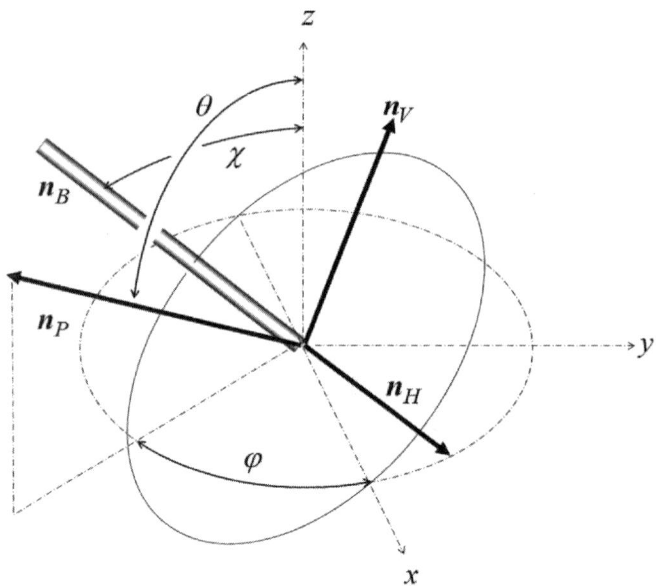


**Figure 2.9** Branch scattering—definition of angles.

The cross section will vary with aspect angle $\varphi$. However, branches are expected to be isotropically distributed. The expected RCS is therefore obtained by circular averaging with respect to $\varphi$.

For the branch radius, length, and axil angle of $a$ = 0.02m, $L$ = 2m, and $\chi$ = 70°, respectively, and with $\eta = 6(1 + 0.3i)$, the following RCS values are obtained for horizontal/vertical polarization.

Comparing Tables 2.1 and 2.3 for horizontally polarized backscattering, the mean RCS for a branch is seen to be below the stem backscatter by 34 to 40 dB. As expected, the vertically polarized backscattering becomes weaker with increasing incidence angle, that is, with shallower incidence, when the ***E*** field and branch directions tend to be orthogonal.

**Table 2.3**
Branch RCS Values for Horizontal/Vertical Polarization

| | 50° | 60° | 70° |
|---|---|---|---|
| 6m | −34.0/−36.4 dBm$^2$ | −33.9/−37.4 dBm$^2$ | −33.8/−38.7 dBm$^2$ |
| 1.5m | −14.3/−19.2 dBm$^2$ | −14.6/−20.7 dBm$^2$ | −14.8/−22.5 dBm$^2$ |
| 0.5m | −10.0/−15.7 dBm$^2$ | −10.4/−17.4 dBm$^2$ | −10.6/−18.6 dBm$^2$ |

Consider the combined scattering effect of many branches. If these are densely packed, they will interact with each other to an extent by multiple scattering, both causing an enhanced amount of immediate backscattering and the continual tapping of power of the impinging field. Neglecting these two effects, the overall RCS contribution from the tree branches is $N \times \langle\sigma\rangle$ with $\langle\sigma\rangle$ being the $\varphi$-averaged RCS and $N$ the number of branches. With a totality of 100 branches, their combined contribution to tree RCS at horizontal polarization will be generally lower than −15 dB below the stem cross section for all the nine wavelength and incidence angle cases considered.

Below the issue of attenuation of the impinging field will be treated. The coupling effect (i.e., degree to which multiple scattering occurs) on backscatter levels can be assessed according to the following argument. As defined, the RCS is an estimate of the total cross section (assuming an isotropic scatterer), so for any two branches (with cross section $\sigma$ and separation $D$), the coupling is given by the ratio $\sigma/\pi D^2$. With the currently assumed density of branches, $100/15 \approx 8$ branches per meter of stem, it can be assumed that within a mean distance of 0.5m, each branch would have 8 other branches. Hence, if the coupling at larger distances is neglected, the degree of coupling becomes $\sigma \times 8/\pi \times 0.5^2 = \sigma + 10$ dB. The coupling is thus indicated negligible for 6m wavelength but not in the cases of shorter wavelengths. However, attenuation to be analyzed below becomes strong in the case where coupling do occur and will obviously act to reduce the backscattering levels. Therefore, also in these cases, the branch RCS contribution is expected not to exceed $N \times \langle\sigma\rangle$, though attenuation will also reduce tree stem and target RCS levels. The end result is that it is not possible to generally conclude whether backscattering contributions can be neglected or not for the shorter wavelengths; it will depend on the detailed circumstances.

Just as for tree stems, also for branches there will be ground interaction (2.214), determined by Fresnel's reflection coefficients. With $L$ being the height above the ground of the individual branches, the exponents in (2.214) can be assumed rapidly and independently oscillating when averaging over a set of branches. The scattering pattern for the individual branch is essentially isotropic in the $\theta$ direction, so $\Phi(\theta,\theta) \approx 1$. The overall branch RCS contribution becomes

$$\Sigma = N\langle\sigma\rangle\left(1 + 4R^2 + R^4\right) \tag{2.217}$$

In contrast to the coherent ground interaction for the tree stem, the branches, being elevated from ground, will interact with ground in such a way that the backscattered power is increased. Typically, the expression in the parenthesis amounts to about 4 dB for 60° incidence and $\eta = 2.2$. Therefore, unless the effects of field attenuation will become dominant, the conclusion is that for the nine cases considered, stem response is dominating over foliage response with a margin always better than 10 dB.

In particular, at steep incidence angles, the propagation path through a forest will have to pass multiple trees, and attenuation may become significant. Knowing the extent to which trees cause attenuation is crucial for predicting radar detection performance for targets concealed in forest.

The concept of wave propagation attenuation will be given the following precise meaning: as will be discussed in subsequent chapters, the principles for focusing the radar signal scattered from ground assume this signal to have been traveling from radar to ground at constant speed (i.e., the speed of light), in a straight line, whereupon it bounces back to the radar in a straight line at the same constant speed. Any part of the signal deviating from this assumption does not contribute to the process of focused imaging of the ground, but will instead increase the noise background in the image. When ground is screened by vegetation, the scattering process determines the fraction of the impinging signal, which is scattered into directions different from the original one and which part of the signal is left to contribute in the ground imaging. The only part of the scattered signal that does so is the forward scattered signal (i.e., the signal in the direction $\vartheta = \pi - \theta$ and $\phi = 0$ in the case of cylinder scattering), which combines with the incident radar signal in the imaging process. Forward scattering occurs with an amplitude of opposite phase to the incident signal, so when combining the incident signal, this will lose in strength (i.e., the object causes a shadow). To what degree this happens can be investigated from the theory derived.

The following is a simple approach to derive plane wave field attenuation due to forward scatter.

Consider the situation of plane wave impinging along the positive $x$-axis and illuminating an infinite cylinder concentric along the $z$-axis. The field in front of the cylinder is $\hat{\boldsymbol{E}}(x,0,0) + \check{\boldsymbol{E}}(x,0,0)$. The sum expresses the attenuation due to the cylinder scatter. It is noted that the field is a combination of a plane wave and a cylindrical wave, that is, the second term has an amplitude decreasing as $1/\sqrt{x}$, whereas the first term modulus is constant. Consequently, the attenuation diminishes with increased distance to the cylinder. Now rather than the single cylinder, assume a "fence" of parallel cylinders distributed along the $y$-axis. Because of the ever expanding cylinder backscatter fields, at sufficient distance, there will be several cylinders, for which their forward scattering components provide contributions to the net field along the $x$-axis, just like the spherical waves from several cylinder segments of length $\Delta L$ contributed to the scattered field of an infinite cylinder, in the derivation of the finite cylinder RCS. By the same argument, the segment of the $y$-axis, for which the scatter contributions have a phase error less than $\pi/2$ effectively determine this field, outside the segment amplitudes starts to rapidly alter sign and will mutually cancel. The width of the contributing segment is $\sqrt{\pi x/k}$, as per the discussion on cylinder scattering. Given a position on the $x$-axis at a distance $x$ from the cylinder fence, the distance to any of the cylinders contributing to the field varies from $x$ to $x + \lambda/4 = x + \pi/2k$. To compute the contribution from the cylinders at the considered position, the mean value $\rho = x + \pi/4k$ may be assumed, in analogous to the cylinder case. If the separation between the cylinders in the fence is $\Delta y$, the number of contributing cylinders is $\sqrt{\pi x/k}/\Delta y$.

Due to the increased number of cylinders contributing, the amplitude loss due to cylinder wave divergence is precisely balanced. The net forward scattered field will in effect constitute a plane wave. The resulting attenuation, determined by the combined field of the impinging plane wave and the forward scattered plane wave from the set of cylinders, has several notable properties. For instance, it will be doubled by two similar fences along the $x$ direction or by packing cylinders at double density. In fact, this method for evaluating attenuation can be applied to cylinders with randomized $x$ and $y$ positions, if contained in a layer of finite depth $D$. It is not required that they are aligned to be parallel, since according to the derivation of cylinder scattering, the forward scattering follow rules similar to specular scattering and thus may be assumed to be insensitive to incidence angle. Therefore, the cylinders may tilt in the $x,z$-plane. If they also tilt in the $y,z$-plane, the two polarization states will mix, as will be discussed subsequently.

Assume a sufficient distance $x$ from the layer of cylinders, so that the contributing number of cylinders is "large" in some sense. Let $\gamma$ be the number of cylinders per surface area (in the $x,y$-plane) within the layer. Thus, $\langle \Delta y \rangle = 1/(\gamma D)$. The signal, having passed the layer of cylinders and at sufficient distance from this, assumes the form

$$\breve{\boldsymbol{E}}_{\text{plane}}(x,0,0) = D\gamma\sqrt{\frac{\pi x}{k}}\breve{\boldsymbol{E}}_{\text{cyl}}\left(x+\frac{\pi}{4k},\pi,0\right) \tag{2.218}$$

For the formula to be fully applicable, the cylinder distribution must be sufficiently thin, in order that the cylinders will not to shadow or otherwise interfere with each other. The attenuation through the layer is

$$\text{attenuation} = \frac{\left|\widehat{\boldsymbol{E}}(x,0,0) + \breve{\boldsymbol{E}}_{\text{plane}}(x,0,0)\right|^2}{\left|\widehat{\boldsymbol{E}}(x,0,0)\right|^2} \tag{2.219}$$

To obtain explicit expressions, consider forward scattering and attenuation, according to (2.157) and (2.194)

$$\begin{aligned} \boldsymbol{M}_k^n &= -kZ_n'(\rho k)\boldsymbol{e}_\phi = -\frac{e^{i\pi/4}}{i^n}\sqrt{\frac{2k}{\pi\rho}}e^{i\rho k}\boldsymbol{e}_\phi \\ \boldsymbol{N}_k^n &= kZ_n(\rho k)\boldsymbol{e}_z = \frac{e^{-i\pi/4}}{i^n}\sqrt{\frac{2k}{\pi\rho}}e^{i\rho k}\boldsymbol{e}_z \end{aligned} \tag{2.220}$$

which implies that

$$\begin{aligned} &\text{TM polarization:} \quad \breve{\boldsymbol{E}}_{\text{TM}} = -\sum_{n=0}^{\infty} i^n b_n \boldsymbol{N}_k^n = -e^{-i\pi/4}\sqrt{\frac{2k}{\pi\rho}}e^{ikx}\boldsymbol{e}_z\sum_{n=0}^{\infty} b_n \\ &\text{TE polarization:} \quad \breve{\boldsymbol{E}}_{\text{TE}} = -\sum_{n=0}^{\infty} i^n a_n \boldsymbol{M}_k^n = e^{i\pi/4}\sqrt{\frac{2k}{\pi\rho}}e^{ikx}\boldsymbol{e}_\phi\sum_{n=0}^{\infty} a_n \end{aligned} \tag{2.221}$$

In the nominator in (2.217), the impinging plane wave is given by (2.154) with $b = \widehat{k}$ and $a = \widehat{k}$ in the respective cases of TM and TE polarizations

$$\begin{aligned} &\text{TM polarization:} \quad \boldsymbol{E}_{\text{TM}} = -k^2 e^{ikx}\boldsymbol{e}_z, \\ &\text{TE polarization:} \quad \boldsymbol{E}_{\text{TE}} = k^2 e^{ikx}\boldsymbol{e}_y \end{aligned} \tag{2.222}$$

Here, $\boldsymbol{e}_y = \boldsymbol{e}_\phi$. It follows that:

$$\begin{aligned} \text{TM attenuation} &= \left|1 + \sqrt{2}\frac{D\gamma}{k^2}\sum_{n=0}^{\infty} b_n\right|^2 \\ \text{TE attenuation} &= \left|1 + i\sqrt{2}\frac{D\gamma}{k^2}\sum_{n=0}^{\infty} a_n\right|^2 \end{aligned} \tag{2.223}$$

A generalized though approximate model, useful for a thicker cylinder layers (with intrinsic shadowing included), is attained if the requirement for sufficient distance in (2.216) is skipped and the formula is considered applicable already at $x = \Delta x$, with $\Delta x$ being the layer thickness. Then $\hat{\boldsymbol{E}}(\Delta x,0,0)$ can be considered the impinging field to a second layer of cylinders and so forth, whereby

$$\begin{aligned}\breve{\boldsymbol{E}}_{\mathrm{TM}}(x+\Delta x,0,0) &= \left(1+\Delta x\gamma a_{\mathrm{TM}}\right)e^{ik\Delta x}\hat{\boldsymbol{E}}_{\mathrm{TM}}(x,0,0);\ a_{\mathrm{TM}} = \frac{\sqrt{2}}{k^2}\sum_{n=0}^{\infty} b_n \\ \breve{\boldsymbol{E}}_{\mathrm{TE}}(x+\Delta x,0,0) &= \left(1+\Delta x\gamma a_{\mathrm{TE}}\right)e^{ik\Delta x}\hat{\boldsymbol{E}}_{\mathrm{TE}}(x,0,0);\ a_{\mathrm{TE}} = i\frac{\sqrt{2}}{k^2}\sum_{n=0}^{\infty} a_n\end{aligned} \tag{2.224}$$

The attenuation may vary greatly with polarization state, that is, attenuation may act as a polarization filter. Polarization effects are taken into account just as for cylinder backscattering, that is, with the polarization state determined from the direction $\boldsymbol{e}_{\mathrm{TE}} = \boldsymbol{e}_P \times \boldsymbol{e}_C / |\boldsymbol{e}_P \times \boldsymbol{e}_C|$, where $\boldsymbol{e}_P$ is the unit direction of propagation and $\boldsymbol{e}_C$ the unit direction of the cylinder. The TE polarization state is when the $\boldsymbol{E}$ field is aligned with $\boldsymbol{e}_{\mathrm{TE}}$, while TM occurs when $\boldsymbol{E}$ is aligned with $\boldsymbol{e}_{\mathrm{TM}} = \boldsymbol{e}_{\mathrm{TE}} \times \boldsymbol{e}_P$. The horizontally and vertically polarized impinging $\boldsymbol{E}$ fields (i.e., $\boldsymbol{E}_H$ and $\boldsymbol{E}_V$) are related to the fields $\boldsymbol{E}_{\mathrm{TM}}$ and $\boldsymbol{E}_{\mathrm{TE}}$ according to

$$\begin{aligned}\cos\psi &= \frac{\left|\boldsymbol{e}_z \bullet \left(\boldsymbol{e}_P \times \boldsymbol{e}_C\right)\right|}{\left|\boldsymbol{e}_P \times \boldsymbol{e}_C\right|} \Rightarrow \\ \boldsymbol{E}_{\mathrm{TM}} &= \boldsymbol{E}_H \cos\psi + \boldsymbol{E}_V \sin\psi \\ \boldsymbol{E}_{\mathrm{TE}} &= -\boldsymbol{E}_H \sin\psi + \boldsymbol{E}_V \cos\psi\end{aligned} \tag{2.225}$$

Note that $\psi$ is an angle between a vector and an axis and as such only assumes values $\leq \pi/2$ (the angle between two vectors can take any value up to $\pi$). No depolarization occurs when the $\boldsymbol{E}$ field is either aligned or transverse to the cylinder. The $\boldsymbol{E}_{\mathrm{TM}}$ and $\boldsymbol{E}_{\mathrm{TE}}$ fields are thus the eigenvectors to the depolarization matrix. For any plane wave field decomposed into its vertical and horizontally polarized components, the attenuation and its associated depolarization become

$$\begin{aligned}&\begin{pmatrix}\hat{\boldsymbol{E}}_{H,x+\Delta x} \\ \hat{\boldsymbol{E}}_{V,x+\Delta x}\end{pmatrix} = e^{ik\Delta x}\left[1+\gamma\Delta x \boldsymbol{a}\left(\boldsymbol{e}_P \times \boldsymbol{e}_C\right)\right]\begin{pmatrix}\hat{\boldsymbol{E}}_{H,x} \\ \hat{\boldsymbol{E}}_{V,x}\end{pmatrix}; \\ &\boldsymbol{a}\left(\boldsymbol{e}_P \times \boldsymbol{e}_C, \gamma\Delta x\right) = \begin{pmatrix} a_{\mathrm{TM}}\cos^2\psi + a_{\mathrm{TE}}\sin^2\psi & \left(a_{\mathrm{TM}} - a_{\mathrm{TE}}\right)\cos\psi\sin\psi \\ \left(a_{\mathrm{TM}} - a_{\mathrm{TE}}\right)\cos\psi\sin\psi & a_{\mathrm{TM}}\sin^2\psi + a_{\mathrm{TE}}\cos^2\psi\end{pmatrix}\end{aligned} \tag{2.226}$$

The formula is valid for $\gamma\Delta x\boldsymbol{a}$ small. It can be applied recursively also letting $\boldsymbol{e}_P \times \boldsymbol{e}_C$ change in each recursive step, allowing the analysis of situations with sets of cylinder with different orientations intermixed. In (2.227), $\boldsymbol{a}(\boldsymbol{e}_P \times \boldsymbol{e}_C)$ has the form $R(\psi)^{-1}\boldsymbol{a}(\boldsymbol{e}_P \times \boldsymbol{e}_C)_{\text{diag}}R(\psi)$, where $R(\psi)$ is the orthogonal matrix of vector basis rotation of the angle $\psi$. As a consequence, in the special case of a cylinder distribution, where $\boldsymbol{e}_P \times \boldsymbol{e}_C$ does not change for an extended interval in the $x$ direction, (2.226) generalizes to

$$\alpha_{\text{TM}} = e^{\sqrt{2}\frac{x\gamma}{k^2}\sum_{n=0}^{\infty} b_n}; \; \alpha_{\text{TE}} = e^{i\sqrt{2}\frac{x\gamma}{k^2}\sum_{n=0}^{\infty} a_n} \Rightarrow$$

$$\begin{pmatrix} \hat{E}_{H,x} \\ \hat{E}_{V,x} \end{pmatrix} = e^{ikx} \begin{pmatrix} \alpha_{\text{TM}}\cos^2\psi + \alpha_{\text{TE}}\sin^2\psi & (\alpha_{\text{TM}} - \alpha_{\text{TE}})\cos\psi\sin\psi \\ (\alpha_{\text{TM}} - \alpha_{\text{TE}})\cos\psi\sin\psi & \alpha_{\text{TM}}\sin^2\psi + \alpha_{\text{TE}}\cos^2\psi \end{pmatrix} \begin{pmatrix} \hat{E}_{H,0} \\ \hat{E}_{V,0} \end{pmatrix} \tag{2.227}$$

valid for any depth $x$.

Subsequently, (2.227) will be applied heterogeneous situations, such as a tree line, with stems and branches being cylinders of different sizes and orientations. Before addressing this general case, in order to gain some explicit insight into the formula, consider the simplest case, that is, (2.227) for perfectly reflecting cylinders with $ak$ so small that lowest order approximations are applicable. From (2.211), the power flux attenuation as a function of $x$ becomes

$$\begin{aligned} &\text{TM attenuation:} \quad S(x,0,0) = e^{-\pi\sqrt{2}\frac{x\gamma}{k}\frac{\pi/2}{\ln^2(0.890535\,ak)+\pi^2/4}} S(0,0,0) \\ &\text{TE attenuation:} \quad S(x,0,0) = e^{-\frac{\pi}{\sqrt{2}}kx\gamma a^2} S(0,0,0) \end{aligned} \tag{2.228}$$

As expected, attenuation for TE polarization is seen to be much smaller than for TM polarization. Certainly, any realistic treatment of attenuation must incorporate the depolarization effects followed by tree branches and stems pointing in different directions, and the possibilities of doing so by means of (7.98) must be exploited.

Consider (2.227) applied to a case of two superimposed sets of cylinders, represented by $\boldsymbol{a}_1 = \boldsymbol{a}(\boldsymbol{e}_P \times \boldsymbol{e}_{C1})$ and $\boldsymbol{a}_2 = \boldsymbol{a}(\boldsymbol{e}_P \times \boldsymbol{e}_{C2})$. In this case

$$\begin{pmatrix} \hat{E}_{H,x+\Delta x} \\ \hat{E}_{V,x+\Delta x} \end{pmatrix} = (1 + \gamma\Delta x\boldsymbol{a}) \begin{pmatrix} \hat{E}_{H,x} \\ \hat{E}_{V,x} \end{pmatrix}; \; \boldsymbol{a} = \boldsymbol{a}_2 + \boldsymbol{a}_1 + \gamma\Delta x\boldsymbol{a}_2\boldsymbol{a}_1 \tag{2.229}$$

The last matrix term is a second-order correction and can be neglected given that it is sufficiently small. Its exclusion means that the mutual shadowing between cylinders, leading to the exponential decay of the field as in (2.228), is neglected. The approximation is thus valid when the two cylinder sets can be assumed not to produce any significant shadow to each other. Consider, for instance, a single line of trees. For simplicity, the trees may be assumed all equal, each occupying a width $\Delta x = D$ along the $x$-axis and separated from each other by $\Delta y = D$. Suppose that there are $N$ branches per tree. Then, $\gamma = N/D^2$ for the branches and $\gamma = 1/D^2$ for the stems. Given that the branches do not shadow each other, the attenuation can be estimated by the matrix formula

$$\begin{pmatrix} \widehat{E}_{H,x+\Delta x} \\ \widehat{E}_{V,x+\Delta x} \end{pmatrix} = \boldsymbol{A} \begin{pmatrix} \widehat{E}_{H,x} \\ \widehat{E}_{V,x} \end{pmatrix}; \boldsymbol{A} = 1 + \frac{1}{D}\boldsymbol{a}_{\text{stem}} + \frac{N}{D}\left\langle \mathbf{a}_{\text{branch}} \right\rangle_{\forall\varphi} \tag{2.230}$$

Keeping to the assumptions used for branch backscatter, $N = 100$ with branch mean radius $a = 0.02$m, stem mean radius $a = 0.2$m, and branch mean length 2m will be used for numerical evaluation. Allowing for some space between the trees, the branch mean length of 2m is converted to mean a mean tree separation of $D = 6$m. Given these parameters together with the previous assumption of stem height 15m, the wood content of the forest will be 755 m$^3$/ha, corresponding to 755 tons/ha of biomass. Such a figure can be reached in reality in areas of extremely dense spruce forest. Biomass figures even higher are rarely met, not even for tropical rain forest. The index of refraction will be assumed to be $\eta = 6(1 + 0.3i)$. The axil and incidence angles will be varied to obtain statistical variability in results.

Being a first-order expansion of an exponential function, the convergence in (2.230) is rapid, meaning that even if the modulus of the matrix coefficients in the second and third terms in $\boldsymbol{A}$ reach a value 0.5, the approximation remains fairly correct. Still, when outside the approximation bounds, (2.227) breaks down rapidly and entirely. As was anticipated, coupling between the branches cannot be neglected for the shorter wavelengths in Tables 2.1 to 2.3. An analysis of approximation errors does show that (with the assumed tree parameters) when basing the mean value of forward branch scatter on entire trees, as in (2.230), the approximation is too coarse. By refining $\Delta x$ increments and iterating (2.227) several times, accuracy is improved. Presently, $\Delta x = D/4$ is selected, realizing the net attenuation across the tree width $D$ by iterating (2.227) four times for the tree plumes, intertwined with one iteration for the tree stem. The matrix $\boldsymbol{A}$ now takes the shape

$$\begin{aligned} A = & \left(1+\frac{N}{4D}\left\langle \boldsymbol{a}_{\text{branch}}\right\rangle_{180^\circ \geq |\phi| \geq 120^\circ}\right)\left(1+\frac{N}{4D}\left\langle \boldsymbol{a}_{\text{branch}}\right\rangle_{120^\circ \geq |\phi| \geq 90^\circ}\right) \\ & \times\left(1+\frac{1}{D}\boldsymbol{a}_{\text{stem}}\right)\left(1+\frac{N}{4D}\left\langle \boldsymbol{a}_{\text{branch}}\right\rangle_{90^\circ \geq |\phi| \geq 60^\circ}\right)\left(1+\frac{N}{4D}\left\langle \boldsymbol{a}_{\text{branch}}\right\rangle_{60^\circ \geq |\phi|}\right) \end{aligned} \tag{2.231}$$

This refined attenuation expression is a valid approximation also for the very dense tree structure that is considered. Figure 2.10 shows the result in the HH and VV polarization cases for different choices of incidence and axil angles. Notable is the strong attenuation of around 1 m. This dip, which shifts with branch radius and index of refraction, corresponds loosely (but not exactly it is at slightly larger $\lambda$s and also more narrow) to the TM backscatter maximum in Figure 2.10. As both branches and incidence tend to the horizontal, the dip becomes stronger at the horizontal polarization, whereas otherwise the VV attenuation is stronger.

Comparing Table 2.3 and Figure 2.10, it is observed that polarization has somewhat different effects for attenuation and backscattering. The explanation seems to be that the latter depends on the diffuse backscatter response (2.208), while the former does not (nor does it depend on stem or branch lengths). The diffuse backscattering response gives extra weight to the situations when the impinging signal hits branches at a near broadside, a situation that is encountered to a high degree for horizontal polarization and given that branches are fairly horizontal. However, the diffuse backscattering component has no relevance to forward scattering and thus to attenuation. At the oblique axil and incidence angles that are considered, it turns out that vertically polarized fields project on branch orientation more effectively than horizontally polarized fields, resulting in that vertical polarization is more vulnerable to attenuation, as is seen in Figure 2.10.

For thicker vegetation layers, the attenuation values in decibels double every 6m. A radar signal path reaching the ground at 60° incidence with the tree height of 15m is 60m long. It would thus amount to the decibel values of Figure 2.10 being multiplied by a factor of 10. Only for wavelengths $\lambda > 5$m will the resulting attenuation be smaller than 10 dB. In essence, this conclusion agrees with experience, though in almost any case of actual target detection, forest is not that dense. Target may more likely be concealed by tree screens rather than being deployed in the midst of next to impenetrable forest.

Hence, the attenuation values arrived at compare well with the overall but not so detailed practical experiences of low-frequency radar. The strong attenuation at around 1m wavelength and, in particular, at vertical polarization well supports the negative experiences regarding target detection using vertical

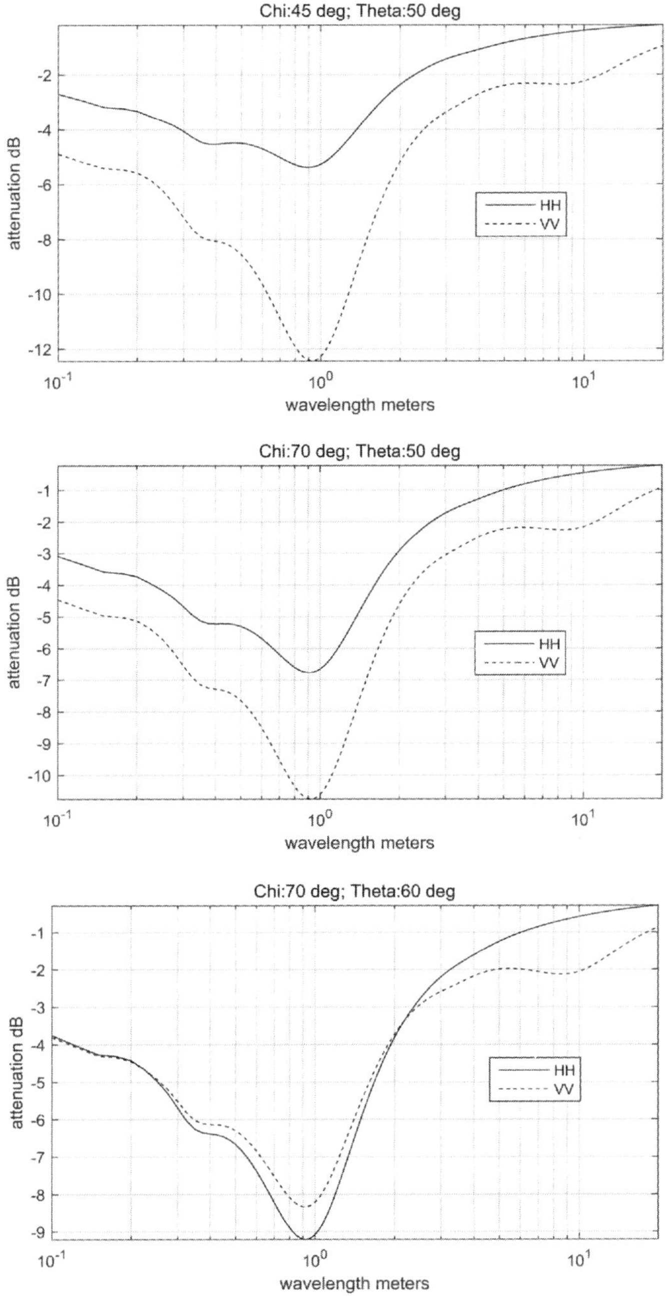


**Figure 2.10** Attenuation caused by a tree line, 6m thick. Trees are composed of branches (given by their axil angle $\chi$) with a radius of 0.02m and vertical stems with a radius of 0.2m. Both the number of stems and branches are given by their surface density in the horizontal plane and correspond to one stem every 6m and 100 branches per stem. The three diagrams correspond to different choices of incidence and axil angle.

polarization at this wavelength. It is also observed that at this wavelength in particular, the attenuation will suppress the stem responses within a forest. It may thus be that backscattering at frequencies of about 300 MHz largely emanates from branches rather than the stems, whereas at low frequencies stem responses continue to dominate.

This treatment of attenuation concludes the development of electromagnetic scattering theory, which is required in this book. The remaining chapters will make reference to what has been concluded here and develop its details for applications and closer comparison with the experimental results.

## References

[1] Landau, L.D., and E.M. Lifshitz, *The Classical Theory of Fields*, Oxford: Butterworth-Heinemann, 2000.

[2] Landau, L.D., and E.M. Lifshitz, *Electrodynamics of Continuous Media*, Amsterdam: Elsevier, 2004.

[3] Sorrentino, R., and G. Bianch, *Microwave and RF Engineering*, Hoboken, NY: John Wiley & Sons, 2010.

[4] Daniels D. J. (Ed.), *Ground Penetrating Radar*, London, IET, 2004.

[5] Valenzula, G.R., "Depolarization of EM Waves by Slightly Rough Surfaces," *IEEE Trans. on Antennas and Propagation*, Vol. AP-15, July 1967, pp. 551–557.

[6] Fung A.K., *Microwave Scattering and Emission Models for Users*, Norwood, MA: Artech House Inc., 2010.

[7] Ulaby, Fa.T., R.K. Moore, and A.K. Fung, *Microwave Remote Sensing*, Vol II, Norwood, MA: Artech House Inc., 1982.

[8] Beckmann, P., and A. Spizzichino, *The Scattering of Electromagnetic Waves From Rough Surfaces*, Norwood, MA: Artech House Inc., 1987.

[9] Stratton, J.A., *Electromagnetic Theory*, Hoboken, NJ: IEEE Press, 2007.

[10] Na, L. "Manipulation of Particles on Optical Waveguides," Thesis, University of Southampton, September 2000.

[11] Knott, E.F., J.F. Shaeffer, and M.T. Tuley, *Radar Cross Section*, Norwood, MA: Artech House, 1993.

[12] Ruck, G.T., et al., *Radar Cross Section Handbook*, Vol I, Berlin, Germany: Plenum Press (also Springer Verlag), 1970.

[13] Balanis, C.A., *Advanced Engineering Electromagnetics*, Hoboken, NY: John Wiley & Sons, 2012.

[14] Lebedev, N.N. "Special Functions and their Applications," Upper Saddle River, NY: Prentice Hall Inc, 1965.

[15] Lösch F., *Tables of Higher Functions*, Leipzig, East Germany: B. G. Teubner Verlagsgesellschaft, 1966.

[16] http://dlmf.nist.gov/.

[17] Olver, F.W.J., et al., *NIST Handbook of Mathematical Functions*, Washington, DC: U.S. Department of Commerce, National Institute of Standards and Technology, Cambridge: Cambridge University Press, 2010.

[18] Stiles, J.M., and K. Sarabandi, "A Scattering Model for Thin Dielectric Cylinders of Arbitrary Cross section and Electrical Length," *IEEE Trans. on Antennas and Propagation*, 44, February 1996, pp. 260–266.

[19] Sarabandi, K., and T.B.A. Senior, "Low-Frequency Scattering From Cylindrical Structures at Oblique Incidence," *IEEE Trans. on Geoscience and Remote Sensing*, Vol. 28, September 1990, pp. 879–885.

[20] Karam, M.A., A.K. Fung, and Y.M. M. Antar, "Electromagnetic Wave Scattering from Some Vegetation Samples," *IEEE Trans. on Geoscience and Remote Sensing*, Vol. 26, No. 6, November 1988, pp. 799–808.

[21] Israelsson, H., et al., "Retrieval of Forest Stem Volume Using VHF SAR," *IEEE Trans. on Geoscience and Remote Sensing*, Vol. 35, No. 1, January 1997, pp. 36–40.

[22] Israelsson, H., et al., "A Coherent Scattering Model to Determine Forest Backscattering in the VHF-Band," *IEEE Trans. on Geoscience and Remote Sensing*, Vol. 38, No. 1, January 2000, pp. 238–248.

[23] Smith-Jonforsen, G., L.M.H. Ulander, X. Luo, "Low VHF-Band Backscatter From Coniferous Forests on Sloping Terrain," *IEEE Trans. on Geoscience and Remote Sensing*, Vol. 43, No. 10, October 2005, pp. 2246–2260.

[24] Smith-Jonforsen, G., L.M.H. Ulander, X. Luo, "A Physical-Optics Model for Double-Bounce Scattering From Tree Stems Standing on an Undulating Ground Surface," *IEEE Trans. on Geoscience and Remote Sensing*, Vol. 46, No. 9, September 2008, pp. 2607–2621.

[25] Forest Service, Forest Products Laboratory: "Dielectric Variations of Wood and Hardboard: Variation with Temperature, Frequency, Moisture Content and Grain Orientation," USDA Forest Service Research Paper FPL 245, U.S. Department of Agriculture, 1975.

# 3

# Meter Wavelength Synthetic Aperture Radar

## 3.1 Introduction

The foregoing chapter demonstrated that electromagnetic meter waves possess significant capability to probe into media, with possibilities to detect targets concealed by vegetation and soil and also to measure ground and vegetation characteristics. The typical application to be discussed is radar operating from an aircraft, bypassing the area under investigation, and exploring it from a distance. The distance may vary from a few hundred meters to tens of kilometers. A number of fundamental questions arise regarding the basic feasibility of such a system. The foremost question is that of resolution; that is, how the backscattering from far away objects can be isolated from each other. A follow-up question concerns what sensitivity of the radar is required, since the backscattered power from the small objects here considered will be very small. This chapter and the next chapter will be organized to address these two questions in turn, starting with this chapter describing the principles of SAR.

The SAR principle is a means to create a two-dimensional radar image from either side of the aircraft operating the radar. The SAR image amplitude is *reflectivity*: the concept of reflectivity was introduced for bare ground in the foregoing chapter. The SAR principle is based on the contention that ground in general, forest covered ground, for instance, scatters as a locally homogeneous random surface, just like bare ground. Assuming randomness, the RCS of some small ground segment will be proportional to the area of the segment. For instance, in the case of forested, but otherwise flat and featureless ground, the RCS of the segment will be the sum of the RCS values of the trees within the segment. The ratio between RCS and area of the ground segment defines the reflectivity within the ground segment. As will be subsequently demonstrated, the reflectivity concept also holds for nonrandom structures. For instance, a horizontal cylinder is attributed an RCS value in the SAR process, though this value is not the maximum broadside RCS, but an average RCS. The former grows with cylinder length squared, while the latter will be demonstrated to be proportional to length. Thus, cylinder RCS is obtained as cylinder reflectivity times cylinder length. The same relation also holds for other linear ground features like house walls, fences, and so on. In all, the concept of reflectivity as RCS per unit area is universally applicable.

As defined in Chapter 2, RCS is determined from the ratio between the impinging plane wave field amplitude and the scattered spherical wave field amplitude. This ratio, when squared, compensated for range attenuation by multiplication with range squared, and multiplied with $4\pi$, yields the RCS. Note that the plane wave assumed is a very good approximation for the actual spherical wave transmitted by the radar, since reflectivity is considered in terms of a small area ground segment, while the distance to the radar is large. In the literature on SAR, also the ratio between the impinging and scattered field, compensated for range attenuation, is referred to as reflectivity. In this book, it will, for the sake of clarity, be referred to as the *reflectivity density*, while the word reflectivity will be reserved to mean the RCS per unit ground area (synonymous to reflectivity, the term *backscattering coefficient* is often encountered in the literature). A further standard convention will be to represent reflectivity density as a complex value, this by retaining only the positive frequency part of the reflectivity density Fourier transform (or the frequencies in a half-plane in the case of a two-dimensional spectrum), which yields a complex valued spatial function after an inverse Fourier transform. The reflectivity for any point-like object in the SAR image is then obtained as $4\pi$ times the modulus squared of the reflectivity density associated with the object. The phase of the reflectivity density value expresses the distance from the flight pass to the point scattering center in (as will be seen) twice the number of wavelengths times $2\pi$. Note that according to these definitions, reflectivity as well as reflectivity density will be dimensionless quantities.

From the viewpoint of target detection, the basic resolution requirement in a SAR image is that an individual target can be isolated from the scattering response, coming from neighboring structures. The radar bandwidth is one parameter, which governs the attainable resolution. However, there are several factors, which cause resolution to depend on frequency in other ways than just bandwidth. As will be analyzed, the position of the target scattering center depends on frequency. Thus, resolution will be limited by the amount of range shift of the scattering center, which occurs within the adopted radar band. Frequency also influences resolution due to the frequency dependence of object backscattering amplitude, which was analyzed in Section 2.7. It may be that an object illuminated by a wideband radar signal is well into the Rayleigh region for a part of the frequency band of the signal. Clearly, the resolution will then be determined by the high-frequency part of the illuminating signal. Alternatively, it may be that obscuring forest attenuates the high-frequency part of the illuminating signal to the extent that resolution is reduced to that of the low frequency response.

These concerns, related to the choice of radar band, are present for any type of wideband radar. Before entering into an analysis of the SAR principles, a discussion is provided on reflectivity density reconstruction from the radar transmitted and received signals in the setting of any radar using large fractional bandwidth. The limitations as to what must be considered maximum useful bandwidth will be detailed. Relying on these results, the chapter thereafter continues by providing a formulation of SAR imaging and SAR image reconstruction principles, when using large fractional bandwidth (in contrast to several standard methods for microwave SAR imaging, where assumptions of small fractional bandwidth are adopted). Details, for example, meter wave radar processing and motion compensation methods, are left to Chapter 5. The SAR principles derived enables translation of the scattering theory of the foregoing chapter into reflectivity predictions for SAR images of structures such a forest, buildings, fencing, and so on. A comparison between these predictions and experimental results are provided. The chapter ends by considering the options for imaging moving objects with meter-wave SAR, given the generalized SAR principles, which have been laid down.

## 3.2 Useful Bandwidth

To achieve its objective of combining discrimination of targets of meter order extension with the capability of penetrating vegetation and even soil, the radar must operate at meter order wavelengths (as was instigated in Chapter 2) and also possess meter order resolution. In other words, the radar must operate with large fractional bandwidth (the term *ultrawideband* (UWB) was introduced in Chapter 1, though UWB may also mean that microwave frequencies are included in the radar signal). Generalizing microwave radar imaging principles to large fractional bandwidth requires care at every stage.

The scattering configuration to be considered henceforward is that of radar illuminating the ground from some position $P$ above it. The radar itself consists of a transmitter and receiver sharing a common antenna. The transmit beam expands as a spherical wave with attenuation $1/r$ ($r$ is the range from radar to target) and so does the scattered beam. Any component of the transmitted $\boldsymbol{E}$ or $\boldsymbol{H}$ field incident on some ground point $Q$ at a distance $r$ from the radar can at that point and time $t$ be represented $C_{TX}f^{TX}(t)/r$, where $f^{TX}$ is the digital complex valued signal fed to the radar antenna through the transmitter and $C_{TX}/r$ is determined by the radar characteristics (including the antenna diagram) relating $f^{TX}$ to the field component hitting $Q$. As in Section 2.6.9, the contribution to the received radar signal from some small region around $Q$ can be represented as $df^{RX}(t) = \hat{\gamma}(Q)f^{TX}(t - 2r/c)dA(Q)/r^2$. Here, $dA(Q)$ is a small area segment at $Q$ and $f^{RX}$ is the digital signal captured by the radar antenna and having passed through the radar receiver. The characteristics of the radar transmission/reception process are thus collected in the dimensionless function $\hat{\gamma}(Q)$, which is the aforementioned reflectivity density. If there is a medium obscuring the path between radar and the point of reflection, $\hat{\gamma}(Q)$ will also include the attenuation resulting. These dependencies aside, $\hat{\gamma}(Q)$ determines the fraction of the signal received at time $t$, which is backscattered from $dA(Q)$. Consider the overall received signal at time $t$ from a range interval $dr$. According to Section 2.6.9, it is made up of contributions from all elements $dA(Q)$ at the given range $r$. In the general case, the elements make up the ring-shaped intersection between a spherical shell (with radius $r$ and thickness $dr$) and the ground. Denoting positions $Q$ in the ring by an angle $\phi$, the signal contribution from this ring can be expressed as

$$\begin{aligned} df^{RX}(t) &= \oint_{|PQ|=r} \hat{\gamma}(Q) f^{TX}\left(t - \frac{2r}{c}\right)\frac{dA(\phi)}{r^2} = \gamma(r) f\left(t - \frac{2r}{c}\right)\frac{dr}{r}; \\ \gamma(r) &= \oint_{|PQ|=r} \hat{\gamma}(Q) d\phi \end{aligned} \tag{3.1}$$

From (3.1), $\gamma(r)$ is the average reflectivity density for a certain range, while $\hat{\gamma}$ is the reflectivity for the individual ground points being averaged over. However, if the radar antenna is highly directive (e.g., in some microwave radar application), it may be that $\gamma(r)$ will be zero outside just a small segment of the ring in (2.1). In this case, $\gamma(r)$ will approach the local reflectivity $\hat{\gamma}(Q)$.

It may also happen that $\hat{\gamma}(Q)$ and $\gamma(r)$ are time dependent. This case will be considered in the next section. If assumed static, the full received signal at time $t$ can be stated as

$$f^{RX}(t) = \int_0^\infty \gamma(r) f^{TX}\left(t - \frac{2r}{c}\right)\frac{dr}{r} \tag{3.2}$$

Depending on its orientation, the radar antenna will sense some particular combination of components of the electromagnetic field, which always will be tangential to the field direction of propagation. In relating $df^{\mathrm{RX}}(t) = \hat{\gamma}(Q)$ $f^{\mathrm{TX}}(t - 2r/c)dA(Q)/r^2$ to (2.112), it will be understood that $f^{\mathrm{RX}}$ and $f^{\mathrm{TX}}$ represent this combination.

The objective of the radar measurement is to determine the reflectivity density using (3.1), combined with precise knowledge of the transmit signal. There are several technical criteria, to be discussed later, affecting the design of the transmit signal. Presently, just assume the transmit signal to be arbitrary but known, and with a nonzero spectral content throughout the radar band $\omega_c - \pi B \leq \omega \leq \omega_c + \pi B$. Since also the received signal is known, a process of deconvolution (in radar jargon *pulse compression*) can be applied to (3.1), whereupon

$$\left.\begin{array}{r} r \geq 0 \Rightarrow \gamma(r) = 0 \\ r \geq 0 \Rightarrow f^{RX}(t) = \displaystyle\int_0^\infty \gamma(r) f^{TX}\left(t - \frac{2r}{c}\right)\frac{dr}{r} \end{array}\right\} \Rightarrow \left(\frac{\gamma(r)}{r}\right)_{2\omega r/c} = \frac{f_\omega^{RX}}{f_\omega^{TX}} \tag{3.3}$$

By an inverse Fourier transform restricted to the band $\omega_c - \pi B \leq \omega \leq \omega_c + \pi B$, where data are known, the *bandlimited estimate* $\overline{\gamma}(r)/r$ of $\gamma(r)/r$ is obtained. Also note that while deconvolution is the appropriate way of pulse compression for the wide-bandwidth signals presently analyzed, for microwave radar, pulse compression is normally obtained as a matched filter. Matched filter pulse compression corresponds to making the substitution $f_\omega^{\mathrm{RX}}/f_\omega^{\mathrm{TX}} \rightarrow (f_\omega^{\mathrm{TX}})^* f_\omega^{\mathrm{RX}}$ sin (2.3).

The interesting applications of (3.1) and (3.2) are when reflectivity density represent local ground conditions. Consider the case above with a highly directive antenna, i.e., at $r$, i.e., $\gamma = \gamma(r)$ represents the conditions in the immediate vicinity of some ground point $Q$. Scattering physics will then determine some complicated functional dependence $\gamma(r) = \boldsymbol{F}[h(Q),\eta(Q),X(Q),Y(Q),\ldots]$ of variations in ground height, index of refraction, and whatever other range-dependent ground features affect the backscattering. The radar measurement determines an interrelation between these. Just as an example, consider the SPM theory, defining reflectivity density from the ground height profile. Write $\gamma(r) = \chi(x)$; $x = r\sin\theta$ to relate to (2.81) and (2.100) in Chapter 2. Thus, for (2.81) and momentarily disregarding bandwidth

$$\begin{aligned} \hat{f}_{\omega}^{RX} &= -i\frac{2\omega}{c}\frac{\cos\theta}{{}_x\sin\theta}R_E h_{2\omega/c} \Rightarrow \\ \gamma(r) = \chi(x) &= \frac{1}{2\pi}\int e^{ik_x x}\hat{f}_{ck_x/2}^{RX}\,dk_x = -R_E\frac{\cos\theta}{\sin\theta}\frac{d}{dx}\left[\frac{1}{2\pi}\int e^{ik_x x}h_{k_x}\,dk_x\right] \end{aligned} \tag{3.4}$$

Proceeding similarly for (2.100)

$$\begin{aligned} &\text{Horizontal polarization: } \gamma(r) = -R_E\frac{\cos\theta}{\sin\theta}\left[\frac{dh}{dx}\right]_{x=r\sin\theta} \\ &\text{Vertical polarization: } \quad \gamma(r) = R_E\frac{\cos\theta}{\sin\theta}\frac{\sin^2\theta+\cos^2\tilde{\theta}}{\cos\left(\theta-\tilde{\theta}\right)^2}\left[\frac{dh}{dx}\right]_{x=r\sin\theta} \end{aligned} \tag{3.5}$$

Apart from reflectivity density $\gamma(r)$, also consider the relevance of reflectivity $|\gamma(r)|^2$. For the SPM theory, the cross-sectional formulas (Section 2.6.10) directly translate into the case of finite bandwidth. The backscattering amplitude and RCS as functions of frequency just become density functions. The Fourier transform identifies backscattering spectral density with reflectivity density. Parseval's theorem identifies cross section with reflectivity for stationary random ground. The result obviously generalizes from the SPM theory to any ground reflectivity distribution, which has a stationary random character.

In (3.1), it was agreed that the reflectivity $\hat{\gamma}(Q)$ lumped together the backscattered fraction of the field impinging on $Q$ with the radar system transfer characteristic, radar system calibration, and the coefficient $k\cos\theta/4\pi$ of (Section 2.6.9). In regard to the cross-sectional formula (Section 2.6.10), it is possible to let calibration compensate for all transfer effects so that RCS becomes $d\sigma = \left|\hat{\gamma}(Q)\right|^2 dA(Q)$. Henceforward, radar system calibration will invariably be assumed conducted in a manner so that this relation is satisfied, thus endowing $\hat{\gamma}(Q)$ with a direct and intuitive meaning.

The bandwidth limitation in (3.2) has the effect of limiting the measurement in terms of the attainable resolution for the range parameter. The effect is easily analyzed for a point reflector at some range $r_0$, that is, $\gamma = \gamma_0\delta(r - r_0)$. From (2.1), follows that $f^{\mathrm{RX}}(t) = \gamma_0 f^{\mathrm{TX}}(t - 2r_0/c)$ according to which $f_{\omega}^{\mathrm{RX}} = e^{-i2r_0/c}\gamma_0 f_{\omega}^{\mathrm{TX}}$. Hence, the band limited point reflector estimate is

$$\begin{aligned}\bar{\gamma}(r) &= \frac{\gamma_0}{\pi c}\int_{\omega_c-\pi B}^{\omega_c+\pi B} e^{i2\omega(r-r_0)/c}\,d\omega = \frac{\gamma_0 e^{i2\omega_c(r-r_0)/c}}{\pi c}\int_{-\pi B}^{\pi B} e^{i2\omega(r-r_0)/c}\,d\omega \\ &= \frac{2\gamma_0 e^{i2\omega_c(r-r_0)/c}}{\pi c}\int_0^{\pi B}\cos[2\omega(r-r_0)/c]\,d\omega \\ &= \frac{\gamma_0 e^{i2\omega_c(r-r_0)/c}}{\pi(r-r_0)}\sin\frac{2\pi B(r-r_0)}{c}\end{aligned} \tag{3.6}$$

Evidently, due to the limited bandwidth, the local point response is spread indefinitely across all ranges. It attenuates inversely proportional to the distance from $r_0$, remaining approximately constant within the first half-period of the sine function. A fair representation of resolution must take this spreading of sidelobes into account. For the typical cluttered scene of natural ground, there will be a large number of scatters, each providing their own sidelobe structure. Their collective sidelobe pattern will effectively limit detection performance unless resolution is sufficient. Preferably, the resolution estimate adopted should be defined in such a way that a fine resolution also guarantees that sidelobes are effectively suppressed outside the resolution cell. The natural resolution definition taking the sidelobe structure into account is

$$\Delta r = \frac{\int_{-\infty}^{\infty}\left|\bar{\gamma}(r)\right|^2 dr}{\left|\bar{\gamma}(r_0)\right|^2} \tag{3.7}$$

Resolution defined according to (3.7) will be referred to as *weighted resolution*. With this definition, there is an optimum resolution entirely dependent on bandwidth, as follows by representing (3.6) in the Fourier domain. According to Parseval's theorem for a signal with bandwidth $B$

$$\Delta r = 2\pi\frac{\int_{2(\omega_c-\pi B)/c}^{2(\omega_c+\pi B)/c}\left|\bar{\gamma}_k\right|^2 dk}{\left|\int_{2(\omega_c-\pi B)/c}^{2(\omega_c+\pi B)/c}\bar{\gamma}_k e^{ikr_0}\,dk\right|^2} \tag{3.8}$$

In (3.7), the well-known Schwarz inequality can be applied. This inequality stipulates that

$$\int \left|f_k\right|^2 dk \int \left|g_k\right|^2 dk \geq \left|\int f_k g_k^* dk\right|^2 \tag{3.9}$$

or any functions $f_k$ and $g_k$. Selecting $f_k = \gamma_k$ and $g_k = e^{-ikr_0}$, the following inequality for resolution follows that:

$$\Delta r \geq 2\pi \left[ \int_{2(\omega_c - \pi B)/c}^{2(\omega_c + \pi B)/c} dk \right]^{-1} = \frac{c}{2B} \tag{3.10}$$

To obtain equality in (3.9), the two functions $f_k$ and $g_k$ must be linearly dependent. Therefore, for equality in (3.10) $\gamma_k$ must be proportional to $e^{-ikr_0}$, which is the Fourier transform of a point reflector. Indeed, applying (3.7) to (3.6)

$$\Delta r = \int_{-\infty}^{\infty} \frac{\sin^2 \frac{2\pi Br}{c}}{\left(\frac{2\pi Br}{c}\right)^2} dr = \frac{c}{2\pi B} \int_{-\infty}^{\infty} \frac{\sin^2 x}{x^2} dx = \frac{c}{2B} \tag{3.11}$$

Hence, as expected, the point reflector satisfies the resolution limit. This limit resolution will therefore be referred to as *point scatterer resolution* (PSR). Real physical objects will now be considered. Their weighted resolution will be found coarser than PSR, as is to be expected according to inequality (3.8).

The cylinder forms an important case in its own, and thus suits well to illustrate the generalization to finite bandwidth. The spectral reflectivity density is obtained, taking the result leading up to (2.206), but before forming the squared modulus and adjusting by multiplication with $4\pi r_0^2$, where $r_0$ is the range to the cylinder center. Shifting the location of the cylinder from the origin to the position $r_0$ means that the phase for the impinging and backscattered signals must both be shifted corresponding to twice that distance. The following expressions are got:

$$\begin{aligned} &\text{TM polarization: } \gamma_\omega = B(\omega)e^{-i2kr_0};\ B(\omega) = \frac{L}{\pi k r_0}\sum_{n=0}^{\infty}(-1)^n \breve{b}_n \\ &\text{TE polarization: } \gamma_\omega = A(\omega)e^{-i2kr_0};\ A(\omega) = \frac{L}{\pi k r_0}\sum_{n=0}^{\infty}(-1)^n \breve{a}_n \end{aligned} \tag{3.12}$$

The expressions in (3.12) deviate from the point reflector response in that the factor $A$ and $B$ also depends on frequency. Consequently, $\gamma(r)$ will not be focused to a single point if bandwidth is large. Depending on the frequency content of the selected radar band, the spreading will be more or less pronounced. Starting in the optical regime, both $A$ and $B$ have their only frequency dependence through a factor $e^{-i2ka}$, meaning that for these frequencies the response is as a point reflector at the position on the cylinder surface closest to the radar. Approaching frequencies in the resonance regime, spreading of $\overline{\gamma}(r)$ starts and becomes ever more pronounced, the frequencies that are included in the radar band will be lower. As long as the spreading $\overline{\gamma}(r)$ stays within the PSR for the bandwidth utilized, there is little noticeable impact on radar performance. Increasing bandwidth further, the spreading will extend beyond the PSR, meaning that increasing bandwidth may not always provide the expected resolution gain.

The spreading of $\overline{\gamma}(r)$ must be investigated for both targets and clutter objects in order to decide on allowable bandwidths. Consider detection of targets located among trees as a guiding example. From the discussion in Chapter 2, the higher frequency limit of the exploited radar band should be selected so that tree elements, smaller than the target, are put in the Rayleigh region. The criterion for achieving this is found in Section 2.7 to be use of TE polarization with $\lambda_{min} = 15a_{stem}$, where $a_{stem}$ is the tree stem radius. As for cylinder-like targets, the requirement chosen for keeping RCS large was $\lambda_{max} = 10a_{targ}$, where $a_{targ}$ is the target characteristic radius. Bandwidth should be selected to keep targets and clutter structures separated within the achievable resolution. Assume that objects are separated by a characteristic distance $2\alpha a_{targ}$, that is, a distance $\alpha$ times larger than their characteristic extension. It is required that $\Delta r \le 2\alpha a_{targ}$. With $\lambda_{max} = 10a_{targ}$, the required separation becomes related to fractional bandwidth as

$$\Delta r = 2\alpha a_{targ} = \frac{1}{2}\frac{1}{\dfrac{1}{\lambda_{min}} - \dfrac{1}{\lambda_{max}}} \Rightarrow \alpha = \frac{2.5}{\dfrac{\lambda_{max}}{\lambda_{min}} - 1} = 2.5\frac{1-\beta}{\beta} \tag{3.13}$$

Here $\beta$ is the fractional bandwidth, defined as $\beta = B/f_{max} = 1 - f_{min}/f_{max} = 1 - \lambda_{min}/\lambda_{max}$. The minimum object separation $\alpha = 1$ implies a wavelength quotient $\lambda_{max}/\lambda_{min} = 3.5$, i.e., $\beta = 0.71$. The quotient $a_{targ}/a_{stem} = 3\lambda_{max}/2\lambda_{min} = 5.25$ follows. For instance, if $a_{stem} = 0.2$ m, the minimum target size providing good RCS is $2a_{targ} = 2.1$ m. For smaller targets, the bottom frequencies of this band will put the target in the Rayleigh regime with weak backscattering. If this is the case, the low end of the transmitted band might be skipped, since the radar image quality in any case will appear as if fractional bandwidth is reduced. The reduced bandwidth results in reduced resolution, which requires increased separation between ground objects if the image is going to distinguish targets. For instance, by relaxing the resolution requirement to $\alpha$ = 3, it follows from (3.8) that wavelength quotient is $\lambda_{max}/\lambda_{min}$ = 3.5/3, whereas $\beta$ = 0.45 and $1.75a_{stem} \approx a_{targ}$. In this case, targets with size $2a_{targ}$ = 0.7 m for $a_{stem}$ = 0.2 m will respond well to the radar signal.

The question that must be analyzed is if scattering center stability is sufficient so that the weighted resolution attained for real physical objects remains sufficiently close to the PSR, to make the arguments regarding fractional bandwidth valid. The term scattering center refers to the apparent position from where target backscattering occurs at a certain frequency. The concept is an approximate one, since its applicability requires that $\gamma_\omega$ in (3.6) has the form $\gamma_{k/c} = A(k)e^{i[\Phi(k)+2k(r-r_1)]}$; here for the $\omega$ variable, the exponential must vary rapidly, with the magnitude $A(k)$ changing only slowly. Whenever this is the case, the integration (3.2) obtains its dominating contribution around the instant when the phase variation with respect to $k$ is at its minimum. There is then a functional dependence between $k$ and the scattering center location $r$, that is

$$\gamma(r) \neq 0 \Leftrightarrow \frac{d}{dk}\left[\Phi(k) + 2k\left(r - r_1\right)\right] = 0 \Leftrightarrow r(k) = r_1 - \frac{1}{2}\frac{d\Phi}{dk} \tag{3.14}$$

For the point scatter, the position of the scattering center is independent of frequency. The scattering center for the cylinder is obtained by a phase unwrapping the complex valued expressions (3.6), followed by differentiation. The results for either polarization and finite and infinite dielectric properties are illustrated in Figure 3.1. As seen, in the optical regime, the scattering center is on the cylinder mantle but shifts to the cylinder interior when frequencies fall into the resonance and Rayleigh regions. At these frequencies, phase variations of $\gamma_\omega$ become slower, whereas amplitude changes just as fast as phase, making the concept of scattering center unreliable. Nevertheless, Figure 3.1 helps in interpreting accurate numerical results, which now will be provided.

Figures 3.2 and 3.3 illustrate the diverse results obtained when applying (3.2) numerically for cylinders, which are either perfectly reflecting or having finite dielectric properties. The spread in the weighted resolution (3.7) is annotated. As seen, the case of a perfectly reflecting cylinder provides resolution very close to PSR (which in fact is expected from the stability of the scattering center in this case as per Figure 3.1). It can thus be concluded that by choosing a frequency band as per the discussion above, a perfectly reflecting target concealed among tree stems yields significant and well-focused backscatter, which would tend to dominate the response from the surrounding trees, thus enabling detection of the target. However, as is shown in Figure 3.3, the same

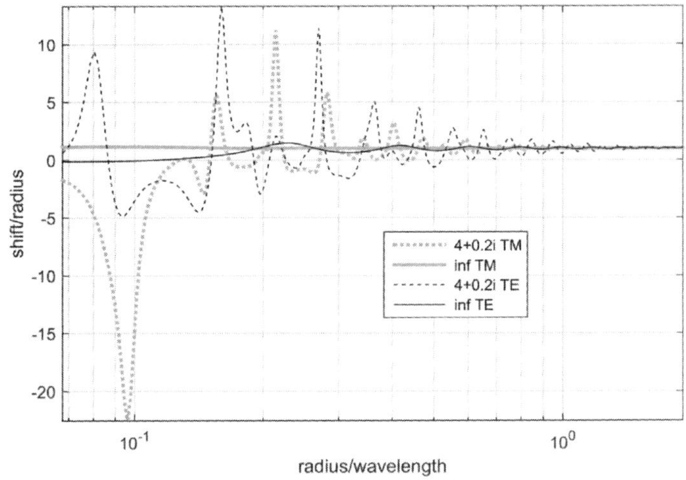


**Figure 3.1** Scattering center shift with frequency. The graph clearly shows that independently of polarization or dielectric properties, the scattering center is fixed at the cylinder mantle throughout the optical regime. Passing into the resonance region, it continues to be stable for vertical polarization and a ideally reflecting cylinder. For horizontal polarization, it becomes slightly variable also in the case of ideal reflection. When dielectric properties are finite, the scattering center becomes very unruly. Because of the unruliness, the scattering center approximation cannot be used for quantitative response estimates though it is an indicator that that significant smearing of the cylinder response is expected.

is not at all true if the target has a moderately low index of refraction. The lack of focus shown is a serious concern. Consider also situations where the target to be detected is a man-made metallic object, which thus can be well focused, but where it is surrounded by similarly sized (and thus high RCS) natural objects, for example, blocks of stone. The index of refraction values for these objects are taken from Figure 2.1. Resonance region scattering properties are insensitive to the shape details, so the scattering behavior of these surrounding objects will be close to that predicted for the cylinder, with Figure 3.3 bearing on their characteristic size. In particular, severely rocky terrain will be poorly imaged by radar operating with large fractional bandwidth within the resonance regime of the typical rock dimensions in the scene, with detection performance for targets hiding in such terrain presumably low. Another area where the scattering center instability will degrade performance is detection of buried nonmetallic objects (plastic mines, for instance) using broadband signals in the near resonance region of these objects.

A partial remedy to this situation is observed, namely, that if the scattering density frequency behavior in the scene imaged can be anticipated,

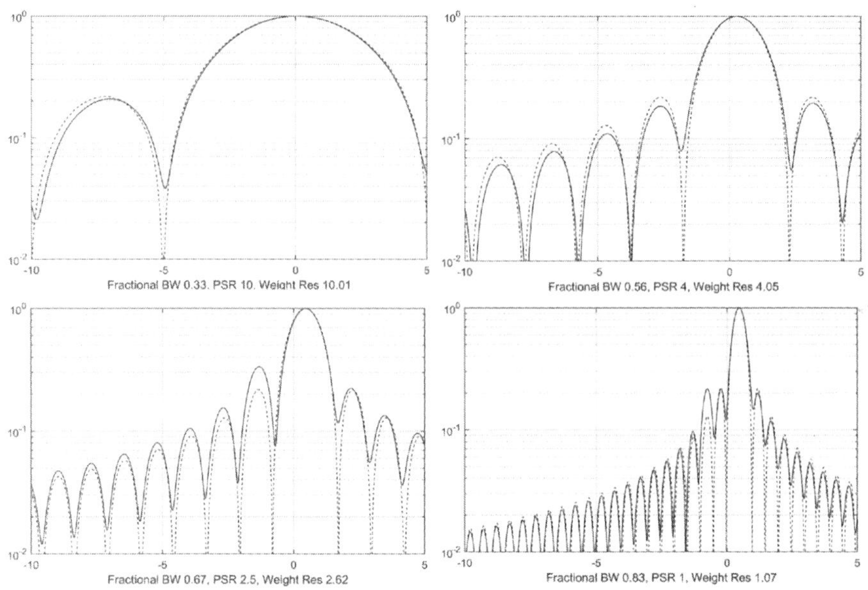


**Figure 3.2** Bandlimited cylinder reflectivity density function (continuous line) and fitted point reflector response (dashed line). The cylinder is perfectly reflecting and centered at the origin. Horizontal scale is range in multiples of object width $2a$ and the vertical scale is peak normed reflectivity density magnitude. The figures covers four cases of different fractional bandwidths, all with $\lambda_{max} = 10a$. The shift of the scattering center from the cylinder center to the cylinder mantle at larger bandwidths is noted. Some increase in sidelobe levels compared to a point reflector are noted, but overall, the response of the reflecting cylinder is well focused in accordance with the stale phase center in Figure 3.1.

scattering center instabilities can be compensated for. This is clear from (3.2). If $f^{TX}_{\omega}$ is substituted for $\gamma^0_{2\omega/c} f^{TX}_{\omega}$, where $\gamma^0_{2\omega/c}$ is the anticipated scattering density of an object that may or may not be located in the scene, the estimated density $\overline{\gamma}_{2\omega/c}$ will represent this object as a point scatterer located at the actual position of the object. To set up $\gamma^0_{2\omega/c}$ requires knowledge of the object shape, size, orientation, and dielectric properties, so the method can only be applied to focus a class of objects that have all these properties in common and preferably only slowly depend on these parameters. Typical clutter objects coming in forms that are quite disparate in their properties can therefore not be refocused in this manner. However, for targets like certain type of plastic mines, the method should be tried. Favorable for detection is when they are sufficiently small to be in Rayleigh region, making the backscattering response independent of shape and orientation. Equally important is that the method

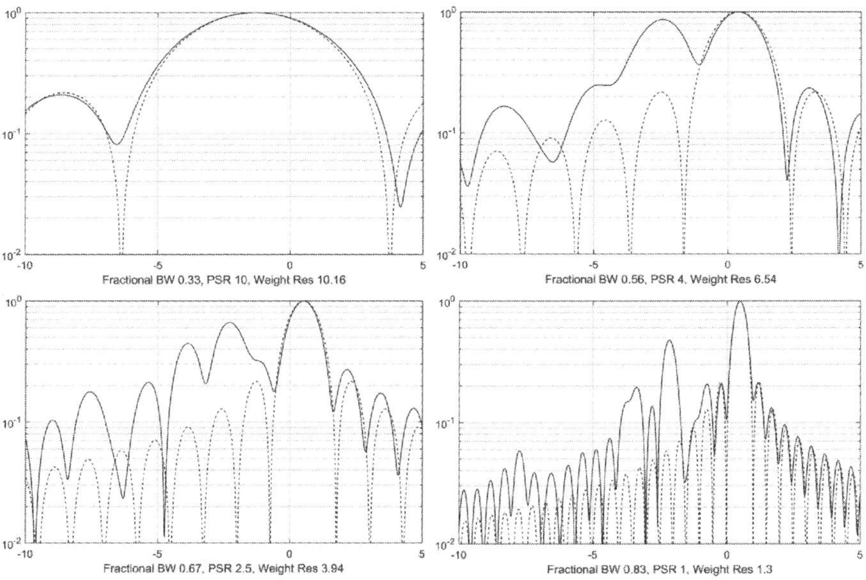


**Figure 3.3** Continuous line represents cylinder with index of refraction $\eta = 4 + 0.3i$, typical for granite (see Figure 2.1)—the figure otherwise complies with Figure 3.2. The scattering center instabilities indicated in Figure 3.1 are here seen as giving rise to significant defocusing. The backscattering response spreads with a large magnitude over a range interval significantly larger than the nominal point scatterer response in all cases but fractional bandwidth 1.5. The overall conclusion is that any reasonable sidelobe suppression is not possible with a resolution better than about five to ten times the object width.

could be applied for detection horizontal underground tunnels, where the diameter of the tunnel is the one parameter that has to be assumed, in order to refocus the image. Typically, detection would be based on running a number of computations with different tunnel diameter settings.

## 3.3 SAR Imaging Principles

A camera resolves the field impinging on its lens into its directional components. In contrast, radar resolves the field impinging on the antenna into temporal components. Since this field is a result of scattering of a field transmitted by the antenna, the temporal dependencies can be resolved into distance, that is, a range dependent reflectivity profile of the backscatter. In its generic form and essentially what will be considered here, radar collects data omnidirectionally, thus not providing any angular resolution at all. However, for either a radar or the camera, by shifting the position of the antenna or lens and repeating the measurement, independent data are obtained, enabling resolution in both angular and distance dimensions. For a camera, with angular resolution being the primary function of the sensor, range resolution is obtained by what is termed *parallax*. With radar, the phenomenon exploited to obtain angular resolution is termed *range migration*, while the image formation technique based on shifting radar position is known as *SAR*. For both optics and radar, it is required that the scene imaged remains stationary during the shift of the sensor.

The alternative to the SAR technique for obtaining angular resolution is a directive radar antenna. For meter-wave radar, the angular resolution attainable from a directive antenna of reasonable dimensions is almost nonexistent. Using the SAR technique, attainable angular resolution will be a fraction wavelength at any distance. Thus, submeter resolution can be attained.

The shifting of the radar position gives a temporal aspect to the collected data. With the range and angular coordinates fixed to the radar, also the reflectivity density will vary in time, even though this time dependence is due to the sensor motion relative to an otherwise fixed scene. Rather than (3.1), a more general, time-dependent relation may be set up between the transmitted and received radar signals and the reflectivity density.

Define the radar range and time of backscatter as the intermediate between a particular instance of transmission $t_1$ and reception $t_2$

$$r = c\frac{t_2 - t_1}{2};\ t = \frac{t_2 + t_1}{2} \tag{3.15}$$

Then instead of (3.1), we have

$$f^{\mathrm{RX}}\left(t_2\right) = \frac{c}{2}\int_0^\infty \gamma(r,t)\, f^{\mathrm{TX}}\left(t_1\right)\frac{dt_1}{r} \tag{3.16}$$

To solve (3.2) with respect to the reflectivity density, note the identity

$$t_2\omega_2 - t_1\omega_1 = \frac{t_2 - t_1}{2}\left(\omega_2 + \omega_1\right) + \frac{t_2 + t_1}{2}\left(\omega_2 - \omega_1\right) \tag{3.17}$$

Thus, performing a Fourier transform of $f^{\mathrm{RX}}$ and inserting the Fourier transform of $f^{\mathrm{TX}}$ in this

$$\frac{\partial(t_1,t_2)}{\partial(r,t)}=\frac{2}{c}\Rightarrow f_{\omega_2}^{\mathrm{RX}}=\frac{1}{2\pi}\int_{-\infty}^{\infty}\left[\frac{\gamma(r,t)}{r}\right]_{\frac{\omega_2+\omega_1}{c},\omega_2-\omega_1} f_{\omega_1}^{\mathrm{RX}}\,d\omega_1 \tag{3.18}$$

Evidently, the temporal development of the reflectivity density has its spectral correspondence in a frequency shift $\omega_2 - \omega_1$ between the transmitted and received signals. The frequency shift is often referred to as *Doppler shift*. This shift can be made measurable by making the transmitted signal periodic (with period $\tau$), in which case it has a discrete Fourier spectrum

$$f_{\omega}^{\mathrm{TX}}=\sum_{n=T(F_c-B/2)}^{T(F_c+B/2)} C_n\delta\left(\omega-\frac{2\pi n}{\tau}\right) \tag{3.19}$$

Here $C_n$ are the known spectral coefficients of the transit signal. In this case

$$2\pi f_{\omega_2}^{\mathrm{RX}}=\sum_{n=T(F_c-B/2)}^{T(F_c+B/2)} C_n\left[\frac{\gamma(r,t)}{r}\right]_{\frac{\omega_2}{c}+\frac{2\pi n}{\tau c},\omega_2-\frac{2\pi n}{\tau}} \tag{3.20}$$

The temporal development of $\gamma(r,t)$ will be assumed band limited, that is, $[\gamma(r,t)/r]_{(\omega_2+\omega_1)/c,\omega_2-\omega_1} = 0$ unless $\omega_2 - \omega_1$ is within a given bandwidth $B_0$ (the *Doppler bandwidth*). Choosing radar transmission periodicity sufficiently small, $2\pi/\tau$ can be made larger than $B_0$. Then for every $\omega_2$, there will be just one index $n = n(\omega_2)$ such that $[\gamma(r,t)/r]_{(\omega_2+\omega_1)/c,\omega_2-\omega_1}$ is nonzero and therefore for every nonzero $f_{\omega_2}^{\mathrm{RX}}$ only one nonzero term in (3.6). It follows that $[\gamma(r,t)/r]_{(\omega_2+\omega_1)/c,\omega_2-\omega_1}$ can be determined from $f_{\omega_2}^{\mathrm{RX}}$. The ratio $B_0/B$, that is, the "slowness" of the temporal development of $\gamma(r,t)$ is set by the maximum time derivative of spatial variations compared to the speed of light. By this rule, most such variations and certainly those to be considered here are extremely slow implying that $|\omega_2 - \omega_1| << |\omega_1|$ within the adopted frequency band. The approximation $\omega_2 + 2\pi n(\omega_2)/\tau \approx 2\omega_2$ will therefore be valid. It agrees with what is referred to as the *start–stop approximation* in radar jargon, that is, it assumes that the time variation of the reflectivity is negligible in establishing the spatial character. An approximate bandlimited estimate $\overline{\gamma}(r,t)$ of $\overline{\gamma}(r,t)$ thereby follows:

$$\left[\frac{\overline{\gamma}(r,t)}{r}\right]_{\frac{2\omega}{c},\Delta\omega}=2\pi\frac{f_{\frac{2\pi n}{\tau}+\Delta\omega}^{\mathrm{RX}}}{C_{n(\omega)}} \tag{3.21}$$

If required (which will not be the case here), the more exact method of using (3.20) directly could be used as an alternative to (3.21). Also note that in (3.21) only frequencies $\omega > 0$ needs to be considered; since $\omega$ represents the radial spectrum of $\gamma(r,t)$ realized by a Fourier Bessel transform, only the positive frequencies are required.

As mentioned, in SAR, the time dependence in $\gamma(r,t)$ is caused by a uniform sensor motion across a stationary underlying reflectivity distribution. The stationary reflectivity density is a feature of the ground, basically expressible as a function $\hat{\gamma}(P)$, where $P$ is any point on the undulating ground surface. The sensor motion is assumed to occur straight and uniformly along the $y$-axis, as per Figure 3.4. With respect to the $y$-axis, the ground surface can be expressed in cylinder coordinates, for example, in terms of $y$, incidence angle $\theta$, and cylinder radius $\rho$ as $\rho = \rho(\theta,y)$, or alternatively $\theta = \theta(\rho,y)$. Correspondingly, the reflectivity density forming the SAR image can be expressed as $\hat{\gamma}(P) = \hat{\gamma}(\theta, y) = \hat{\gamma}(\rho, y)$. The last representation, referred to as *slant range* representation, is essential for the SAR technique. In adopting this, it must be understood that it is not universally valid. The representation is certainly ambiguous regarding the ground to the right and left of the sensor path; there will be two angles $\theta_1 > 0$ and $\theta_2 < 0$ representing two distinct ground points, for which the reflectivity contributions become juxtaposed, when representing them with coordinates $\rho,y$. For this reason, the SAR imaging process must be limited to one side of the sensor path, and it will be required that the radar system is designed to avoid or suppress responses from the other side. Restricting imaging to one side is still not a sufficient condition to avoid image anomalies. For one, if the ground rises too steeply with increasing $\theta$, it may happen that $\rho$ does not increase or may even diminish. Also in this case, $\theta = \theta(\rho,y)$ becomes an ambiguous representation of the ground surface. Secondly, situations may arise where the ground drops with increasing $\rho$, making $\theta$ diminishing. In this case, $\rho = \rho(\theta,y)$ is an ambiguous representation of the ground surface. In both cases, SAR imaging breaks down. The first case (known as image *layover*) encompasses specular reflection from a ground segment, effectively disabling any resolution based on the SAR method within that segment. The second case is that of *shadowing* a certain ground segment, obviously disabling SAR resolution within the segment. The requirement for nonanomalous SAR imaging is thus that the functions $\rho = \rho(\theta,y)$ and $\theta = \theta(\rho,y)$ are one to one and each other's inverse.

Assume that the sensor moves with constant speed $v$ along the $y$-axis. Since $\gamma(r,t)$ represents the collective reflectivity density, stemming from the radar range $r$ at time $t$, it will be the superposition of all backscatter responses from the intersection between the ground and a sphere of radius $r$, centered

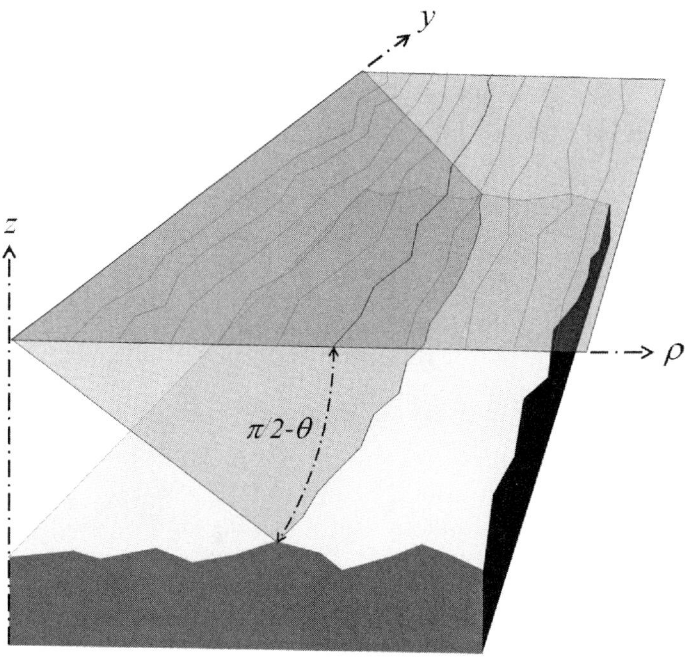


**Figure 3.4** SAR imaging geometry. The sensor path constitutes the $y$-axis. The reflectivity density is basically defined on the undulating ground surface. It can be represented as a function of position along the $y$-axis and slant range (the distance from the sensor path to the ground point) if 1) data is restricted to a half space bounded by a plane through the sensor path and 2) for this half space there is a one-to-one map $\rho = \rho(\theta, y)$ with inverse $\theta = \theta(\rho, y)$, defining the ground surface. Consequently, the reflectivity density function can be represented $\hat{\gamma} = \hat{\gamma}(\rho, y)$. The geometry conventions serve to represent the data received by the radar as decomposed into components, each providing the backscattering from the intersection of the ground surface with a sphere, centered on any sensor position $vt$along the $y$-axis, and with any radius $r$. The ground points of the intersection are given by coordinates $\rho = r\cos\phi$, $y = vt + r\sin\phi$, for $-\pi/2 \leq \phi\pi/2$.

on the sensor position $y = vt$ at time $t$. The sphere may be split up into circles with their axes coinciding with the $y$-axis, each circle with the radius $\rho = r\cos\phi$ and its center at the $y$-position $vt + r\sin\phi$. Considering just one side of the sensor path, such a circle meets the ground at the point $\theta(\rho,y) = \theta(r\cos\phi, vt + r\sin\phi)$. Since the reflectivity at this point is $\hat{\gamma}(\rho, y)$, the following relation is obtained from (3.1):

$$\gamma(r,t) = \int_{-\pi/2}^{\pi/2} \hat{\gamma}(r\cos\phi, vt + r\sin\phi)\, d\phi \tag{3.22}$$

For the subsequent analytical treatment, it will be convenient to restate (3.22) slightly, by allowing $\hat{\gamma}(\rho,t)$ to extend to negative $\rho$ and assuming symmetry $\hat{\gamma}(\rho,t)=\hat{\gamma}(-\rho,t)$. Thereby, the integral in (3.22) can be extended over the full circle, that is

$$\gamma(r,t)=\frac{1}{2}\oint\hat{\gamma}(r\cos\phi,vt+r\sin\phi)d\phi \tag{3.23}$$

Note, however, that $\hat{\gamma}(\rho,t)$ only has physical significance for $\rho > 0$.

In (3.24), the function $\gamma(r,t)$ is determined from radar data according to (3.21), while $\hat{\gamma}(\rho,y)$ is the yet unknown SAR image. Bandwidth limitations will reflect on the resolution attainable for $\hat{\gamma}(\rho,y)$ as will be analyzed in the sequel. To determine $\hat{\gamma}(\rho,y)$, (3.22) may be viewed as an integral equation. Mathematically, it belongs to the realm of *integral geometry* and has a linear counterpart (straight lines instead of circles) in the *Radon transform*, forming the basis of X-ray tomography. The equation can be solved formally by spectral decomposition, starting by performing a Fourier transformation with respect to time[1]

$$\begin{aligned}\gamma(r)_{\Delta\omega}&=\frac{1}{2}\oint\int_{t=-\infty}^{\infty}\hat{\gamma}(r\cos\phi,vt+r\sin\phi)e^{-i\Delta\omega t}\,dtd\phi\\&=\frac{1}{2v}\oint e^{i\left(\frac{\Delta\omega}{v}\right)r\sin\phi}\hat{\gamma}(r\cos\phi)_{\frac{\Delta\omega}{v}}\,d\phi\end{aligned} \tag{3.24}$$

Expand $\hat{\gamma}(r\cos\phi)_{\omega/v}$ with respect to its Fourier transform. The following expression is obtained:

$$\gamma(r)_{\Delta\omega}=\frac{1}{4\pi v}\int_{-\infty}^{\infty}\left\{\oint e^{i\left[\left(\frac{\Delta\omega}{v}\right)r\sin\phi+k_\rho r\cos\phi\right]}d\phi\right\}\hat{\gamma}_{k_\rho,\frac{\Delta\omega}{v}}dk_\rho \tag{3.25}$$

Put $\boldsymbol{k}=(\omega/v,k_\rho)$ and $\boldsymbol{r}=(r\sin\phi,r\cos\phi)$. The exponent can be restated as

$$\left(\frac{\Delta\omega}{v}\right)r\sin\phi+k_\rho r\cos\phi=\boldsymbol{r}\bullet\boldsymbol{k}=rk\cos\psi \tag{3.26}$$

[1] For functions $f(x,y)$ of two parameters, the Fourier transform with respect to the first parameter is denoted by $f_u(y)$; the Fourier transform with respect to the second is denoted by $f(x)_v$, and the transform with respect to both is denoted by $f_{u,v}$.

where $\psi$ is the angle between $\boldsymbol{k}$ and $\boldsymbol{r}$. Note the well-known Bessel function integral representation

$$J_0(z) = \frac{1}{2\pi}\oint e^{iz\sin\psi}\, d\psi$$

It follows that:

$$\gamma(r)_{\Delta\omega} = \frac{1}{2v}\int_{-\infty}^{\infty} J_0\left(r\sqrt{k_\rho^2 + \left(\frac{\Delta\omega}{c}\right)^2}\right)\hat{\gamma}_{k_\rho,\frac{\Delta\omega}{v}} dk_\rho \tag{3.28}$$

Integration with respect to $k_\rho$ is now replaced by integration with respect to $k = \sqrt{k_\rho^2 + (\Delta\omega/c)^2}$, with values $0 < k < \infty$. The substitution is one to one, only when the integral in (3.28) is restricted to either $k_\rho > 0$ or $k_\rho < 0$. Since $\hat{\gamma}(\rho,t)$ was assumed symmetric for $\rho$ positive or negative, so is $\hat{\gamma}_{k_\rho,\omega/v}$ symmetric, that is, either case gives the same result. Hence

$$\gamma(r)_{\Delta\omega} = \frac{1}{v}\int_0^{\infty} J_0(rk)\hat{\gamma}_{k_\rho,\frac{\Delta\omega}{v}} dk_\rho = \frac{1}{v}\int_0^{\infty} J_0(rk)\hat{\gamma}_{\sqrt{k^2-\left(\frac{\Delta\omega}{c}\right)^2},\frac{\Delta\omega}{v}} \frac{kdk}{\sqrt{k^2 - \left(\frac{\Delta\omega}{c}\right)^2}} \tag{3.29}$$

In Chapter 2, the Fourier–Bessel theorem was used in (2.146) to subdivide an arbitrary cylinder coordinate function in cylinder coordinate particular solutions to the wave equation. The same theorem in its form

$$f_k^{FB} = \int_0^{\infty} J_0(rk)f(r)rdr \Leftrightarrow f(r) = \int_0^{\infty} J_0(rk)\, f_k^{FB} kdk \tag{3.30}$$

now implies the following spectral representation of $\hat{\gamma}(\rho, y)$:

$$\hat{\gamma}_{k_\rho,\frac{\Delta\omega}{v}} = v\left|k_\rho\right|\int_0^{\infty} J_0\left(r\sqrt{k_\rho^2 + \left(\frac{\Delta\omega}{v}\right)^2}\right)\gamma(r)_{\Delta\omega} rdr \tag{3.31}$$

Here, extension to negative $k_\rho$ is based on the assumed symmetry of $\hat{\gamma}_{k_\rho,\omega/v}$. The modulus sign can be removed by relating the right-hand side of (3.28) to the *Hilbert transform* $\hat{\gamma}^H(\rho, y)$ of $\hat{\gamma}(\rho, y)$ with respect to $\rho$. A Fourier transform of a Hilbert transformed function corresponds to multiplication with the sign function, precisely changing $|k_\rho|$ to $k_\rho$. Also setting $k_y = \omega/v$, (3.28) can be expressed entirely in spatial wave numbers

$$\hat{\gamma}^H_{k_\rho,k_y} = \frac{v}{2} k_\rho \int_0^\infty J_0\left(r\sqrt{k_\rho^2 + k_y^2}\right)\gamma(r)_{vk_y} r dr \tag{3.32}$$

In other words, up to multiplication with the spatial frequency $k_\rho$, the two-dimensional Fourier Hilbert transform of the SAR image equals the measured reflectivity density, Fourier transformed with respect to time, and Fourier–Bessel transformed with respect to range.

It should be remarked that (3.32) constitutes an exact solution to the SAR image reconstruction problem, in contrast to the several approximate solutions that are established, and in use for microwave SAR applications (see [1]). The accuracy of the solution becomes of increasing importance when resolution approaches the wavelength limit. Thus, the exact solutions are essential for meter-wave SAR, operating at large fractional bandwidth. Formulating the SAR reconstruction problem as an integral equation (3.24) was initially done by the author in the mid-eighties. Its formal solution, (3.32) was soon thereafter provided by the late mathematician Lars–Erik Anderson (see [2] and [3]). At the same time, but independently, an identical problem was treated by Fawcett [4], motivated by seismological applications. Well known to the radar community is the later derivation of this SAR inversion formula by Rocca et al. [5]. The publication [6] deals with a version of the SAR inversion problem, where the ground reflectivity density is interpreted as the differential of a ground height function, as follows from the SPM theory according to (3.5). Depending on the validity of this interpretation, it is possible to reconstruct ground topography from the reflectivity density. However, ground is rarely so bare and uninhibited that the SPM theory holds to the required degree. Thus, [6] proved a dead end.

In the early stages of the Swedish CARABAS development (3.32) was used for actual radar processing. It has, however, a fundamental shortcoming in the assumption of a uniform and rectilinear sensor motion. Actually, as will be analyzed subsequently, deviations from the uniform motion must be carefully registered and compensatory actions inserted in the SAR image formation. Even so, the method in its formulations (3.24) and (3.32) is a key for understanding SAR in the case of near wavelength resolution and will here be used repeatedly.

It is possible to reformulate the method in a way that clearly indicates how to compensate for any sensor motion nonuniformity. This reformulation will now be provided. Define a function $\chi(\rho,y)$ from the measured reflectivity density $\gamma(r,t)$ as follows:

$$\chi(\rho,y)=\int_{-\infty}^{\infty}\gamma\left[\sqrt{\rho^2+(y-vt)^2},t\right]dt \tag{3.33}$$

Evidently, for the given SAR image position $\rho,y$, the integral superimposes all measured reflectivity densities that contain that position. These measurements form a one-dimensional set, with one density value $\gamma[\sqrt{\rho^2+(y-vt)^2},t]$ contributing for each point of time $t$. Integral (3.34) is commonly referred to as a *backprojection* integral. The function $\chi(\rho,y)$ will be closely related to the SAR image amplitude $\hat{\gamma}(\rho,y)$. To see the connection, make the substitution $y''=y-vt$ as follows:

$$\chi(\rho,y)=-\frac{1}{v}\int_{-\infty}^{\infty}\gamma\left[\sqrt{\rho^2+y''^2},\frac{y-y''}{v}\right]dy'' \tag{3.34}$$

By Fourier transforming $\chi$ with respect both variables and then transforming $\gamma$ with respect to $y$

$$\begin{aligned}\chi_{k_\rho,k_y}&=-\frac{1}{v}\int_{y=-\infty}^{\infty}\int_{\rho=-\infty}^{\infty}\int_{y''=-\infty}^{\infty}\gamma\left(\sqrt{\rho^2+y''^2},\frac{y-y''}{v}\right)e^{-i(k_\rho\rho+k_yy)}dyd\rho dy''\\&=-\frac{1}{v}\int_{\rho=-\infty}^{\infty}\int_{y''=-\infty}^{\infty}e^{-i(k_\rho\rho+k_yy'')}\gamma\left(\sqrt{\rho^2+y''^2}\right)_{vk_y}d\rho dy''\end{aligned} \tag{3.35}$$

Similar to the substitution in (3.25), put $\boldsymbol{k}=(k_y,k_\rho)$ and $\boldsymbol{r}=(y'',\rho)$, whereby the exponent equals $rk\cos\psi$. Also noting that $d\rho dy''=rdrd\psi$

$$\begin{aligned}\chi_{k_\rho,k_y}&=-\frac{1}{v}\int_0^{\infty}\left(\oint e^{-ikr\cos\psi}d\psi\right)\gamma(r)_{vk_y}rdr\\&=-\frac{1}{2\pi v}\int_0^{\infty}J_0\left(r\sqrt{k_y^2+k_\rho^2}\right)\gamma(r)_{vk_y}rdr\end{aligned} \tag{3.36}$$

The expression is very similar to (3.32). Comparing (3.32) and (3.36), note the relation $f'_u=iuf_u$ between the Fourier transform of a function $f$ and the transform of its derivative $f'$. Thus

$$\hat{\gamma}^{H}_{k_\rho,k_y} = -2\pi v^2 k_\rho \chi_{k_\rho,k_y} \Leftrightarrow \hat{\gamma}^{H}(\rho,y) = i\pi v^2 \frac{\partial}{\partial \rho}\chi(\rho,y) \qquad (3.37)$$

Apart from a differentiation operation, the SAR image can evidently be obtained by the backprojection operation (3.33).

Note the convention of representing the reflectivity function by positive frequencies with respect to range, that is, presently with respect to $\rho$. Since the Hilbert transform only affects negative frequencies $k_\rho$, this representation remains unaffected by the Hilbert transform, which thus has no practical impact. Some insight into the appearance of the differentiation occurring in (3.37) is gained from Figure 3.5.

The natural generalization of the backprojection integral is

$$\overline{\chi}(P) = \frac{1}{\langle v \rangle} \int_{-\infty}^{\infty} \gamma\left(\left|Q_t P\right|, t\right) ds \qquad (3.38)$$

Here $Q_t$ is sensor position at time $t$ and $P$ any point on the ground. Integration is suitably conducted with respect to sensor path length $ds$. By division

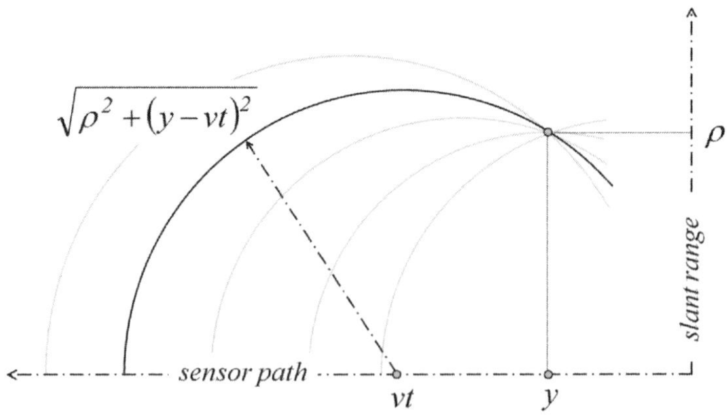


**Figure 3.5** Backprojection algorithm (3.20). For $\chi = \chi(\rho,y)$, the integral superimpose all amplitudes $\gamma(r,t)$, which contains the ground point $\theta = \theta(\rho,y)$. Since these amplitudes stem from spheres (represented by circle in slant range), which only intersect each other at the position $\theta = \theta(\rho,y)$, any particular response outside this position will be contained in just one of the spheres and thus one amplitude $\gamma(r,t)$. However, the response from $\rho,y$ is contained in all $\gamma(r,t)$, thus enhancing this particular response in comparison to all others. Since the remaining responses provide a single contribution to $\rho,y$, the method adds a constant background level to the focused responses. The differentiation in (3.33) removes this.

with mean velocity the estimate $\overline{\chi}$ agrees with (3.33) in the case of a uniform path. Note that in contrast to (3.33), the expression may not be the solution of any integral equation of the type (3.22) or (3.24). Whereas such a problem and an exact solution also could be set up for a circular path, nonlinear sensor paths make SAR image focusing dependent on knowledge of the ground topography, that is, this knowledge is required in order to utilize (3.38) for a sensor path of arbitrary shape. The suitable procedure is therefore to attempt a straight course for the sensor and use (3.38) for this nearly straight path. The defocusing due to ground topography will be negligible, so assuming the ground to be flat yields a focused image. Another concern is processing efficiency. Fast Fourier transforms can be used to approximate SAR processing based on the spectral domain inversion formula (3.32). Numerically efficient algorithms for SAR image formation are thereby obtained. In contrast, the spatial domain methods based on (3.34) or (3.38) may result in impractical computational burdens. As will be returned to in Chapter 5, a remedy is what is known as fast factorized backprojection (FFBP), making spatial domain processing efficiency become on par with that of the spectral domain inversion. Hence, FFBP turns out as the preferable meter wave SAR processing method.

The exact results will now be restricted to the case of signals of finite bandwidth. This will be done under the assumption valid for SAR that in (3.32),$\gamma(r)_{vk_y} = 0$ for ranges that are short compared to the radar wavelengths. Then, the Bessel function in (3.32) can be substituted by its asymptotic representation $J_0(z) \approx \sqrt{2/\pi z}\cos(z - \pi/4)$, whereupon

$$\hat{\gamma}_{k_\rho,k_y} = \frac{v}{\sqrt{2\pi}}|k_\rho|\int_{r_0}^{\infty}\frac{\cos\left(r\sqrt{k_\rho^2+k_y^{\;2}}-\dfrac{\pi}{4}\right)}{\sqrt{r\sqrt{k_\rho^2+k_y^2}}}\gamma(r)_{vk_y}\,rdr \tag{3.39}$$

Note that as will presently be convenient, the Hilbert transform has been removed by the substitution $k_\rho \to |k_\rho|$. Given that the data $\gamma(r)_{vk_y}$ are band limited within a radar frequency band $\omega_c - \pi B \leq \omega \leq \omega_c + \pi B$

$$\hat{\gamma}_{k_\rho,k_y} = \frac{v}{(2\pi)^{3/2}}|k_\rho|\int_{\omega=\omega_c-\pi B}^{\omega_c+\pi B}\int_{r_0}^{\infty}\frac{\cos\left(r\sqrt{k_\rho^{\;2}+k_y^{\;2}}-\dfrac{\pi}{4}\right)}{\sqrt{r\sqrt{k_\rho^{\;2}+k_y^{\;2}}}}e^{ir\kappa}\gamma_{\kappa,vk_y}\,rdrd\kappa \tag{3.40}$$

Here, according to (3.21), $\gamma_{\kappa,vk_y}$ is obtained from the radar data according to $\kappa = 2\omega/c$, $vk_y = \Delta\omega$. Since in (3.38) the exponential and the cosine oscillates rapidly in comparison with $r/\sqrt{rk}$ , the integral will not provide any net amplitude unless the phases in the integrand balance each other. It follows that:

$$\begin{aligned}\hat{\gamma}_{k_\rho,\Delta\omega/v} &\propto \int_{\omega_c-\pi B}^{\omega_c+\pi B}\left[\int_{-\infty}^{\infty} e^{ir\left(2\frac{\omega}{c}-\sqrt{k_\rho^2+\left(\frac{\Delta\omega}{v}\right)^2}\right)}dr\right]\gamma_{2\frac{\omega}{c},\Delta\omega}\,d\omega \\ &= \int_{\omega_c-\pi B}^{\omega_c+\pi B}\delta\left(2\frac{\omega}{c}-\sqrt{k_\rho^2+\left(\frac{\Delta\omega}{v}\right)^2}\right)\gamma_{2\frac{\omega}{c},\Delta\omega}\,d\omega\end{aligned} \tag{3.41}$$

The interpretation is as follows: for $k_\rho$ and $\Delta\omega$ given $\hat{\gamma}_{k_\rho,\Delta\omega/v}$ is nonzero only if $2\omega_c - 2\pi B \leq c\sqrt{k_\rho^2 + (\Delta\omega/v^2} \leq 2\omega_c + 2\pi B$. Hence, the SAR image spectrum forms a circular annulus with $2\omega_c - 2\pi B$ and $2\omega_c + 2\pi B$ as inner and outer radii. Moreover $k_\rho = \sqrt{(2\omega/c)^2 - (\Delta\omega/v)^2}$ for any given transmit frequency $\omega$.

To investigate the effect of a finite integration time, or equivalently finite SAR path length, assume a single point source at slant range $\rho_0$, conveniently located at $y = 0$

$$\hat{\gamma}(\rho,y) = C\delta(\rho-\rho_0)\delta(y) \tag{3.42}$$

where $C$ is the strength of the reflector. First, calculate radar data $\gamma(\rho,t)$ for this type of distribution. From (3.22)

$$\gamma(r,t) = C\int_{-\pi/2}^{\pi/2}\delta(r\cos\phi-\rho_0)\delta(vt+r\sin\phi)\,d\phi \tag{3.43}$$

Note that since $\gamma(r,t)$ is dimensionless, $C$ has the physical dimension Length$^{-2}$. Decompose $\gamma(\rho,t)$ into Doppler frequencies $\Delta\omega$ by the Fourier transform $\gamma(r)_{\Delta\omega}$ with respect to the time parameter. Setting the reflectivity to zero outside the limited SAR integration time, an estimate $\bar{\hat{\gamma}}(r)_{\Delta\omega}$ of $\hat{\gamma}(r)_{\Delta\omega}$ is obtained as

$$\bar{\hat{\gamma}}(r)_{\Delta\omega} = \begin{cases}\dfrac{C}{v}\displaystyle\int_{-\pi/2}^{\pi/2}\delta(r\cos\phi-\rho_0)e^{-i\left(\frac{\Delta\omega}{v}\right)r\sin\phi}\,d\phi \Leftarrow r\sin|\phi| \leq vT \\ 0 \Leftarrow r\sin|\phi| > vT\end{cases} \tag{3.44}$$

Integration with respect $\phi$ is performed by making the substitution $\phi \rightarrow r\cos\phi$, whereupon it is found that

$$\bar{\hat{\gamma}}(r)_{\Delta\omega} = \begin{cases} \dfrac{2C}{v} \dfrac{e^{-i\left(\frac{\Delta\omega}{v}\right)\sqrt{r^2-\rho_0^2}}}{\sqrt{r^2-\rho_0^2}} \Leftarrow \rho_0 \le r \le \sqrt{\rho_0^2+(vT)^2} \\ 0 \text{ elsewhere} \end{cases} \tag{3.45}$$

Since $\rho_0$ is large compared to wavelength, it can be demonstrated that a certain Doppler frequency $\omega$ is coupled to a certain position on the $y$-axis. By insertion of the asymptotic representation $J_0(z) \approx \sqrt{2/\pi z}\cos(z-\pi/4)$ in (3.31)

$$\bar{\hat{\gamma}}_{k_\rho,k_y} = \sqrt{\frac{2}{\pi}} C\left|k_\rho\right| \int_{\rho_0}^{\sqrt{\rho_0^2+(vT)^2}} \frac{e^{-i\frac{\pi}{4}} e^{i\left(rk-k_y\sqrt{r^2-\rho_0^2}\right)} + e^{i\frac{\pi}{4}} e^{i\left(-rk+k_y\sqrt{r^2-\rho_0^2}\right)}}{\sqrt{rk}\sqrt{r^2-\rho_0^2}} r\,dr \tag{3.46}$$

Change integration variable $r$ to $y = \sqrt{r^2-\rho_0^2}$, with the integration (3.60) carried out symmetrically across the interval $-vT \le y \le vT$

$$\bar{\hat{\gamma}}_{k_\rho,k_y} = \sqrt{\frac{2}{\pi}} C\left|k_\rho\right| \left[ e^{-i\frac{\pi}{4}} \int_0^{vT} \frac{e^{i(rk-k_y y)}}{\sqrt{rk}} dy + e^{i\frac{\pi}{4}} \int_{-vT}^{0} \frac{e^{-i(rk-k_y y)}}{\sqrt{rk}} dy \right] \tag{3.47}$$

Just as when considering range response in connection with (3.41), since range wavelength ratio is large, the exponentials in the integrand oscillates rapidly in comparison to the denominator. Due to the rapid oscillations, the contribution of the integral, for any given pair of values $k_\rho, k_y$, only occurs when the terms in the phase expressions balance out, that is, when the derivative of the phase expression $\Delta\Phi = k\sqrt{y^2+\rho_0^2} - k_y y$ is zero (the so-called *stationary phase approximation*). Using $k_\rho, k_y$ as suffix for the $\Delta\Phi$-derivative at this instance

$$\left(\frac{d\Delta\Phi}{dy}\right)_{k_\rho,k_y} = \frac{ky}{\sqrt{y^2+\rho_0^2}} - k_y = 0 \Rightarrow y = \frac{k_y\rho_0}{\left|k_\rho\right|} \tag{3.48}$$

Because $y/\rho_0 = k_y/|k_\rho|$ and $|y| \le vT$ for the nonzero part of $\bar{\hat{\gamma}}_{k_\rho,k_y}$, the condition $|k_y| \le vT\sqrt{k^2-k_y^2}/\rho_0$ applies for $\hat{\gamma}_{k_\rho,k_y}$ to be nonzero. In the $k_\rho,k_y$-plane, the condition defines a circular sector with an opening angle $2\Phi = 2\tan^{-1}(vT/\rho_0)$ referred to as the *integration angle*. The restriction of the spectral support thus defined combines with the result (3.41), which was that the spectral support is zero outside an annulus in the SAR image frequency plane.

By the combined requirements of finite integration time and limited bandwidth, the nonzero part of $\bar{\hat{\gamma}}_{k_\rho,k_y}$ becomes restricted to an annulus sector set by $2(\omega_c - \pi B)/c$ as inner and $2(\omega_c + \pi B)/c$ as outer radii, while $\pm\tan^{-1}(vT/\rho_0)$ are the angular limits of the sector with respect to the $k_\rho$-axis. This region in the $k_\rho,k_y$-plane will be denoted $\Omega$ and referred to as the *spectral support* of the SAR image.

Though the results regarding the spectral confinement of $\bar{\hat{\gamma}}_{k_\rho,k_y}$ are approximate in the sense of relying on asymptotic and stationary phase approximations, they nevertheless state that for the practical derivation of $\bar{\hat{\gamma}}_{k_\rho,k_y}$ or $\hat{\gamma}(\rho, y)$ from time and band limited data, the exact expressions (3.31) or (3.33) can be used, with input data $\gamma_{k\rho}(t)$ set to zero outside the frequency band and integration time limits. Doing so, values $\hat{\gamma}_{k_\rho,k_y}$ inside $\Omega$ are obtained accurately, whereas values outside will be close to zero.

In the study above of the one-dimensional range response, (3.8) provided an expression to determine point scatter resolution from reflectivity spectral density. With the spectral representation, $\bar{\hat{\gamma}}_{k_\rho,k_y}$ of SAR image point scatterer follows a generalization of this result to resolution attainable by SAR. The expressions conventionally used for microwave SAR assumes (sometimes implicitly) small fractional bandwidth, and are not valid presently. In fact, for small fractional bandwidth, the spectral support is rectangular and parallel to the $k_\rho,k_y$-axes. Accordingly, resolution can be factorized in two components $\Delta\rho$ and $\Delta y$. As was just established, the wideband SAR the support is not rectangular and such a factorization is not possible. Rather, the natural generalization of (3.7) is as area resolution

$$\Delta A = \frac{\int_{-\infty}^{\infty}\int_{-\infty}^{\infty}\left|\bar{\hat{\gamma}}(\rho,y)\right|^2 d\rho dy}{\left|\bar{\hat{\gamma}}(\rho_0,0)\right|^2} \tag{3.49}$$

By complete analogy with (3.10)

$$\Delta A \geq \frac{(2\pi)^2}{\int_\Omega dk_\rho\, dk_y} \tag{3.50}$$

with equality only for the bandwidth and time truncated point scatterer. In the case of SAR, according to the above analysis

$$\Omega = \left\{ k_\rho = k\cos\phi, k_\rho = k\sin\phi; \frac{\omega_c - \pi B}{c} \le \frac{k}{2} \le \frac{\omega_c + \pi B}{c}; |\phi| \le \tan^{-1}\left(\frac{vT}{\rho_0}\right) \right\} \tag{3.51}$$

The area of $\Omega$ and thus the aerial resolution are readily calculated as

$$\int_\Omega dk_\rho\, dk_y = \frac{16\pi\omega_c B\Phi}{c^2} \Rightarrow \Delta A = \frac{\pi c^2}{4\omega_c B\Phi} = \frac{\pi c^2}{4\omega_c B\tan^{-1}\left(\frac{vT}{\rho_0}\right)} \tag{3.52}$$

This resolution formula stems from CARABAS II work around 1996 [7]. It is valid also in the diffraction limit of indefinitely long integration time and large bandwidth. In the limit of infinite integration time $\Delta A \rightarrow c^2/2\omega_c B$. Also assuming that the frequency band starts at zero, the resolution limit becomes $\Delta A = \lambda_{\min}^2/2\pi$. For SAR operating with 100 MHz as the maximum frequency, the finest possible resolution is therefore around 1.5 m$^2$, while the resolution limit becomes around 1.5 cm$^2$ in the entirely unrealistic case of a 10-GHz SAR with 100% fractional bandwidth. A typical choice of radar parameters is to select bandwidth and center frequency to agree, that is, $\omega_c = 2\pi B$, whereas integration time is set to contribute to resolution with the same amount as bandwidth, that is

$$\Delta A = \frac{c^2}{4B^2} = \frac{\pi c^2}{4\omega_c B\Phi} = \frac{\pi c^2}{8\pi B^2\Phi} \Rightarrow \Phi = \frac{1}{2} \Rightarrow vT \approx \frac{\rho_0}{2} \tag{3.53}$$

Consequently, in this case, SAR path length and stand-off range to the target agrees. In the case of short SAR paths lengths, typical for microwave SAR

$$\Delta A = \rho_0 \frac{\lambda_c}{4vT} \frac{c}{2B} \tag{3.54}$$

The formula expresses the anticipated factorization into range resolution $c/2B$ and the angular resolution $\lambda_c/4vT$, given the SAR path length $2vT$. At the distance $\rho_0$, given the attained angular resolution, the resolution along the $y$-axis becomes $\rho_0\lambda_c/4vT$ and surface resolution thus $c\rho_0\rho\lambda_c/8vTB$.

An important observation in connection with (3.50) is that the best resolution is attained by a flat response function. Calibration of the radar and the various signal processing stages should aim at satisfying this requirement.

With a given center frequency $\omega_c$, bandwidth $B$, and integration angle $2\Phi$, the SAR response from a point scatterer at location $\rho, y = \rho_0, 0$ can be computed as

$$\hat{\gamma}(\rho, y) = \frac{1}{c^2\pi^2} \int_{\omega_c - \pi B}^{\omega_c + \pi B} \int_{-\Phi}^{\Phi} e^{i2\omega\frac{(\rho_0 + \rho\cos\phi + y\sin\phi)}{c}} \omega \, d\omega \, d\phi \tag{3.55}$$

To further investigate the effect of integration angle, again consider the situation of an essentially square-shaped resolution cell, that is, when integration angle and bandwidth contributes equally to resolution. As in (3.57), with fractional bandwidth $\beta = 2\pi B/(\omega_c + \pi B)$

$$\Delta A = \frac{c^2}{4B^2} = \frac{\pi c^2}{4\omega_c B\Phi} \Rightarrow \Phi = \frac{\pi}{\omega_c} B = \frac{\beta}{2 - \beta} \tag{3.56}$$

Since $\beta \le 1$, it follows that $\Phi \le 1 \approx 60°$ for the shaped resolution cell. With larger integration angle, angular resolution becomes finer than achievable range resolution. For a given areal resolution, conditioned by (3.56)

$$\omega_c = \pi \frac{c}{2\sqrt{\Delta A}} \frac{2 - \beta}{\beta} \tag{3.57}$$

According to (3.56) and (3.57), all quantities $\omega_c$, $B$, and $\Phi$ depend on $\Delta A$ and $\beta$, that is, for a fixed resolution area, they only depend on fractional bandwidth. The situation is illustrated in Figure 3.6, where the point response has been plotted for fractional bandwidths $\beta = 1$ and $\beta = 0.1$. The latter case provides an essentially square shape of the spectral support, whereas the former case has an evident circular segment shape. There is an apparent difference how the sidelobes are distributed, whereas there is little difference in sidelobe levels.

Resolution formula (3.56) is very general. For instance, the SAR method may be applied to a circular aperture rather than a linear. The slant range view of Figure 3.4 must then be abandoned with the SAR image instead, given in Cartesian coordinates $x$ and $y$ for a flat ground plane, integral (3.22) is performed in the ground plane (with $r$ meaning ground projected radar range) rather than in the slant range plane. As in (3.48), a certain position of the sensor with respect to the target collects spectral data corresponding to plane waves in the direction between the sensor and the target. Hence, for a circular aperture, the spectral support of the data collected forms a full ring with width to radius ratio given by the fractional band width $\beta = 2\pi B/\omega_{max}$. The resolution is calculated by obtaining the area of the ring

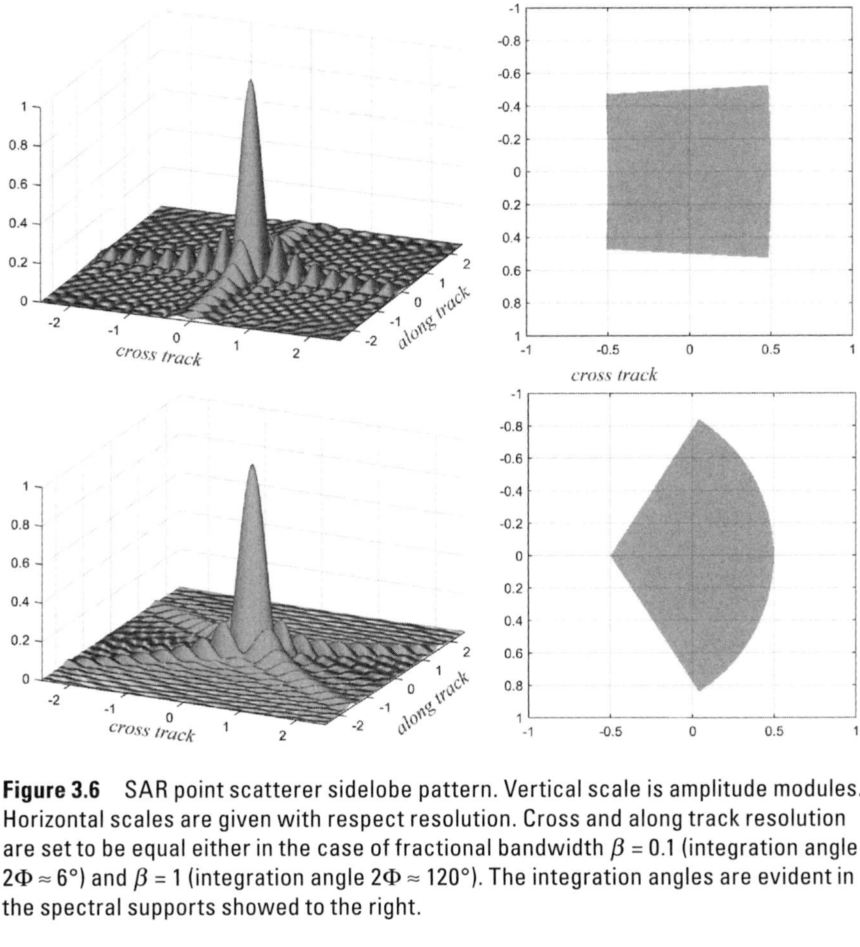


**Figure 3.6** SAR point scatterer sidelobe pattern. Vertical scale is amplitude modules. Horizontal scales are given with respect resolution. Cross and along track resolution are set to be equal either in the case of fractional bandwidth $\beta = 0.1$ (integration angle $2\Phi \approx 6°$) and $\beta = 1$ (integration angle $2\Phi \approx 120°$). The integration angles are evident in the spectral supports showed to the right.

$$\int_{\Omega} dk_x \, dk_y = 2\pi k_c k_B = \frac{4\omega_c B}{c^2} \Rightarrow \Delta A = \frac{\pi^2 c^2}{\omega_c B} \tag{3.58}$$

Inserting the relation $\omega_c = \omega_{\max}(1 - \beta/2)$

$$\Delta A = \frac{1}{\beta(2-\beta)} \frac{\lambda_{\min}^2}{4\pi} \leq \frac{\lambda_{\min}^2}{4\pi} \tag{3.59}$$

Adopting a circular aperture, high resolution can be achieved using a smaller bandwidth than is the case for a linear aperture. The penalty is enhanced sidelobe levels. As is exemplified in Figure 3.7, a considerable amount of the

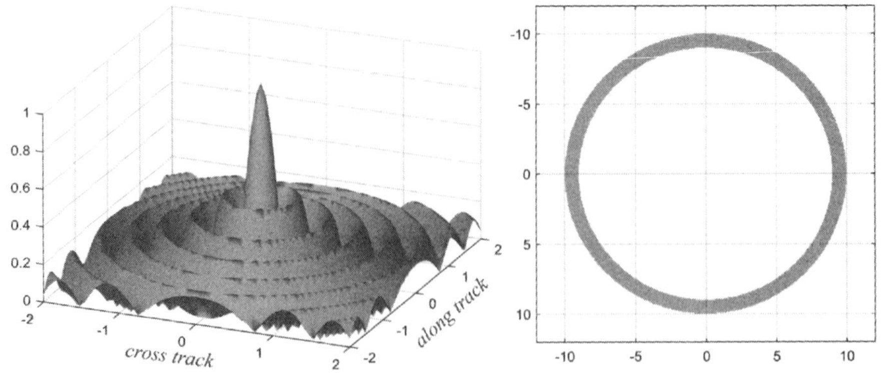


**Figure 3.7** Example of point scatter response for circular, that is, 360° integration angle, and SAR aperture with fractional bandwidth $\beta = 0.1$. The spectrum can be considered as sparse compared to its extension in the Fourier plane. As is expected in such a case, sidelobes are considerably more significant than in the cases in Figure 3.5.

backscattered energy will not contribute to point like response but be spread in a considerable sidelobe structure surrounding the object.

Recall the argument in connection with (3.38) that based on the assumption of a rectilinear sensor path, the reflectivity function can be obtained without any requirement of knowing the ground topography. While the SAR processing for a circular path can also be based on (3.38), accurate ground topography data is required in this case.

For extended objects, the scattering center position depends on direction of incidence of the impinging field. This effect must be taken into account, when exploiting large SAR integration angles and thus in particular in the case of meter-wave SAR. Assume that the scattering center for an extended object is located at some SAR image position $P$ during a part $A$ of the aperture, while for another part $A'$, the scattering center has shifted to $P'$. The attainable resolution when imaging the object is set by the condition that the scattering center shift during $A$ equals the resolution attained for $A$. Such a resolution limitation affects, for instance, a large (i.e., radius $a >> $ wavelength) vertical reflecting cylinder standing on the horizontal ground. Symmetry dictates that the scattering center for a particular sensor position is located at that part of the cylinder wall, which is momentarily closest to the sensor. For a circular aperture, SAR resolution formula (3.49) therefore implies that $\Delta A = 2\pi a \times c/2B$, that is, surface resolution equals bandwidth limited range resolution times the cylinder circumference. For a limited integration angle, only a correspondingly sized segment of the cylinder will appear in the SAR image, while the surface resolution area will diminish by the corresponding amount. An account of circular aperture imaging for CARABAS II is provided in [8].

A case reminiscent of circular aperture cylinder imaging is the imaging from a linear aperture of objects extending linearly in the ground plane. Important cases are how straight walls, for example, the sides of a building, overhead electric cables, underground cables, piping, and tunnels respond to the meter-wave SAR imaging process. Among these, cables and piping fall into the category of cylindrical objects that are thin compared to wavelength and will only produce significant backscatter at horizontal polarization. For cylinder diameter in the resonance regime, there is little factual difference in cylinder scattering between horizontal and vertical polarizations, though polarization dependencies arise due to fading and refraction, caused by the vicinity to the ground. In the resonance regime, the difference in scattering behavior from a flat vertical wall and a horizontal cylinder of comparable dimension is relatively marginal. Hence, considering SAR imaging of horizontal cylinders approximate conclusions can also be drawn regarding SAR imaging of linear structures, such as building walls.

The geometry to be considered is that of a cylinder, horizontally positioned with respect to a flat plane representing the ground surface. The ground is transversed by a horizontal radar path at height $h$ above the plane, the path represented by the $y$-axis. Denote by $\hat{y}$ the vertical projection on ground of the $y$-axis. The cylinder is assumed to have its midpoint at $y = \hat{y} = 0$ and slant range $\rho$. In the course of time, the range to the cylinder midpoint is $r_t = \sqrt{\rho^2 + v^2t^2}$ , that is, as previously the radar is assumed to pass through the origin $y = 0$ at time $t = 0$. The orientation of the cylinder is given by "misalignment" angle $\psi$, between its direction $\boldsymbol{n}$ and the $\hat{y}$-axis.

As the radar moves along, the cylinder will be imaged from a continually changing angle. Denote by $\theta_t$ this angle, precisely understood to be the angle between the cylinder direction $\boldsymbol{n}$ and the direction from the cylinder midpoint to the current radar position. From (2.208), the maximum backscattering will occur when the radar is positioned broadside to the cylinder, that is, at a time $t = \tau$ when $\theta_\tau = \pi/2$. Since the radar moves straight and uniformly, the velocity vector in the reference frame of the cylinder is noted to be

$$\boldsymbol{v} = v\boldsymbol{e}_y = v\cos\psi\boldsymbol{n} + v\sin\psi\boldsymbol{e}_z \times \boldsymbol{n} \tag{3.60}$$

The radar position in the reference frame of the cylinder is in the course of time

$$\boldsymbol{r}_t = r_t\cos\theta_t\boldsymbol{n} + r_t\sin\theta_t\sin\chi_t\boldsymbol{e}_z \times \boldsymbol{n} - r_t\sin\theta_t\cos\chi_t\boldsymbol{e}_z$$

Forming $\boldsymbol{r}_t - \boldsymbol{r}_0 = \boldsymbol{v}t$ it follows that:

$$\cos\theta_t = \frac{\rho\cos\theta_0 + vt\cos\psi}{r_t};\ r_t = \sqrt{\rho^2 + v^2t^2} \tag{3.62}$$

Expand this expression around the instance of maximum return $t = \tau$, that is, $\theta_\tau = \pi/2$. Due to (3.62), $\rho\cos\theta_0 = v\tau\cos\psi = 0$, so

$$\cos\theta_t \approx \frac{v(t-\tau)\cos\psi}{r_\tau} \tag{3.63}$$

Note that $\cos\theta_0$ is given by the projection of $\boldsymbol{r}_0$ on ground followed by projection on $\boldsymbol{n}$, that is, $\cos\theta_0 = \sqrt{\rho^2 - h^2}\sin\psi/\rho_0$. Thereby, $\tau$ and $r_\tau$ are obtained from the cylinder orientation and location as

$$\begin{aligned} \tau &= \frac{1}{v}\tan\psi\sqrt{\rho^2 - h^2} \\ r_\tau &= \sqrt{\rho^2 + v^2\tau^2} = \frac{\sqrt{\rho^2 - h^2\sin^2\psi}}{\cos\psi} \end{aligned} \tag{3.64}$$

From (3.63), it is possible to relate the angular backscattering response (2.208) to the SAR imaging geometry. Just as was demonstrated for the point scatter range response in (3.11), the amplitude given by (2.208) remains essentially unity within an interval $-\Delta\Phi_1 \le \phi - \phi_\tau \le \Delta\Phi$ (where $\phi_\tau - \tan^{-1}(vt/\rho_0)$), whereas it is effectively zero outside it. The backscattering width $2\Delta\Phi$ will be determined below. Adapting (2.208) to the SAR imaging case is done by noting $k = 2\omega/c$

$$\gamma_{k_\rho,k_y} = \gamma_{\frac{2\omega\cos\phi}{c},\frac{2\omega\sin\phi}{c}} = B(\omega)\frac{\sin\left(\frac{\omega L}{c}\cos\theta_t\right)}{\frac{\omega L}{c}\cos\theta_t}e^{-i2\omega\rho\cos\phi} \tag{3.65}$$

It will not be necessary to specify $B(\omega)$ in order to determine the bandlimited SAR resolution for the cylinder; just assume that it remains approximately constant within the radar band. Subsequently computing cylinder reflectivity $B(\omega)$ will be specified for various cases of a cylinder above or below ground.

The time dependence in (3.65), that is, the radar position on the $y$-axis, has been demonstrated in (3.46) to translate directly into the spectral domain by (3.47) stating that the spectral amplitudes of the image are picked up along the aperture according to the relation $vt/\rho = k_y/|k_\rho|$. Thus, the cylinder response spectral amplitudes are obtained according to

$$\phi = \tan^{-1}\left(\frac{k_y}{\left|k_\rho\right|}\right) \Rightarrow \phi - \phi_\tau = (t-\tau)\left[\frac{d}{dt}\tan^{-1}\left(\frac{vt}{\rho}\right)\right]_{t=\tau} = \frac{v}{\rho}\frac{t-\tau}{1+\left(\frac{v\tau}{\rho}\right)^2} \quad (3.66)$$

Combining with (3.63) and (3.64)

$$\cos\theta_t = \cos\psi\frac{\sqrt{\rho^2+v^2\tau^2}}{\rho}\left(\phi-\phi_\tau\right) = \frac{\sqrt{\rho^2-h^2\sin^2\psi}}{\rho}\left(\phi-\phi_\tau\right) \quad (3.67)$$

Due to the property of (3.65) of being effectively zero for all but angles $\theta_t$ close to $\pi/2$, while being approximately constant for those, the integral mean value theorem yields

$$2\Delta\Phi \approx \int_{-\infty}^{\infty}\frac{\sin\left[\frac{\omega L}{c}\frac{\sqrt{\rho^2-h^2\sin^2\psi}}{\rho}\left(\phi-\phi_\tau\right)\right]}{\frac{\omega L}{c}\frac{\sqrt{\rho^2-h^2\sin^2\psi}}{\rho}\left(\phi-\phi_\tau\right)}d\phi = \pi\frac{c}{\omega L}\frac{\rho}{\sqrt{\rho^2-h^2\sin^2\psi}}$$

(3.68)

From (3.68), the spectral support $\Omega_1$ of (3.65) and thus SAR resolution follows. Indeed

$$\Omega_1 = \left\{\begin{array}{l} k_\rho = k\cos\phi \\ k_y = k\sin\phi \end{array};\begin{array}{c} \frac{\omega_c-\pi B}{c} \le \frac{k}{2} \le \frac{\omega_c+\pi B}{c} \\ \phi_\tau - \frac{\pi}{2}\frac{c}{\omega L}\frac{\rho}{\sqrt{\rho^2-h^2\sin^2\psi}} \le \phi \le \phi_\tau + \frac{\pi}{2}\frac{c}{\omega L}\frac{\rho}{\sqrt{\rho^2-h^2\sin^2\psi}} \\ \phi_\tau = \cos^{-1}\frac{\rho\cos\psi}{\sqrt{\rho^2-h^2\sin^2\psi}} \end{array}\right\}$$

(3.69)

Comparing to the full spectral support $\Omega$ of the SAR image, and depending on the degree of misalignment $\psi$, it may happen that $\Omega_1$ is entirely within the interior of $\Omega$, or it may happen that it is only partially so or even not at all. In all cases, SAR resolution is set by the area of $\Omega_1 \cap \Omega$. Presently, assuming that $\Omega_1 \subset \Omega$, the resolution is calculated as

$$\int_{\Omega_1} dk_\rho \, dk_y = \frac{4}{c^2} \int_{k=2(\omega_c-\pi B)}^{k=2(\omega_c+\pi B)} \int_{-\Phi_1}^{\Phi_1} \omega \, d\Delta\phi \, d\omega$$
$$= 8\pi^2 \frac{B}{Lc} \frac{\rho}{\sqrt{\rho^2 - h^2 \sin^2 \psi}} \Rightarrow \Delta A_{PSR} = \frac{Lc}{2B} \frac{\sqrt{\rho^2 - h^2 \sin^2 \psi}}{\rho} \tag{3.70}$$

The results are the same for any mean frequency $\omega_c$, a fact that formally depends on that the product $\Phi_1 \omega$ is independent of $\omega$. The astonishing conclusion is that the PSR attained is limited from above (and in most imaging situations very close) to bandwidth limited range resolution times the actual extension of the linear structure (a result similar to the case of the large cylinder discussed above). In the case that platform altitude is non-negligible compared to stand off range, resolution is noted to be finer, given that the cylinder is oriented at some nonzero angle compared to the radar path. The effect is partly explained in that the slant range representation of the cylinder will compress in the $\rho$–direction by an amount $\sin\psi\sqrt{\rho^2 - h^2}/\rho$, but will also affect the width of the cylinder in the SAR image. It is noted that formula (3.57) is valid only if the angular span through so that the radar transverses the backscattering beam is small. Cases beyond the applicability of (3.70) arise when $\psi$ approaches $\pi/2$ and $h$ approaches $\rho$.

In Chapter 2, developing the reflectivity concept, it was demonstrated how a locally homogeneous random ground surface decorrelated when illuminated at a particular frequency, leading to a pointwise defined reflectivity, with the interpretation of cross section per unit area. In Section 3.2,it was mentioned that reflectivity concept extends from bare ground to homogenous terrain in general. Forested terrain is of course a prime example of such terrain, though requirements on randomness generally turn out modest and even cultural ground areas (cities and the like, organized by man) are mostly sufficiently random for the concept of cross section per unit area to apply. Moreover, bandwidth can be finite rather than monochromatic, which was the assumption in Chapter 2. Hence, with reflectivity density obtained from SAR processing and with appropriate calibration

$$\sigma = \left| \int_{\Xi} \hat{\gamma}(\rho, y) d\rho dy \right|^2 = \int_{\Xi} \left| \hat{\gamma}(\rho, y) \right|^2 d\rho dy \Rightarrow \sigma_0 = \left| \hat{\gamma}(\rho_0, y_0) \right|^2$$
$$= \lim_{\Xi \to \rho_0, y_0} \frac{\int_{\Xi} \left| \hat{\gamma}(\rho, y) \right|^2 d\rho dy}{A(\Xi)} \tag{3.71}$$

where $\Xi$ is some ground region containing the ground points $\rho_0$ and $y_0$, and $A(\Xi)$ the surface area of that region. The validity of (3.71) does become critical when SAR resolution is finer than the extension of individual objects, since the homogenous random character of the reflectivity density is then broken. The cases of significant cylinder sidelobe structures, discussed in Section 3.2 exemplifies that correlation of the reflectivity density extends beyond the resolution set by bandwidth and consequently that (3.71) is not fully correct in such cases. As for the foregoing analysis of SAR imaging of objects linear in the ground plane, it was demonstrated that physical size, that is, actual object length, and thus SAR image area agree with the achieved SAR resolution. For the reflectivity concept to apply to these objects as well, their cross section must increase linearly with their length, that is, image area. Deriving explicit formulas for the SAR reflectivity of horizontal cylinders, it will now be shown that this is indeed the case.

The cylinder RCS of Chapter 2 can be generalized to finite bandwidth and time variable situations, by realizing that RCS expresses backscatter energy compared to impinging energy. Since energy distributed over different frequencies or time intervals just adds, the scattering cross section for a signal distributed over different time intervals or frequencies will be the average of each of these RCS values

$$\sigma = \frac{1}{4\pi TB} \int\limits_{\omega_c-\pi B}^{\omega_c+\pi B} \int\limits_{-T}^{T} \sigma_{\text{monochromatic}}(\omega,t)\,dt\,d\omega \tag{3.72}$$

For (3.72) to make sense, it is required that time changes in $\sigma_{\text{monochromatic}}(\omega,t)$ are slow compared to the cycle period $2\pi/\omega$. The stationary phase conditions, which are valid for ranges much larger than the wavelength, guarantee this to be the case. The connection to SAR reflectivity is made through Parseval's theorem applied to (3.71)

$$\sigma = \int\limits_{0}^{\infty} \int\limits_{-\infty}^{\infty} \left|\hat{\gamma}(\rho,y)\right|^2 d\rho\,dy = \int\limits_{\Omega} \left|\hat{\gamma}_{k_\rho,k_y}\right|^2 dk_\rho\,dk_y \tag{3.73}$$

The spectral representation of reflectivity in (3.73) must be equated with (3.72) in the case of a SAR image that is empty but for the response from the cylinder target. Inserting the stationary phase relations $vt/\rho = k_y/|k_\rho|$ and $k = 2\omega/c$ into (3.73)

$$\sigma = \frac{4}{c^2} \int\limits_{\Omega} \left|\hat{\gamma}_{k_\rho,k_y}\right|^2 d\phi\,\omega\,d\omega = \int\limits_{\Omega} \left|\hat{\gamma}_{k_\rho,k_y}\right|^2 \frac{4}{c^2} \frac{\rho v\omega}{\rho^2+v^2t^2}\,dt\,d\omega \tag{3.74}$$

Using (2.206) modified by (2.208) for $\sigma_{\text{monochromatic}}$, it follows that:

$$\left|\hat{\gamma}_{k_\rho,k_y}\right|^2 \frac{4}{c^2}\frac{\rho v\omega}{\rho^2+v^2t^2} = \frac{1}{BT}\frac{L^2c^2}{\pi^2\eta^2\omega^2}\left|\sum_{n=0}^{\infty}(-1)^n\breve{b}_n\right|^2\left[\frac{\sin\left(\frac{\omega L}{c}\cos\theta_t\right)}{\frac{\omega L}{c}\cos\theta_t}\right]^2 \tag{3.75}$$

Here, integration time can be substituted for integration angle according to $vT = \rho\tan\Phi$. Assume $\Omega_1 \subset \Omega$ and revert to the previous approximation of a first-order expansion around $t = \tau$. Using the approximation technique of (3.68), the SAR cross section is obtained from (3.73)

$$\sigma = \frac{\sqrt{\rho^2 - h^2\sin^2\psi}}{\rho\tan\Phi\cos\psi}\frac{L}{\pi B}\int_{\omega_c-\pi B}^{\omega_c+\pi B}\frac{c^3}{\eta^2\omega^3}\left|\sum_{n=0}^{\infty}(-1)^n\breve{b}_n\right|^2 d\omega \tag{3.76}$$

Note that (3.76) is never divergent since $\Phi \geq \Delta\Phi$ because $\Omega_1 \subset \Omega$ in order than (3.76) should be valid. The RCS is seen to be independent of slant range $\rho$ up to the particulars, when imaging a cylinder askew with the SAR path and from a certain elevation. The maximum value for $\sigma$ is obtained when $\Phi = \Delta\Phi$, in which case (3.68) implies that

$$\sigma = \frac{\rho^2 - h^2\sin^2\psi}{\rho^2\cos\psi}\frac{2L^2}{\pi^2 B}\int_{\omega_c-\pi B}^{\omega_c+\pi B}\frac{c^2}{\eta^2\omega^2}\left|\sum_{n=0}^{\infty}(-1)^n\breve{b}_n\right|^2 d\omega \tag{3.77}$$

The reduction of cross section, which occurs when integration angle is increased above $2\Delta\Phi$, is expected, since for an ever-increasing integration time, the misalignment of the cylinder backscattering beam and radar position will be ever more influential. The reflectivity along the extended cylinder is found by dividing (3.76) with SAR resolution (3.70)

$$\sigma_0 = \frac{\sigma}{\Delta A_{PSR}} = \frac{1}{\tan\Phi\ \cos\psi}\frac{2}{\pi}\int_{\omega_c-\pi B}^{\omega_c+\pi B}\frac{c^3}{\eta^2\omega^3}\left|\sum_{n=0}^{\infty}(-1)^n\breve{b}_n\right|^2 d\left(\frac{\omega}{c}\right) \tag{3.78}$$

As expected, the reflectivity is independent of the length of the cylinder. In fact, the SAR (i.e., average) RCS increases in proportion to cylinder length, whereas the peak RCS increases with cylinder length squared.

Reflectivity formula (3.78) has important applications for determining the detectability of wire fences, overground and underground cables, pipelines, and tunnels; all these target categories are normally constructed with the objective of being segment wise straight and are thus well represented by horizontal straight cylinders. The different target categories are characterized by different wavelength to circumference ratios, as well as different dielectric properties of target and the surrounding medium. Ideal scatterers (representing perfectly conducting cables, metal wire fences, and metal pipes) as well as cavities (unlined tunnels) will be considered. For a perfectly conducting cable, the diameter might be just a few millimeters, whereas a tunnel might have a diameter of, say, 2m. Wavelengths might range from submeter to several meters. The circumference wavelength ratios to be considered may thus vary from, say, $10^{-3}$ to 10. A radar fractional bandwidth of $\beta = 2\pi B/(\omega_c + \pi B) = 2/3$ makes the bandwidth and center frequency agree, and is what is recommended and also practiced in the CARABAS systems. Integration angle $\Phi = 1/2$ makes integration time and bandwidth contribute equally to the SAR resolution and is also the recommended choice, for which experimental data also are available. The results of reflectivity computations under these conditions for buried cylinders are provided in Figure 3.8. Figure 3.9 shows the results for buried and above-ground thin straight wires. For both figures, fading and refraction effects have been incorporated, adopting the models of Chapter 2.

As for incorporating refraction effects, the following arguments have been applied. With circumference comparable to or larger than the wavelength, the target response for tunnels will approach the optical regime, implying that horizontal polarization is no longer obvious in order to obtain the best possibilities for detection. Since polarization has no effect on the angular behavior of cylinder scattering, reflectivity for vertical polarization is obtained (according to (2.206) merely by replacing $\breve{b}_n$ for $\breve{a}_n$ in (3.77)). Reflectivity values representing tunnels will be given here for both polarizations. Since the reflectivity is independent of cylinder length, an extended cylinder can be understood as composed of multiple short segments. Each of these only scatters weakly. In Section 2.4, it was demonstrated that coupling between the ground surface and buried targets is negligible for weakly scattering objects. Accordingly, the ground surface affects the tunnel response by two-way reflection losses in the ground surface. The resulting overall RCS reduction was expressed in (2.193) for the weak scattering of buried spheres. Since the weak scatting assumption still applies, so does (2.193). A slight simplification is introduced in that the use of a complex index of refraction and consequently complex subsurface propagation angles $\widehat{\theta}$ has little impact on the resulting reflectivity values.

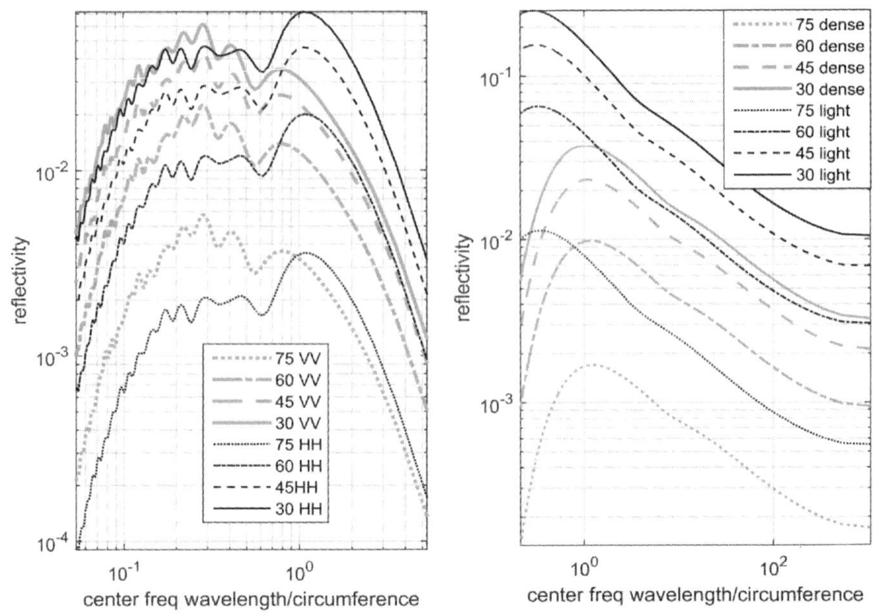


**Figure 3.8** Reflectivity for cylinder buried at depth to top equal to radius and 75°, 60°, 45°, and 30° incidence angles. Right: the case of a perfectly conducting cylinder (e.g., metallic wire) target for two cases of dielectrically lighter and denser ground medium, i.e., $\eta = 2.2 + 0.07i$ and $\eta = 4.4 + 0.24i$. Left: the case of a cavity (tunnel) target for vertical and horizontal polarizations and the light ground medium.

Thus, these are based on Snell's law in its real form $\eta_{re}\sin\hat{\theta} = \sin\theta$, whereas $\hat{\theta} = \hat{\theta}_0$ and (2.193) assumes the form

$$\begin{aligned}
&\text{Horizontal:} \quad \sigma_0 = e^{-\frac{4kD\eta_{im}}{\cos\hat{\theta}}}\left(\frac{\tan\hat{\theta}}{\tan\theta}\right)^2\left[1-\left|\frac{\sin\left(\theta-\hat{\theta}\right)}{\sin\left(\theta+\hat{\theta}\right)}\right|\right]^2\sigma_0^{\text{buried}} \\
&\text{Vertical:} \quad \sigma_0 = e^{-\frac{4kD\eta_{im}}{\cos\hat{\theta}}}\left(\frac{\tan\hat{\theta}}{\tan\theta}\right)^2\left[1-\frac{\tan\left(\theta-\hat{\theta}\right)}{\tan\left(\theta+\hat{\theta}\right)}\right]^2\sigma_0^{\text{buried}}
\end{aligned} \tag{3.79}$$

For wire targets, it is noted that their responses will be deeply in the Raleigh regime for the wavelengths considered, so horizontal polarization is a requirement for their detection. Again, there will be refraction losses according to (2.193). Figure 3.8 depicts the situation for buried metallic wires and also thicker cylinders (typically representing pipelines) at horizontal polarization.

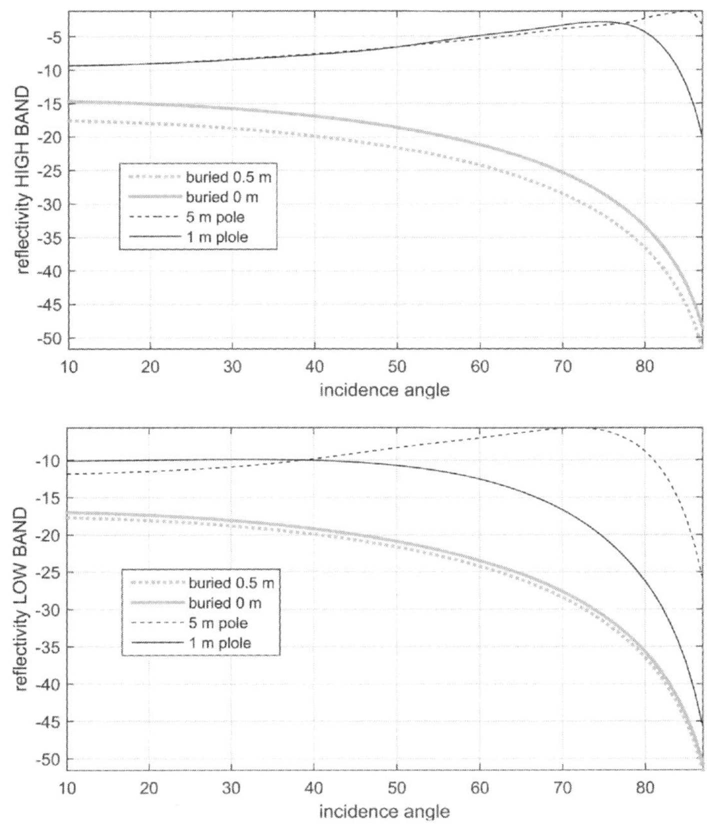


**Figure 3.9** Reflectivity examples for thin horizontal cylinder. Light soil conditions $\eta = 2.2 + 0.07i$ are assumed. By "high" and "low," the bands 30–80 MHz and 135–360 MHz are understood. The cylinder is a thin perfectly reflecting wire of 0.005m thickness, located at poles 1m or 5m above ground or buried right below ground surface (depth to top equal to radius) or at 0.5m depth.

The results for underground tunnels use a different scale for wavelength to circumference ratio. For both plots, depth to the cylinder center is assumed to equal cylinder diameter (thus, thin wires are very shallowly buried). In the case of a lossless ground medium, the result is independent of the burial depth. For true lossy media, an increased burial depth may mean a sharp decrease in reflectivity compared to the diagrams. It is noted that steep incidence angles is favorable for obtaining reasonably high reflectivity values. Tunnels are noted to require wavelengths of the order of the tunnel circumference in order to obtain significant reflectivity values. Any shorter wavelength reduces reflectivity due to medium internal losses, even in the case of the low-loss medium on which the diagram is based.

For overground horizontal cylinders, mostly wire fences and electric wires on poles, the phenomenon of ground fading becomes an important factor. Ground fading was analyzed at length in Chapter 2 and was quantitatively stated in the form of formula (2.214), which is directly applicable to reflectivity. Ground fading is effective when the cylinder is close to the ground, fencing being the typical example. For wires placed well above the ground, electric wires on poles being the typical example, the fading process reverses into an enhancement of the backscattered signal. The square of the gain factor in (2.214) determines the resulting effect on RCS and reflectivity. As numerical examples, consider backscattering from a metallic wire with a diameter $2a$ = 0.005m in the cases of location 1m and 5m above ground. For comparison, a case of the wire being buried at zero and 0.5m depth is also considered. The computations are carried out for two frequency bands 30 to 80 MHz and 135 to 360 MHz (these are the bands essentially realized in the following CARABAS III experimental system), implying wavelengths at the center frequencies of 5.5m and 1.2m, respectively. A light soil ground with $\eta = 2.2 + 0.07i$ is selected to keep contrasts between overhead and buried target responses at a minimum. Figure 3.9 depicts the results in (3.22), indicating that wires above the ground are strong targets. As expected, fading becomes pronounced for wires close to the ground, at the lower band. For cables buried at 0.5m depth, the reflectivity is similar at the higher and lower bands. The somewhat higher wire cross section at the higher band becomes balanced by attenuation losses being higher at the higher band.

## 3.4 Comparison with Collected SAR data

To acquire experimental data enabling comparison with theoretical predictions is a major challenge for meter-wave radar. Many of the theoretical predictions made above, relate to cross section values. For microwave RCS, data can be obtained from mounting objects on a calibrated pedestal and using a directive antenna, combined with range gating, and preferably conducting the experiment in an anechoic camber. For meter-wave SAR, the near impossibility of forming a directive antenna or creating an anechoic chamber calls for different methods. The objectives of such measurements are anyway to verify that the here adopted method of generic and simplified geometry models are valid for real objects and real situations. A comparison between theory and experiments is best conducted by field tests adopting scenes where ground truth is well known or controllable.

Calibration of SAR data can be relative or absolute. With relative calibration, a reference object is identified in the scene and SAR image amplitudes computed with respect to this. With absolute calibration, the SAR images are generated in the scale of reflectivity or equivalently RCS by knowing in absolute terms the radar system transfer characteristics. Though attractive, this latter alternative is difficult to achieve in practice since very many parameters affect the SAR image amplitude. Some of these parameters are the result of random circumstances like turbulence and wind gusts, or which parts of the radar band have to be suppressed due to interference from communications signals. While these variations are remedied by compensatory actions in the SAR processing (further discussed in Chapters 4 and 5), the amplitude scale in the resulting SAR will be affected in ways difficult to foresee and prevent absolute calibration. It should be stressed that the meter-wave SAR image is very stable in all other respects. Hence, once calibrated the meter-wave SAR image will be noticeably more robust than a microwave SAR image for variations in heading, altitude of SAR path, the particular transmit waveform, and so on. Equalizing the mean SAR image amplitude between meter-wave SAR images of a particular piece of scenery collected at different occasions, small targets entering into or vanishing from the scene can be detected. This detection technique, called *change detection*, will be returned to and discussed in detail in Chapter 6.

For relative calibration, the reference object (calibration target) is suitably artificial and purposely deployed for a particular radar measurement. There are a number of design requirements on such a target, foremost:

1. Its reflectivity should not change in the course of time (mechanical robustness required).
2. It should not be directive, that is, it should be as independent as possible of its orientation with respect to the SAR path.
3. It should yield a strong and point like response across the several octaves of relevant radar frequencies and for either polarization.
4. Its reflectivity should have as weak coupling to the environment, as is possible.
5. It must be possible to calibrate over the relevant frequencies by a suitable method of accurate electromagnetic modeling.
6. It should be compact/transportable/easy to assemble–disassemble to make logistics manageable.

Recall the brief history of Swedish meter wave SAR development provided in the introduction (Chapter 1). The Swedish Defense Research Agency FOI had developed the original design for meter wave SAR in the form of the 25- to 85-MHz CARABAS I radar installed in Sabreliner Business Jet (see Figure 1.1 and for a written account [9]). When the CARABAS I tests commenced around 1991, the U.S. defense research organization MIT Lincoln Laboratory joined. A primary task for the team was to establish baseline performance, that is, resolution, sensitivity, and so on. As for calibration targets, a number of innovative designs were evaluated initially, all considered conceivable for the unconventional frequency band. Loop and log periodic structures, both active and passive, were all tried. It was found that the by far most satisfying structure was a large trihedral, positioned with one side flat on ground (see Figure 3.10). Its size was selected to be a compromise between transportability and reflectivity. Microwave trihedrals are normally designed to be many wavelengths large, but imply dimensions of tens of meters for the 25- to 85-MHz band. The dimensions selected provided 5m sides. The design allowed the reflector to be disassembled and transported as a set of manageable flat packages.

The standard narrow band trihedral RCS cross-sectional formula (where $L$ is the length of trihedral)

**Figure 3.10** A 5m corner reflector deployed in tropical rain forest in Panama in an early test (in the nineties). The purpose of the test was to measure foliage attenuation of the low frequency radar signal, by operating the SAR aircraft at the other side of the vegetation screen toward which the reflector is directed. The modular (but necessarily rigid) design of the corner reflector in the form of 1m × 1m aluminum sections is evident.

$$\sigma = \frac{4\pi}{3}\frac{L^4}{\lambda^2} \tag{3.80}$$

is translated into its broad band correspondence by averaging spectral density contributions. Hence

$$\begin{aligned}\sigma &= \frac{4\pi L^4}{3}\frac{1}{2\pi B}\int_{\omega_c-\pi B}^{\omega_c+\pi B}\frac{1}{\lambda^2}d\omega = \frac{L^4}{6\pi^2c^2B}\int_{\omega_c-\pi B}^{\omega_c+\pi B}\omega^2 d\omega \\ &= \frac{L^4}{18\pi^2c^2B}\left[\left(\omega_c+\pi B\right)^3-\left(\omega_c-\pi B\right)^3\right]\end{aligned} \tag{3.81}$$

Applying (3.81) to the band 25 to 85 MHz yields $\sigma = 95$ m$^2$. However, (3.80) assumes a wavelength small compared to the trihedral size, so the formula is not appropriate at this band for a 5m trihedral. In particular, when the trihedral is electrically small, it will couple to its surroundings. Since the triangular side of the trihedral is not large compared to the wavelength, a significant part of the backscattering response is subjected to the fading mechanisms, governed by incidence angle, ground dielectric properties, and polarization, as discussed in Section 2.7.

For fading, consider Figure 2.7. Fading for horizontal polarization is seen depend relatively slowly on incidence angle and dielectric properties, though it is noted that for objects not shallow, it may imply an amplification of the backscatter. Using numerical electromagnetic solvers (see [10] and [11] for details), the RCS $\sigma = 250$ m$^2$ was derived and has become the routinely used average for horizontal polarization, approximately valid for all practical incidence angles and types of surrounding ground at 25 to 85 MHz. Fading exhibits a much a stronger incidence angle variation for vertical than for horizontal polarization, and it is an absolute necessity to take this variation into account when estimating vertical polarization RCS. While $\sigma = 250$ m$^2$ is assumed for horizontal polarization at the incidence angles commonly used, for vertical polarization, this value becomes significantly reduced. The incidence angle dependent fading difference between the polarizations is expressed by (2.214).

Examples of calibrated meter-wave SAR imagery are provided in Figure 3.11. The imagery was obtained in 2015, by means of the CARABAS III system, operating in either of the bands 27.5 to 82.5 MHz (low band) or 137.5 to 357.5 MHz (high band). Polarization is selectable between horizontal and vertical. The system will be described in more detail in connection with hardware design principles, accounted for in Chapter 4. The SAR images depict the same 350m × 600m mixed open and forested land. Images are georectified, that is, map true with north pointing upward. The radar views the scene from a path inclined 45° to the orientation of the images at the top left, that is, path heading is either SW or NE. The distance from the path to the trihedral is 1000m and path altitude is 250m above ground. The incidence angle becomes 75° and is thus more shallow than the foliage penetration "rule of thumb" (discussed in Chapter 2) that surveillance range should be shorter than twice the platform altitude. Still, the image set is useful for comparing prediction and measurement. The forested land is concentrated to the upper right corner of Figure 3.11 and consists mainly of mature coniferous trees. In the midsection of the images in the figure, there are a few scattered wooden/stone family houses and garages/sheds, as well as isolated mostly deciduous capital trees (generally oak). To the left of the figure, there is a farm with larger barns and stable wooden buildings. The circular structure seen at the right of the figure is a manure deposit, adjacent to the stable, and encircled by a 1.5m high concrete wall. Surrounding the farm, there are farm fields and paddocks, on occasion lined with wire fences.

The top image of Figure 3.11 is high-band horizontal polarization, the middle image is low-band horizontal polarization, and the bottom image is low-band vertical polarization. As expected, the line structures caused by wire fences are well visible in the high band image (those seen are aligned to give specular reflection within the SAR integration angle, as explained in Section 3). Again, according to expectations, they are not visible in the low-band vertically polarized image and just barely perceptible in the low-band horizontally polarized image.

To assess the actual reflectivity values in the three images, the trihedral reflectivity is determined first, based on that the trihedral RCS is assumed to be $\sigma$ = 250 m$^2$ for the low band and horizontal polarization. For vertical polarization, (2.214) implies that fading is about 9 dB stronger at vertical than at horizontal polarization, with ground dielectric properties having no significant influence. Thus, RCS $\sigma$ = 31 m$^2$ will be assumed for low-band vertical polarization. For the high band, (2.195) is assumed sufficiently correct, yielding $\sigma$ = 1900 m$^2$. To determine the trihedral RCS in the SAR images, integration must be conducted over some region $\Xi$ in the image sufficiently large to encompass the reflector sidelobes but so small that the RCS is not unduly influenced by objects in the vicinity of the reflector. Selecting $\Xi$ as a 10m × 10m square around the trihedral, as illustrated in Figure 3.10, seems the suitable compromise. The RCS in the SAR image is estimated as

$$\sigma = \int_{\Xi} \left|\hat{\gamma}(\rho, y)\right|^2 d\rho dy = A_{\text{pixel}} \sum_{\Xi} \left|\hat{\gamma}(n\Delta\rho, m\Delta y)\right|_{n,m}^2 \tag{3.82}$$

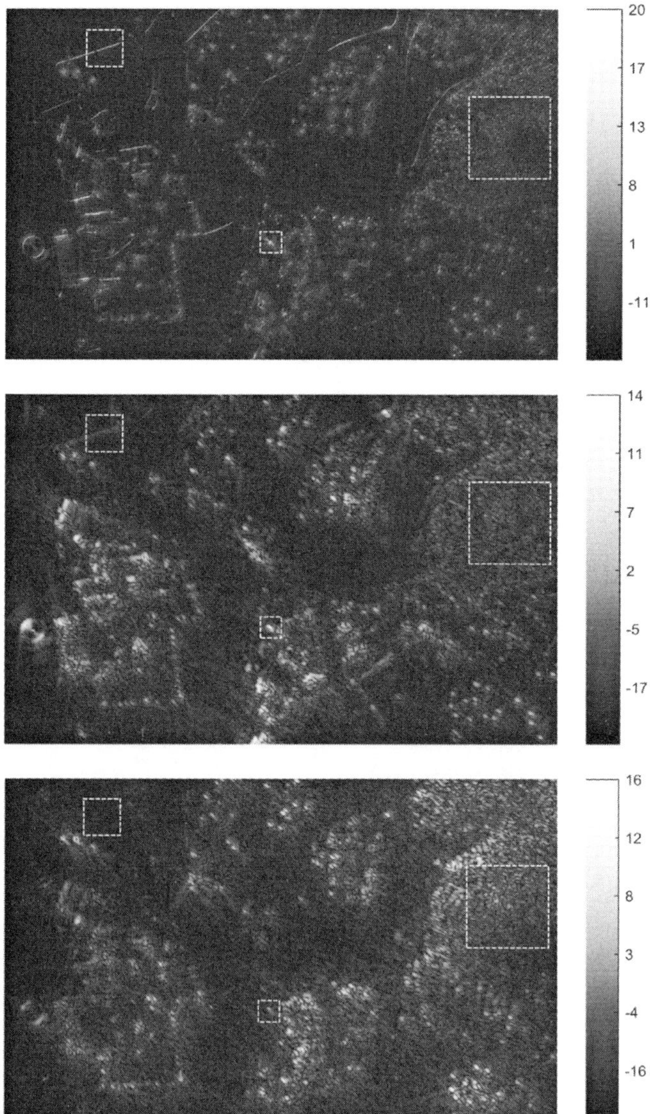


**Figure 3.11** SAR images of an about 350m × 600m large agricultural area. Topmost: horizontally polarized using the band 137.5 to 357.5 MHz. Middle: horizontally polarized using the band 27.5 to 82.5 MHz. Bottom: vertically polarized using the same band. The dotted squares are (a) a 10m sample region for the trihedral reference (square in the middle), (b) an 80m sample region for forest (square on the right), and (c) a 40m sample region for cattle fence reflector (square on the upper left). The images are given in reflectivity decibel values, according to the gray scale on the right. The high band image is saturated for the trihedral response.

where $A_{pixel}$ is the pixel area (0.5m × 0.5m and 0.2m × 0.2m is the low- and high-band SAR images). Trihedral resolution is determined according to (3.49) with integration restricted to Ξ.

Scaling the uncalibrated reflectivity density to agree with the known trihedral RCS, the forest and fence reflectivities can be estimated. For the forest, a suitable homogenous sample is obtained by considering an 80m × 80m square, seen to the left of the SAR images. A fence sample is captured by a 40m × 40m square seen to the right of the images. Mean reflectivity values for the forest are obtained by calculating the overall RCS within the 80m × 80m square by (3.82) and dividing by the area. For the fence, the maximum reflectivity is obtained for each north–south line in the square, whereupon the set of maxima is averaged across the east–west direction. Table 3.1 is then obtained by these reflectivity values.

Compare first with the theoretical point scatter resolution given by (3.52). For all three images, the integration angle is $2\Phi = 60°$, whereby (3.52) yields the high- and low-band resolution values of 0.41 and 7.44 $m^2$, respectively. The differences between theoretical and measured resolution depends on a several tuning factors in operating the radar system and when performing the SAR image processing (as will be described in the sequel). Some degradation of the theoretical figures are to be expected, making the actually measured resolution 0.67 $m^2$ a good indication of that the system works as intended. The longer wavelengths at low band tend to make tuning errors smaller, implying that the theoretical resolution value should be easier to achieve in this case. However, as it happened for the low-band imagery in Figure 3.11, the flying conditions when capturing the data were difficult with very strong cross winds upsetting the path and orientation for the helicopter radar platform. The slightly larger than normal resolution degradation at horizontal polarization is affected by these circumstances. The inferior resolution figure for vertical polarization is most likely due to an enhanced vulnerability to imaging errors in this case.

As for forest reflectivity, to make a comparison with theoretical predictions more complete, a richer statistical material, obtained in the cooperation

**Table 3.1**
Mean Reflectivity Values Obtained by Calculating the Overall RCS

| | Trihedral | | | |
|---|---|---|---|---|
| | **Resolution** | $\sigma_0$ | **Forest** $\sigma_0$ | **Fence** $\sigma_0$ |
| High Horizontal | 0.67 $m^2$ | 35 dB | –8.1 dB | 4.9 dB |
| Low Horizontal | 9.7 $m^2$ | 14 dB | –15 dB | –14 dB |
| Low Vertical | 16 $m^2$ | 2.9 dB | –7.7 dB | NA |

between FOI and MIT Lincoln Laboratory mentioned above, will also be invoked. Figure 3.12 is a compilation of reflectivity and attenuation measurement results obtained from several campaigns in the mid-nineties using the NASA AIRSAR system for microwave frequencies, the UWB SRI FOLPEN radar for P-band, and the CARABAS I radar for the 25- to 85-MHz band. The methodology for obtaining reflectivity with low radar frequencies was as described here. Attenuation was determined by comparing responses from trihedral in the open and put behind tree screens (as in Figure 3.10). Test sites included Panama, in order to obtain conditions representative of the tropical forest belt and Maine in the US representative of semiboreal forest surrounding the northern hemisphere. A fuller account of these experiments is found in [12]–[14].

The cumulative probability, in Figure 3.12 (top), is understood as the fractional number of tests finding the attenuation below the given value. By exceedance, probability in Figure 3.12 (bottom) is understood as the fractional number of tests finding the reflectivity level higher than the given

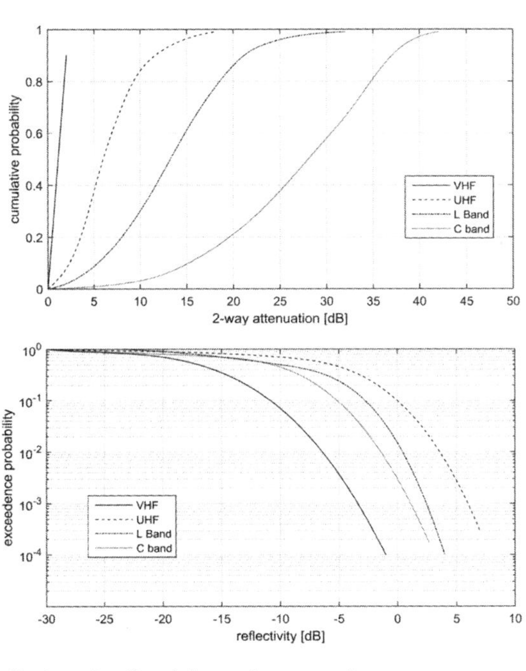


**Figure 3.12** Compilation of reflectivity and attenuation measurement results obtained from several campaigns in the mid-nineties using the NASA AIRSAR system for microwave frequencies, the UWB SRI FOLPEN radar for P-band, and the Swedish CARABAS I radar for the 25- to 85-MHz band. Test sites include Panama, in order to obtain conditions representative for the tropical forest belt, and Maine in the US, representative of semiboreal forest surrounding the northern hemisphere.

value. Median attenuation and reflectivity are thus obtained from 0.5 probabilities. The median reflectivity values are found to be −17 and −5.7 dB for very high frequency (VHF) and ultrahigh frequency (UHF), respectively. The median attenuation values are 1.1 and 6.0 dB, respectively. Comparing with the theoretical predictions given in Figure 2.10 (also recalling that Figure 2.10 represents the case of very dense forest), the overall correlation to the measured VHF/UHF attenuation can be concluded as acceptable. A precise comparison is not sensible since Figure 3.12 combines very disparate sets of data. In particular, the resolution of data in Figure 3.12 is too coarse to see any tendency of attenuation being particularly pronounced at around 1m wavelength, which is predicted in Figure 2.10. It thus remains unsettled whether the nonmonotonic character of the graphs in Figure 2.10 depends on the fact that the model adopted becomes inaccurate at higher frequencies or if actually branches remains the dominating source of attenuation, making the attenuation graphs true also at 500 MHz. If that is the case, foliage penetration radar might do better with the center frequency at 500 MHz than at 300 MHz, though a firm conclusion to that end cannot be made unless finer tree elements than branches are included in the modeling. That the attenuation decreases and eventually becomes marginal for frequencies lower than 300 MHz, as in Figure 2.10, is in full compliance with Figure 3.12.

In Chapter 2, analysis was provided, showing that the main tree RCS contribution generally comes from the tree stem. Incorporating ground fading, tree stem cross section was calculated with the results provided in Tables 2.1 and 2.2 for typical trees, 15m high and with a 0.2m stem radius. Presently, for the same tree size, assuming the incidence angle to be 75° and redoing the calculation at the CARABAS III high- and low-band center frequencies (wavelengths: 1.4m and 5.5m), the RCS values become 25.8 dBsm for high-band horizontal polarization and 4.9 and 13.6 dBsm for low-band horizontal and vertical polarizations, respectively. For the attenuation example in Chapter 2, a tree separation of just 6m was assumed, which (as was pointed out) represents very dense forest. Though dense, the forest in Figure 3.11 is not extremely so and a tree separation of 9m will be assumed. It follows that there are −19 dB trees per square meter, that is, disregarding attenuation, the reflectivity values become 6.8, −14.1, and −5.4 dB for high-band horizontal polarization, low-band horizontal polarization, and low-band vertical polarization, respectively. Attenuation (according to Figures 2.10 and 3.12) will be strong for the high band, though certainly below the very high attenuation values of Figure 2.10. Very likely branches will become an influential backscattering source, just considering the RCS estimates for branches in Table 2.3, and the branch RCS for a tree with 100 branches is 15 dBsm, that is, a reflectivity contribution of −3 dB, disregarding attenuation. According to Table 3.1, the attenuation should result in a combined reflectivity of −8.1 dB, that is, it must be limited to a few decibels for branches and of the order 15 dB for the tree stems, figures that seem reasonable in view of the near horizontal incidence angle (for the attenuation plot in Figure 3.12, incidence angles have been considerably steeper). Note that in Figure 3.12, the mean UHF forest reflectivity is −5.7 dB, that is, higher than what is estimated in Table 3.1, supporting that there is excessive attenuation at the near horizontal incidence in the images in Figure 3.11.

Given the coarseness with which the trihedral RCS for low-band vertical polarization was settled, the measured reflectivity value of −7.7 dB agrees reasonably well with the prediction −5.4 dB.

Comparing the theory and experiment for the fence, it should be noted that this is positioned on a farm field with wet agricultural soil. An index of refraction $\eta = 4.5 + 0.3i$ seems reasonable in this case according to Figure 2.1. The actual fence is for keeping cattle and consists of an electric tape 0.025m wide, about 1.25m above ground. The tape is expected to scatter like a cylindrical wire with diameter equaling the width of the tape. Hence, $ak = 5.6 \times 10^{-2}$ for the high band and $ak = 1.4 \times 10^{-2}$ for the low band. The ground range from the radar path to the fence is 800m, which, given an altitude of 250m, means an incidence angle of 73°. The fence is oriented at $\Psi = 30°$ with respect to the aircraft path. According to (3.65), this leads to a minor enhancement of the reflectivity values. With the assumed parameters, (3.65) implies $\sigma_0 = -18$ dB and $\sigma_0 = 2.9$ dB for the low and high bands, respectively. Comparing with Table 3.1, the measured values thus become underestimated by 2 and 4 dB for the high and low bands, respectively. For the low band, the discrepancy may well be because signal levels are close to the noise threshold, with the noise adding to the fence reflectivity, thus enhancing the measured reflectivity. For the high band, explanations must be more speculative, possibly the trihedral RCS is not as large as $\sigma = 1900$ sqm, which would result in lower measured reflectivity values not only for the fence but also the forest. The latter is already on the low side compared to likely values in Figure 3.12. On the other hand, at the very shallow incidence angles, it may be that forest attenuation results in unusually low reflectivity (i.e., lower than has been arrived at in Table 3.1).

The arguments regarding reflectivity values exemplifies that the interpretation of SAR measurement results can be difficult. There is a large number of influencing factors, in particular at low frequencies, with ground structures partially transparent and perhaps in the resonance regime. The theoretical models and predictions, which have been developed in this chapter and the foregoing chapter, are unavoidably partial and approximate, while the nature they model is indefinably variable and complex. The aim of the comparison with experiments has been to demonstrate that even so, the models provide a good theoretical understanding of actual SAR imagery. Particularly important is that the theory developed indicates how radar imagery is best acquired to achieve the given measurement objectives, by a physics-based selection of imaging parameters, such as frequency polarization and incidence angle.

## 3.5 SAR Imaging of Moving Objects

As was discussed in Section 3.3 for SAR, the time dependence in the measured reflectivity distribution $\gamma(r,t)$ is assumed to be entirely due to the sensor moving uniformly with respect to a reflectivity distribution $\hat{\gamma}(P)$, which is tied to the ground points $P$ in a fixed and time independent manner. The two functions $\gamma(r,t)$ and $\hat{\gamma}(P)$ can in this case be interconnected through the integral (3.22), representing the radar measurement.

While the ground reflectivity distribution being stationary is a reasonable assumption for ground in general, it is obviously not universally valid. Changes in reflectivity might occur on a long-term basis with targets entering or vanishing between repeated registrations, which are the situations exploited in change detection. Movements that are actually ongoing during SAR registration will also be commonly encountered. On many occasions, wind may cause movements (e.g., making trees sway). Wind movement can be very noticeable in microwave radar imaging. For meter-wave SAR, it is less of a concern, since branches being the tree elements moving most have little influence on the reflectivity distribution, as was analyzed in Chapter 2. Another source for reflectivity nonstationarity is vehicle targets on the move. In principle, detection of targets moving ought to be a simpler than static target detection; the very movement constitutes a signature that distinguishes a target from the surrounding stationary clutter. Furthermore, the high degree of concealment possible for static targets cannot be attained for targets en route. Thus, targets too small to be detectable when stationary ought to be detectable when on the move. This assumption turns out to be valid. The impact of target motion on SAR imaging will now be analyzed in detail.

A first demonstration is to show that as long as a target movement remains linear, the SAR imaging technique will focus the radar response. Assume the radar platform moving in the $y$ direction at constant altitude $z = h$, with a ground speed vector $\boldsymbol{v}$. On the flat ground (the $x$,$y$-plane), there is a single point reflector, linearly moving with a constant velocity vector $\boldsymbol{V} = (\dot{x},\dot{y})$(to be referred to as the "mover"). The coordinates $\rho_0 = \sqrt{x_0^2 + h^2}, y_0$ represent the mover location at the center time $t = 0$ of the SAR integration. Since the reflector is the only object proving reflectivity in the SAR image, (3.8) becomes

$$\gamma(r,t) = \int_{-\pi/2}^{\pi/2} \delta\left(r\cos\phi - \rho_0 - \dot{\rho}t\right)\delta\left(vt + r\sin\phi - y_0 - \dot{y}t\right)d\phi \tag{3.83}$$

Here, $\dot{\rho} = \dot{x}x(t)/\sqrt{x(t)^2 + h^2}$ with $x(t) = x_0 + \dot{x}t$. With $\dot{x}$ constant, $\dot{\rho}$ will as a consequence change in course of time. Since the fractional change of $x(t)/\sqrt{x(t)^2 + h^2}$ during integration time will be small, the $\dot{\rho}$ change will be small. Hence, in the sequel (and as a good approximation) $\dot{\rho}$, just as $\dot{x}$, will be considered constant.

Performing the integration, we obtain

$$\begin{aligned} \gamma(r,t) &= -\int_{-\pi/2}^{\pi/2} \frac{\delta\left(r\cos\phi - \rho_0 - \dot{\rho}t\right)\delta\left(vt + r\sin\phi - y_0 - \dot{y}t\right)}{r\sin\phi} d(r\cos\phi) \\ &= -\frac{\delta\left[\sqrt{r^2 - \left(\rho_0 + \dot{\rho}t\right)^2} - y_0 + \left(v - \dot{y}\right)t\right]}{\sqrt{r^2 - \left(\rho_0 + \dot{\rho}t\right)^2}} \end{aligned} \tag{3.84}$$

It follows that $\gamma(r,t)$ is zero unless range $r$ and time $t$ are interrelated:

$$r(t) = \sqrt{\left(\rho_0 + \dot{\rho}t\right)^2 + \left(y_0 - vt + \dot{y}t\right)^2} \tag{3.85}$$

Note that (5.3) is the range between mover and radar platform. Therefore, performing a rotation around the $z$-axis in the rest frame of the platform, $r(t)$ will not change, nor will radar data. Performing the same rotation for all times of a SAR registration, an original linear mover motion becomes rotated into another linear mover motion, heading in the rotated direction. The SAR data for this motion is the same as the first one. Indeed, there is an entire class of linear mover configurations, given by arbitrarily selecting the angle of rotation, which are indiscernible in SAR data.

Most interestingly, there is for any state of motion of the ground mover a particular rotation, which will take the mover to move along the $y$-axis, by which the geometry of the mover radar configuration will coincide with that of the standard SAR configuration of the radar platform moving along the $y$-axis with respect to stationary ground. The mover will thus become focused by SAR processing, though speed parameter in the SAR processing must be adjusted from platform ground speed to platform speed in the mover rest frame (or depending on sign conventions, the mover speed in the platform rest frame). This property of SAR of also focusing targets in linear motion seems not always that well understood within the SAR community. It is explicit in [16] and [17], while traceable in [15] and [18]. It is better recognized in recent publications like [19]–[21].

In order to obtain the explicit expressions for SAR image parameters for moving targets, the angle between the mover velocity in the rest frame of the platform and the $y$-axis is noted to be

$$\psi_0 = \sin^{-1}\frac{\dot{\rho}}{\sqrt{(v-\dot{y})^2+\dot{\rho}^2}} \tag{3.86}$$

This clearly is the angle for which a rotation take the mover to move along the $y$-axis, that is, to assume the standard SAR configuration. Use polar coordinates $\rho_0 = r_0\cos\phi_0$ and $y_0 = r_0\sin\phi_0$. The mover path after rotation through the angle $-\psi_0$ but in the rest frame of the platform takes the form

$$\begin{bmatrix}\rho(t)\\ y(t)\end{bmatrix} = \begin{bmatrix}\cos\psi_0 & -\sin\psi_0\\ \sin\psi_0 & \cos\psi_0\end{bmatrix}\begin{bmatrix}r_0\cos\phi_0 - \dot{\rho}t\\ r_0\sin\phi_0 - (v-\dot{y})t\end{bmatrix} \tag{3.87}$$

or

$$\begin{bmatrix}\rho(t)\\ y(t)\end{bmatrix} = \frac{1}{\sqrt{\dot{\rho}^2+(v-\dot{y})^2}} \times\begin{bmatrix}-\dot{\rho}r_0\sin\phi_0 + (v-\dot{y})r_0\cos\phi_0\\ \dot{\rho}r_0\cos\phi_0 + (v-\dot{y})r_0\sin\phi_0 - \left[\dot{\rho}^2+(v-\dot{y})^2\right]t\end{bmatrix} \tag{3.88}$$

Comparing to SAR imaging, applied to stationary ground, it is seen that any linear motion is mapped by the SAR process into the single point $\rho(0),y(0)$. The SAR processing speed parameter is seen to be $w = \sqrt{\dot{\rho}^2+(v-\dot{y})^2}$, that is, the relative speed between radar and mover (or the mover speed in the radar rest frame). In particular, it may happen that a mover has the same relative speed to the radar as the relative speed between radar and the ground. In this case, the mover will appear as a focused object in the ground SAR image. With the platform ground speed set to $v$ and the mover velocity vector to $\boldsymbol{V} = (\dot{x},\dot{y})$ the condition for such incidental focusing is

$$v = \sqrt{\dot{\rho}^2+(v-\dot{y})^2} \Rightarrow \dot{y} = v \pm \sqrt{v^2-\dot{\rho}^2} \tag{3.89}$$

Hence, if $\sqrt{\dot{\rho}^2 + \dot{y}^2} << v$, it follows that $\dot{y} \approx \dot{\rho}^2/2v << \dot{\rho}$, that is, for slow ground movers, (3.89) becomes satisfied approximately, if their heading is close to perpendicular to the aircraft path. In this case, it is not possible to discern, by the requirement that objects are focused in the SAR image, if they are static or moving. The conclusion might be regarded as a serious weakness in the SAR principle, since such focused moving objects will appear displaced from their true trajectory. Their actual position or even their trajectory cannot be resolved by the SAR principle alone. These aspects of SAR imaging of moving objects will be explained closer.

In the radar rest frame, the shortest distance between radar and mover is when the mover velocity is at right angles with the direction from the radar to the mover. In an arbitrary frame, this translates into the condition for minimum distance, which occurs when the relative speed vector is orthogonal to the vector pointing from the radar to the mover. In (3.88), this situation happens at a time $t_0$ for which $y(t_0) = 0$. Hence

$$t_0 = \frac{\dot{\rho} r_0 \cos\phi_0 + (v - \dot{y}) r_0 \sin\phi_0}{\dot{\rho}^2 + (v - \dot{y})^2} \tag{3.90}$$

The mover minimum range is given by the $\rho$ coordinate in (3.88). The mover path length from SAR image center time to the time of minimum distance becomes

$$t_0 \sqrt{\dot{\rho}^2 + (v - \dot{y})^2} = \frac{\dot{\rho} r_0 \cos\phi_0 + \left(v - \dot{y}_{mvr}\right) r_0 \sin\phi_0}{\sqrt{\dot{\rho}^2 + (v - \dot{y})^2}} \tag{3.91}$$

that is, it is the $y$ coordinate in (3.88) for $t = 0$. These results can be summarized in a somewhat different interpretation of a SAR image, than just assuming it to correspond to coordinates of stationary ground (see Figure 3.13). As illustrated, SAR image across track range, the $\rho$ coordinate, can more generally be understood as the minimum range for any imaged reflector, stationary or moving. The $y$ coordinate is the path length from the position of the reflector at center time of the SAR image to the position of minimum range. With this understanding of SAR imaging, linearly moving targets with any heading and velocity will appear focused and detectable in one and the same SAR image. The image will, however, not provide the full set of motion parameters, allowing the proper location of the movers on the stationary ground, which requires further knowledge. The association between SAR images and ground locations in preceding sections was only possible because the targets were stationary. To obtain true positioning of moving targets, SAR imaging must be enhanced to provide further mover data. Suitable methods of mover positioning will be discussed below.

Consider the ground range $r$ and azimuth angle $\phi$ at the ground point, which occupies the same position as the mover in the SAR image

$$
\begin{aligned}
r &= \sqrt{\rho(0)^2 + y(0)^2} \\
\phi &= \sin^{-1}\frac{y(0)}{\sqrt{\rho(0)^2 + y(0)^2}}
\end{aligned}
\tag{3.92}
$$

Since (3.88) amounts to a rotation through $-\psi_0$ at $t = 0$, the difference between actual location and the location indicated in the SAR images can be stated as $r = r_0$; $\phi = \phi_0 - \psi_0$. As said, the misalignment $\psi_0$ cannot be determined from the SAR process. For the mover being a ground vehicle and the radar platform an aircraft, there is a considerable platform speed advantage over the mover.

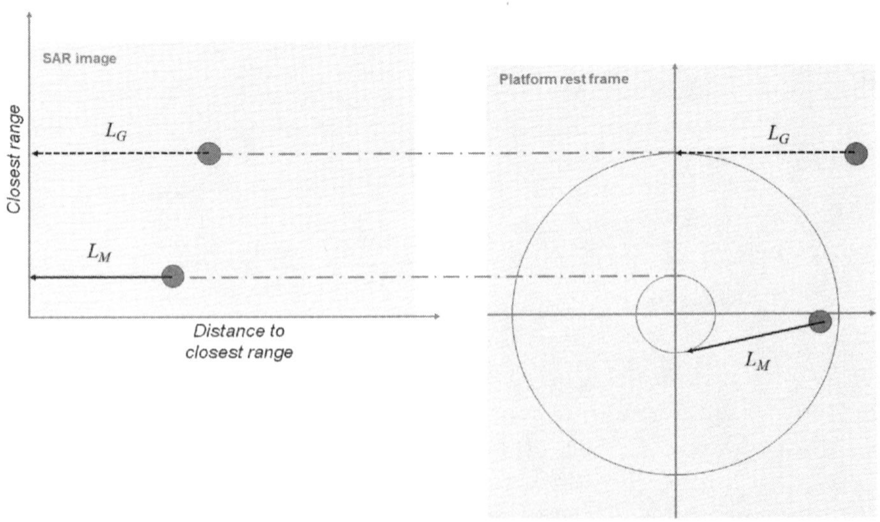


**Figure 3.13** Left: nature of SAR image representation of moving targets. Right: the platform moves to the right with respect to the ground, so in the platform rest frame, the ground moves to the left. The figure illustrates that ground point imaging is a mere special case of imaging of linear movers. All linear motions are tangents to their minimum range in the radar platform rest frame. Their SAR image azimuth position is given by the time lapsed from the zero time of SAR image formation to the instant minimum range is reached.

The misalignment $\psi_0$ will then be small. A first-order expansion in $\dot{x}/v$ and $\dot{y}/v$ may be used for (3.86), whereupon the following set of expressions relates known mover motion parameters to SAR image parameters:

$$\begin{cases} r = r_0 \\ \phi = \phi_0 - \psi_0 \\ \psi_0 = \dfrac{\dot{\rho}_0}{v} \\ w = v - \dot{y}_0 \end{cases} \tag{3.93}$$

The misalignment given by (3.93) is well known in microwave SAR; for obvious reasons, the apparent displacement of targets is called the *train off the track* effect. It opens for obvious ways of deceiving SAR-based targeting techniques. According to the previous conclusion, if a SAR surveillance aircraft is approaching a deployed ground vehicle, just by letting the vehicle move slowly in a direction at right angles to the radar path, the SAR image will wrongly position the vehicle, while focusing of the vehicle will remain. The angular pointing error $\psi_0$, may be small but with a many kilometer standoff range, it will correspond to a displacement in the $y$ direction of many tens of meters, invalidating the nominally high accuracy of SAR targeting.

Limitations in the applicability of SAR focusing of movers arise due to target movement nonlinarites. Target nonlinearities can be either a target speed magnitude change or a heading change. For the target response in the SAR image, the nonlinarites will impose an along track smearing $\delta y$ and/or cross track smearing $\delta\rho$. Here, it will be assumed that target speeds are slow compared to the platform, that is, $V/v$ is small, where $V$ designates mover ground speed. Therefore, in the platform rest frame, the target trajectories will head in directions close to the negative $y$-axis.

Along the track, smearing relates to mover speed magnitude change $\delta V$ under the integration time $T$ according to

$$\delta y = \delta VT \tag{3.94}$$

For a heading change, there will be smearing both in the along and across track directions. The across-track smearing in slant range is obvious and due to the fact that the minimum range of the linear motion changes according to the heading change, that is

$$\delta\rho = \frac{\delta V}{v}|y| \tag{3.95}$$

However, also the distance to minimum range changes with a heading change, which leads to along track smearing. In the platform rest frame, the direction to minimum range is at right angles to the direction of the mover, so a heading change gives rise to an equal change for this direction. Since the mover direction is almost equal the negative $y$ direction, the $y$ displacement thereby becomes

$$\delta y = \frac{\delta V}{v}\rho \tag{3.96}$$

As has been discussed, for large SAR integration angles the image spectral support will not be rectangular and resolution cannot be factorized into along and across track resolution. For movers, motion linearity limits integration angle, so factorization can be allowed. Note that in (3.54), integration time is set to $2T$. Presently, integration time is just $T$, so the along and across the track, resolution expressions become

$$\Delta y = \frac{\rho\lambda_c}{2Tv};\ \Delta\rho = \frac{c}{2B} \tag{3.97}$$

where $\lambda_c$ is center frequency wavelength and $B$ bandwidth. In (3.94) to (3.96) it must be required that $\delta\rho$ and $\delta y$ are smaller than $\Delta\rho$ and $\Delta y$, respectively. Combining with (3.94) to (3.97), criteria are obtained for this to be the case (here assuming $\cos\phi \approx 1$)

$$\begin{aligned} T &\leq \frac{\lambda_c}{2\delta V} && \text{(a)} \\ |y| &\leq \frac{vc}{2B\delta V} && \text{(b)} \\ \rho &\geq \frac{2T^2 v\delta V}{\lambda_c} && \text{(c)} \end{aligned} \tag{3.98}$$

Apart from these relations, there is a functional, albeit loose, relationship between speed uncertainty $\delta V$ and integration time $T$. Always, during a sufficiently short integration time $T$, speed fluctuations $\delta V$ will be indefinitely small, since accelerations are limited by target inertia. However, considering seconds of integration time, limitations on speed fluctuations will be due to the fact that speeds and directions are controlled by intent. A typical example would be a vehicle or even a human or an animal, purposefully moving along a road or a path. Roads and paths are invariably created as the shortest course between two points, under loose constraints of minimizing energy expenditure. Energy preservation also dictates that target speed will not change unnecessarily. It thus seems appropriate that with an associated high probability, $\delta V$ remains small also for extended intervals of time. However, according to condition (3.98a), the allowable speed fluctuation $\delta V$ decreases as $T$ increases, Thus, there is a maximum integration time beyond which condition (3.98a) will be violated.

As an example, some experiments were made in connection with recent CARABAS III development in logging human walk. A fit test person was equipped with a GPS data logging device. Tests were made with the person conducting walks with and without a 25-kg back pack, and also at different walking speeds. It is required that the probability remains reasonably high for the mover to be focused in the SAR image according to condition (3.98a), say larger than 50%. From the collected walk data, this probability was computed by dividing the east and west speed data into segments of duration $T$ and computed $\delta V$ as the standard deviation of the speed for each such interval. From the set of segments, the cumulative probability for $\delta V$ being less than any given value was plotted for $T$ variable. Assuming wavelength $\lambda_c$ = 1m (approximately the mean wavelength for CARABAS III high band; see Section 3.4), it was found that the 50% percentile of sufficiently linear segments of walk corresponded to $T$ = 8.5s of integration time and accordingly $\delta V \leq$ 0.07 m/s. Indications are that for most types of movers and wavelengths of meter order, the integration time around 10s would satisfy the assumption of mover linearity. In the recent test conducted with CARABAS III (see Chapter 6), integration times were extended to 30s, sometimes resulting in significant target defocusing, but in other cases in the target remaining relatively well focused. The traditional designation for SAR detection of moving ground targets is GMTI. For microwave GMTI systems integration time might be just a fraction of a second (for the Saab NORA X-band system, integration times above 0.5s gave no resolution improvement [22]). Comparison with microwave processing is returned to briefly in Chapter 6.

Inserting (3.98a) into (3.98b) and (3.98c), this restricts the imaging area for the mover in range and azimuth. As implied

$$
\begin{aligned}
x_{\min} &= Tv \\
y_{\max} &= \frac{f_c}{B} Tv
\end{aligned}
\tag{3.99}
$$

Here $f_c$ is the center frequency. For the platform speed $v$ = 25 m/s as in the CARABAS III experimental system, $Tv = 8.5 \times 25 = 220$m. The closest possible GMTI imaging range is thus 220m, which is not much of a restriction. More restrictive is the fact that the broadside field of view $2|y|_{\max}$ is limited by the center frequency to bandwidth ratio according to (3.98b). For instance, if $B$ = 55 MHz is selected, then $f_c/B = 4$ and the total azimuth field of view is $2|y|_{\max} = 1700$m. At a platform speed of 25 m/s, this distance will take 68s to traverse, meaning that about 8 successive registrations for 8.5s each could be made of the same moving target.

Object-sized resolution is ideal for detecting concealed objects, but the integration time and bandwidth limitations may prevent this level of resolution to be reached for meter waves. In the above example, assume $r$ = 5 km. Then $\Delta y$ = 14m and $\Delta x$ = 2.7m, that is, the area resolution is $\Delta A$ = 38 m$^2$. If the mover to be detected is a walking human being, the resolution of 0.67 m$^2$, which was achieved according to Table 3.I, would be desirable (indeed, humans standing erect are detectable in the CARABAS III SAR imagery at this level of resolution). The moving target resolution is thus 18 dB above the optimal value. For detecting a vehicle of ordinary size, a resolution of 10 m$^2$ may be suitable, in which case achievable resolution is only 6 dB above optimal value. Increasing the radar range, resolution can be maintained by increasing platform speed.

The established remedy to reduce the influence of clutter for moving targets is the so-called *displaced phase center array antenna* (DPCA). Microwave GMTI relies (and is implicit in the so-called STAP technique, see [23]) on this method, which is also is applicable to meter waves. For the extended integration time in this case, the DPCA principle amounts to obtaining two SAR images, with respect to a forward and a rear antenna phase center. Assume the antennas separated by the along track distance $D$. Correspondingly, the ground surface in the two SAR images will be separated by a shift $D$ in the along track coordinate, whereas the cross-track coordinate remains unshifted. For a mover, in the platform rest frame, denote the angle between mover and platform heading by $\hat{\phi}$ and designate by ± the front and rear antenna SAR image coordinates. The difference in closest range and along track coordinate, that is, distance to closest range, becomes

$$\left.\begin{aligned} \rho_+ - \rho_- &= D\sin\hat{\phi} \\ y_+ - y_- &= D\cos\hat{\phi} \end{aligned}\right\} \Leftarrow \hat{\phi} = \sin^{-1}\frac{\dot{\rho}}{\sqrt{\dot{\rho}^2 + (v - \dot{y})^2}} \tag{3.100}$$

To the lowest order in $V/v$

$$\begin{aligned} \rho_+ - \rho_- &= D\frac{\dot{\rho}}{v} \\ y_+ - y_- &= D \end{aligned} \tag{3.101}$$

It follows that the cross range shift is to the first-order proportional to mover ground speed. The along track image displacement is (approximately) insensitive to object ground speed, that is, it will be the same when imaging static and slowly moving objects. Matching the coordinates of the front and rear images to the stationary ground (i.e., clutter data), subtraction between SAR images will suppress the clutter. However, the mover data will not cancel, since the expected cross range displacement give rise to a phase shift. The fractional mover reflectivity density loss in the process of clutter cancellation becomes

$$\Delta_{\text{mover}} = e^{i\frac{4\pi}{\lambda_c}\frac{D\dot{\rho}}{2v}} - e^{-i\frac{4\pi}{\lambda_c}\frac{D\dot{\rho}}{2v}} = 2i\sin\left(\frac{2\pi}{\lambda_c}\frac{D\dot{\rho}}{v}\right) \tag{3.102}$$

The level of suppression of the clutter is limited by transfer function dissimilarities, as well independent noise entering into the forward and rear antenna channels. Leaving the details on this subject to Chapter 6, denote the suppression level by $\gamma_{\text{clutter}}$. The mover target-to-clutter ratio after clutter suppression becomes according to (3.97) and (3.98)

$$\Gamma_{TC} = \frac{\left|\Delta_{\text{mover}}\right|^2 \sigma_{\text{mover}}}{\left|\Delta_{\text{clutter}}\right|^2 \sigma_0 \Delta x \Delta y} = \frac{\sigma_{\text{mover}}}{\left|\Delta_{\text{clutter}}\right|^2 \sigma_0} \frac{2vT}{r\lambda_c}\frac{2B}{c}\sin^2\left(\frac{2\pi}{\lambda_c}\frac{D\dot{\rho}}{v}\right) \tag{3.103}$$

An optimal situation arises when the argument of the sine function is $\pi/2$, that is, $v = v = 4\dot{\rho}D/\lambda_c$. Assuming this situation, since it is required for detection that $\Gamma_{TC} \geq 1$, detection requires that clutter reflectivity is limited according to

$$\left|\Delta_{\text{clutter}}\right|^2 \sigma_0 \leq \sigma_{\text{mover}} \frac{4vTB}{rc\lambda_c} \tag{3.104}$$

Further considering the case with the mover being a walking human, (3.103) determines the possibilities for such a mover to become detectable. Model a human as a vertical cylinder with height 1.7m and diameter 0.3m (i.e., 1m waistline) by means of the RCS formulas of the foregoing chapter (including the important ground interaction). The index of refraction of human (or animal) tissue should not be very different from that of vital wood, so $\eta = 6 + 0.31i$ is assumed in accordance with Chapter 2. A mean value $\sigma_{\text{mover}} \approx 3$ dBsm is obtained for horizontal polarization, incidence angle around 60°, and wavelength in the interval 1m to 1.5m. Assume the walker to move at ground speed $\dot{x} = 1.5$ m/s, that is, $\dot{\rho} = 1.3$ m/s, and use the above the experimental system values $B = 55$ MHz, $\lambda_c = 1$m, $r = 5$ km, $v = 25$ m/s, and $T = 8.5$s.

For the example, the optimal antenna displacement is $D = v\lambda_c/4\dot{\rho} = 5$m. Then (3.104) implies $|\Delta_{\text{clutter}}|^2\ \sigma_0 \le -13$ dB. To the extent that the forest reflectivity is due to tree stems returns (with branch returns being negligible), attenuation of the signals in the forest canopy affects the mover RCS and forest reflectivity to a similar degree and thus does not enter into this analysis. For 15m high tree stems of 0.4m diameter and (as typical for dense forest) with tree stems 9m apart, the reflectivity before inclusion of attenuation is $\sigma_0 = 4$ dB, so a level of suppression better than $|\Delta_{\text{clutter}}|^2 = 17$ dB is required. In tests with the CARABAS III experimental system, clutter suppression levels of the order 17 dB have indeed been obtained with human movers in light forest plainly detectable. Due to size constraints, CARABAS III antenna displacement is only $D = 1.5$m implying a 6.5-dB reduction in mover reflectivity compared to ideal antenna separation. In the tests, this reduction was balanced by the relatively moderate clutter level of the light forest. For a full account of test results, see Chapter 6.

The principles presented for mover detection seem less suitable at wavelengths of several meters, even though these provide more favorable ratios between forest reflectivity and target RCS for larger targets such as vehicles. The obstacle, in particular, is that the required forward/rear antenna separation increases beyond the installation constraints of the moderately sized airborne platforms, which are suitable for this type of radar.

As is practiced in microwave GMTI, mover position can be uniquely determined if at least three along track phase centers are used. The basic principle is again to obtain SAR images from each of the phase centers and by interferometric evaluation of these determine the phase difference for corresponding points in the SAR images. As seen in (3.102) and (3.103), with two phase centers, only the residual strength of the mover target after clutter suppression can be determined. With three or more phase centers displaced, by preforming clutter cancellation pairwise between the images, it is also possible to estimate $\dot{\rho}$ from such data. Consider three phase centers separated by the distance $D$. Obtain two clutter cancelled images, the first from the middle and the front phase centers, and the second by the middle and the rear phase

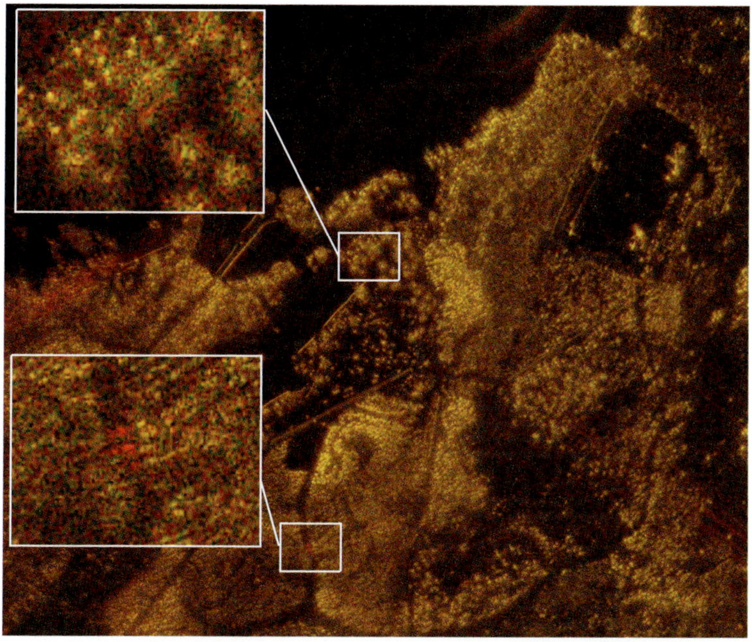

**Color Plate 1** Reference/update color blend for CARABAS III high band. Image size is 1 km × 1 km. The dark topmost part is a lake, the cleared rectangular area to the right is an antenna measurement range.

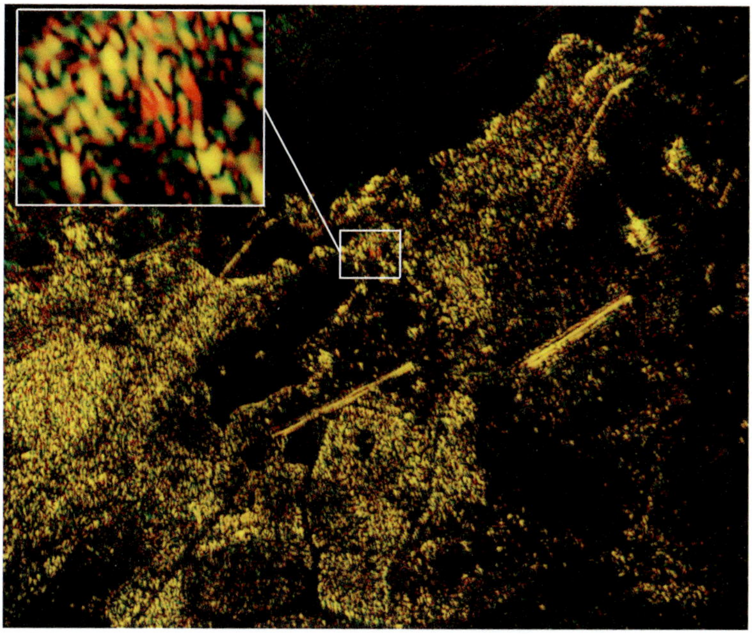

**Color Plate 2** Reference/update color blend for CARABAS III low band. The strong reflection at the measurement range comes from a tower. The linear structures in the image are power lines oriented within SAR integration sector.

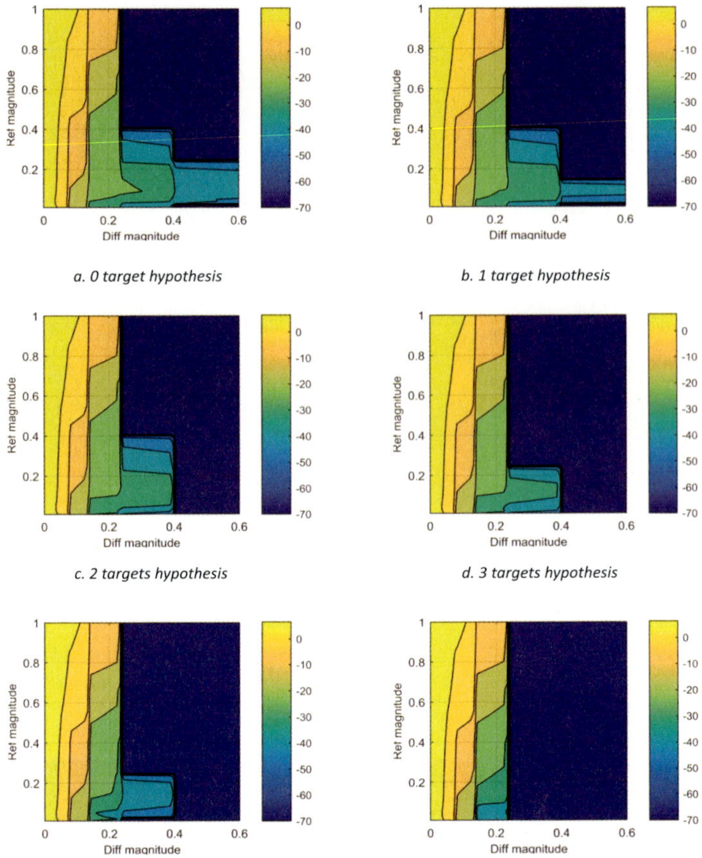


*a. 0 target hypothesis*

*b. 1 target hypothesis*

*c. 2 targets hypothesis*

*d. 3 targets hypothesis*

*e. 4 targets hypothesis*

*f. 5 & 6 targets hypotheses*

**Color Plate 3** Clutter magnitude statistics based on gradual excision of the most probable target nominees. Note that the statistics become stable only after excision of the 5th strongest nominee, even though only the four strongest become assigned high probabilities of being targets. Color scale in dB.

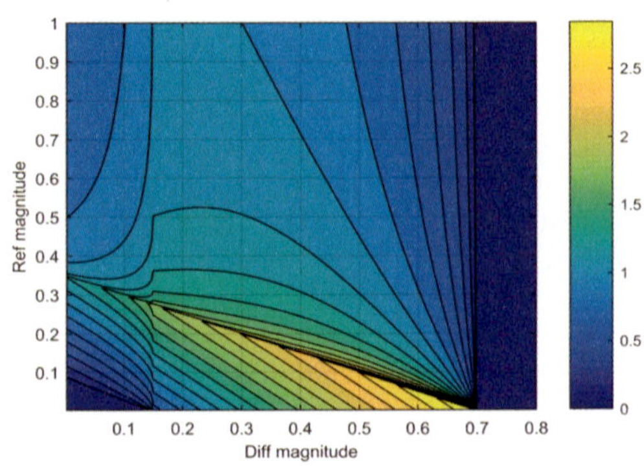


**Color Plate 4** Conditional target magnitude probability density $p\langle a_U | x_T, a_R \rangle$. It is assumed that $a_{min} = 0.15$ and $a_{max} = 0.7$. Color scale is linear.

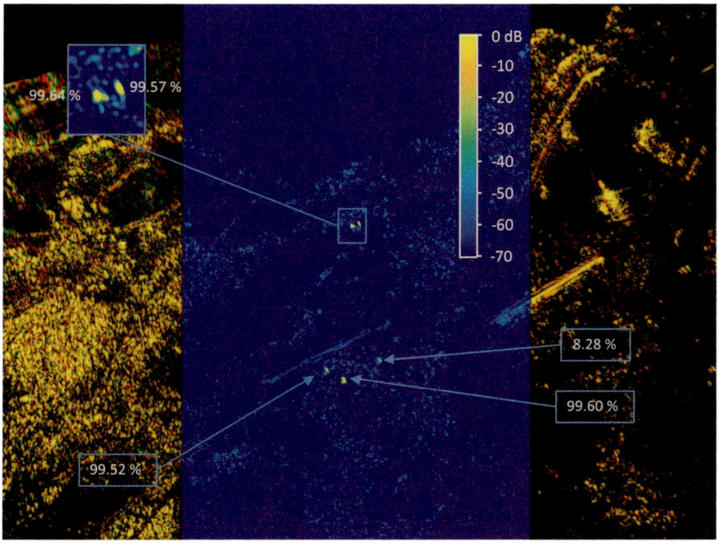


**Color Plate 5** Positions and probabilities for the five most likely target nominees. The 6th target nominee identified (target probability 5.68%) is not the 6th and weakest target but a side lobe in the immediate vicinity of the 2nd strongest target.

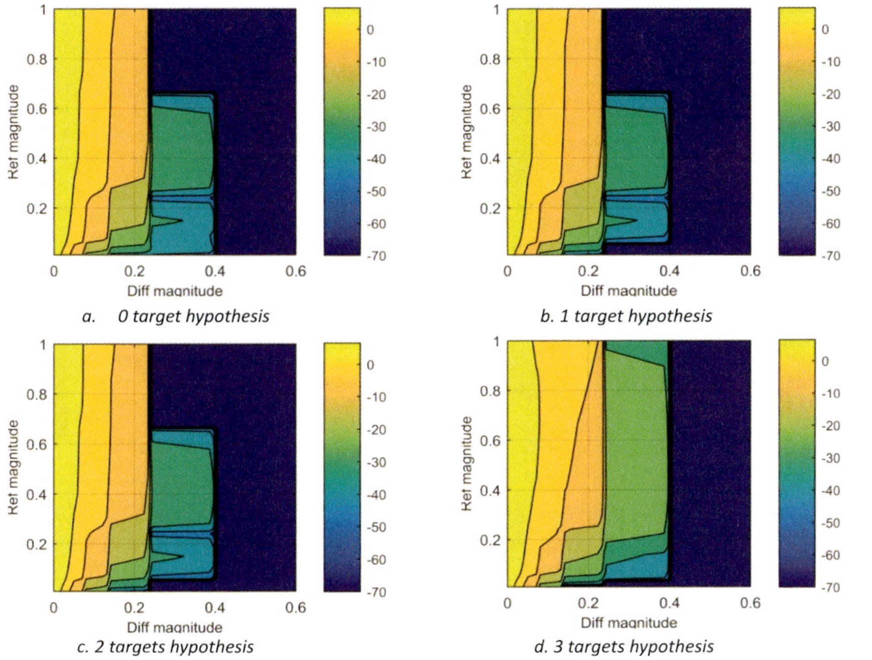


**Color Plate 6** Gradual target excision clutter change statistics for Roxtuna. There is a prevalence of cultural clutter returns characterized by fluctuating responses, causing the statistics. The Bayesian scheme effectively discriminates such clutter changes from changes due to target presence.

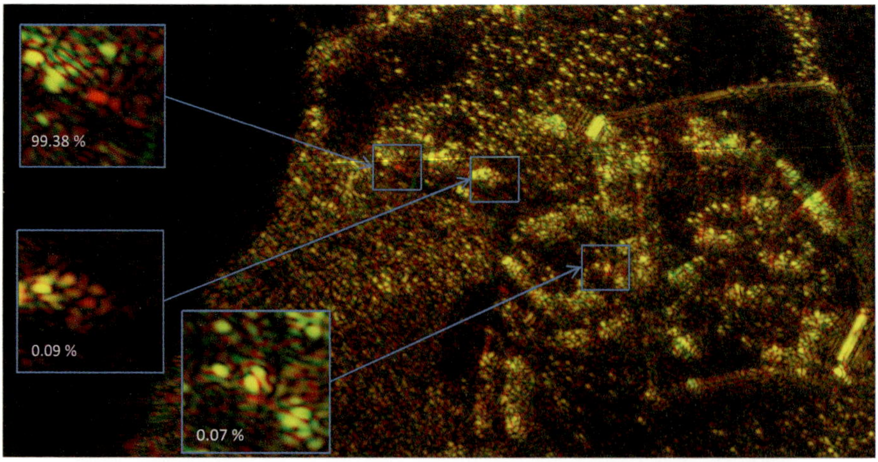


**Color Plate 7** Positions and probabilities for the three most likely target nominees at the Roxtuna site. Only the actual target has significant probability. The red indications, in the middle of the lower image half, turn out with zero target probabilities. Evidently they are too weak to represent vehicle targets.

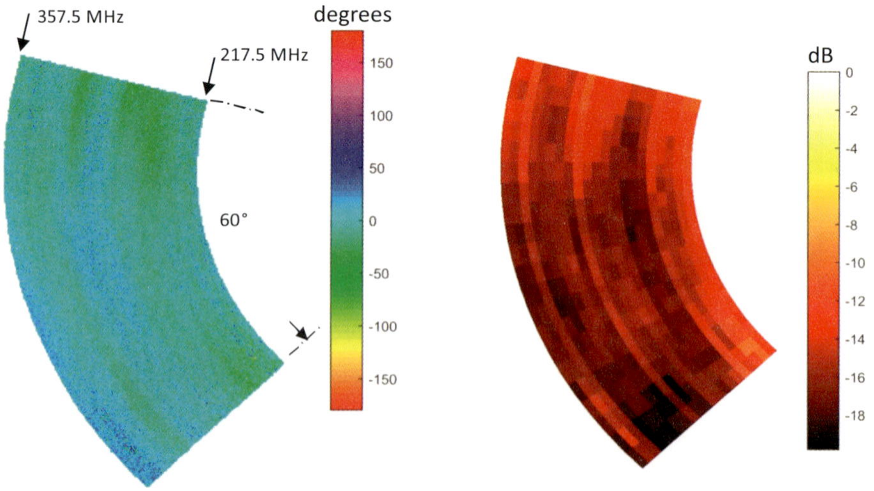


**Color Plate 8** LEFT: The phase of the quotient between forward aft image complex spectra for P9. RIGHT: Covariance eigenvalue quotient for the 441 spectral segments across which the inverse Fourier transform along the minimum and maximum eigendirections provide the mover and stationary SAR images of P9.

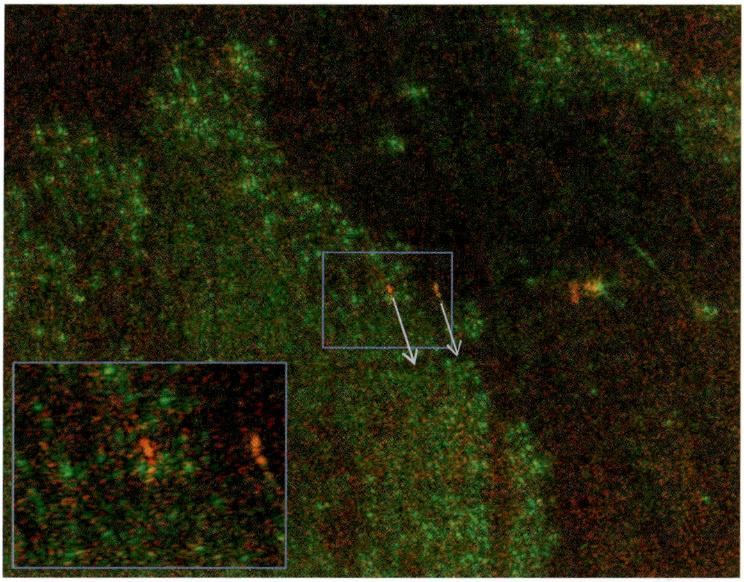

**Color Plate 9** Mover image is red, stationary clutter image is green. Movers are offset their actual position walking along a forest path, with the offset due to their walking speed of 5 km/h (cf. P11). Offset is at right angles to the antenna beam, which tilts from the platform southerly course (down) to compensate easterly winds. No discernable difference in strength of the mover responses are seen, even though they have reached different depths into the forest.

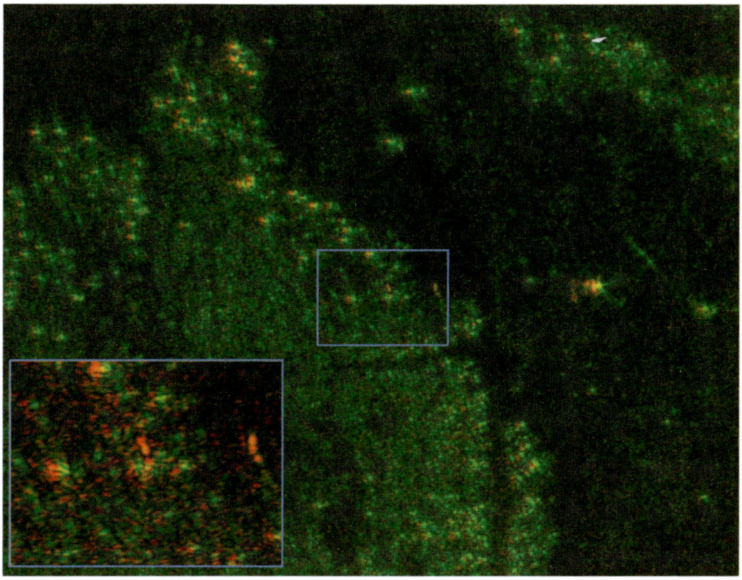

**Color Plate 10** Image based on the same data as P9, but eigenvalue decomposition has been performed over the entire spectrum rather than for the 421 segments. As a result equalization becomes imperfect and stationary clutter leaks into the mover image causing false alarms—some stronger than the mover footprints.

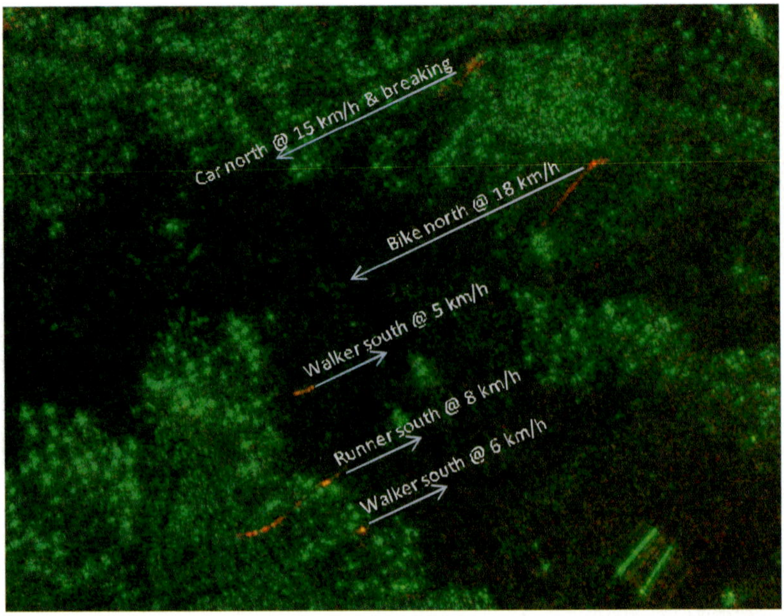


**Color Plate 11** Several types of movers along a forest lane. There is no tree cover but severe stationary clutter. The different speeds and headings are seen in the different displacements. The degree of defocusing reflects the variations in speed for each target during the 30s of integration time.

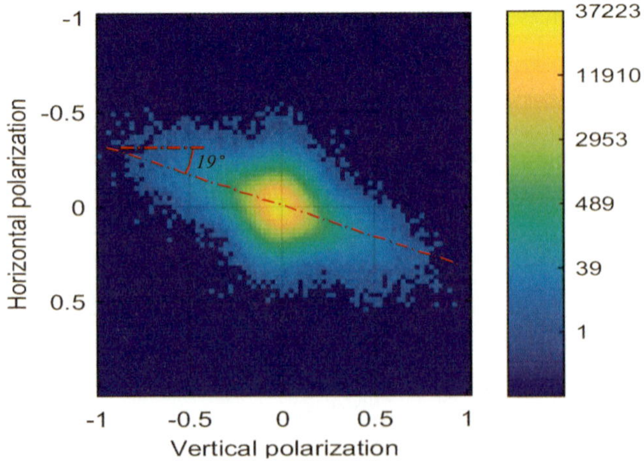


**Color Plate 12** Correlation statistics in number of pixels per amplitude bin for Color Plate 13. The statistics reasonably well fulfill the predicted mean ratio of 19° between horizontal and vertical amplitudes for the dominating surface returns.

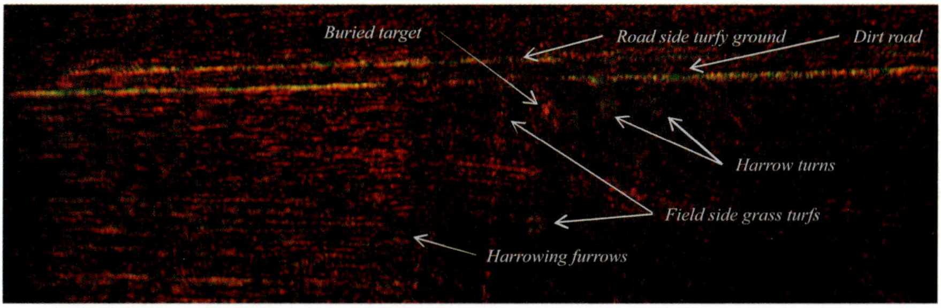


**Color Plate 13** Green is the vertically polarized, red is the horizontally polarized 150 × 50 m SAR image, obtained at 65° incidence. The buried target is evident in horizontal polarization, while visibility is essentially nil for vertical polarization.

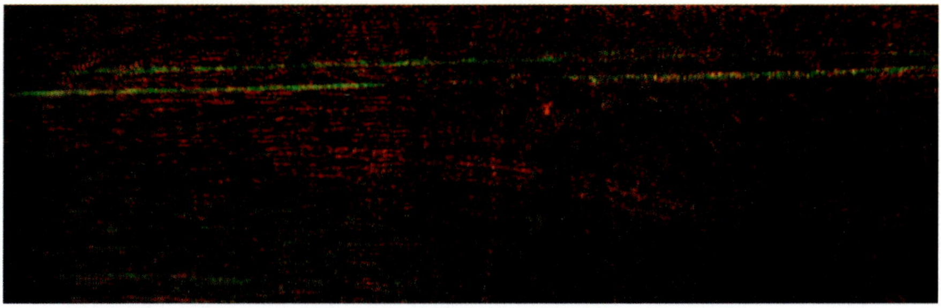

**Color Plate 14** Covariance separation of surface (green color) and subsurface responses (red color) in Color Plate 13. The road side turfy ground and the harrowing furrows are identified as pure surface responses, i.e. has obtained significantly stronger greenish tint.

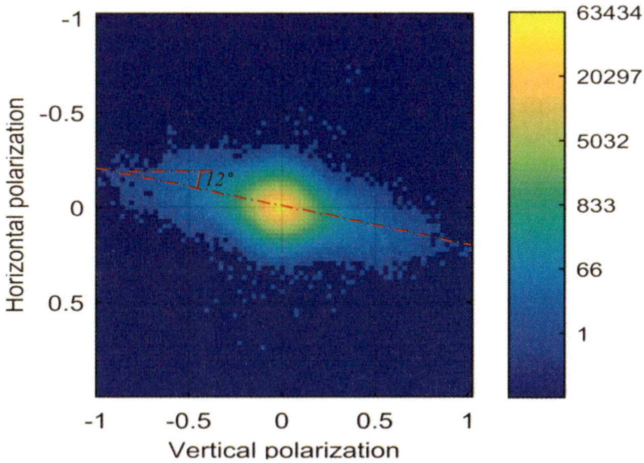


**Color Plate 15** Correlation statistics for image below. The higher subsurface contrast at the near grazing incidence follows from the weaker co-variation of horizontal amplitude with vertical. The measured correlation agrees well with the theoretically predicted mean ratio of 12°.

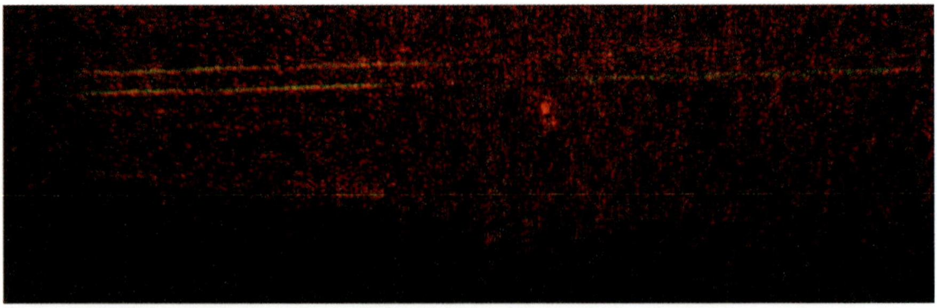

**Color Plate 16** The same type of image as in P13 but for 77.5° incidence. Surface responses are predictably weaker for horizontal polarization.

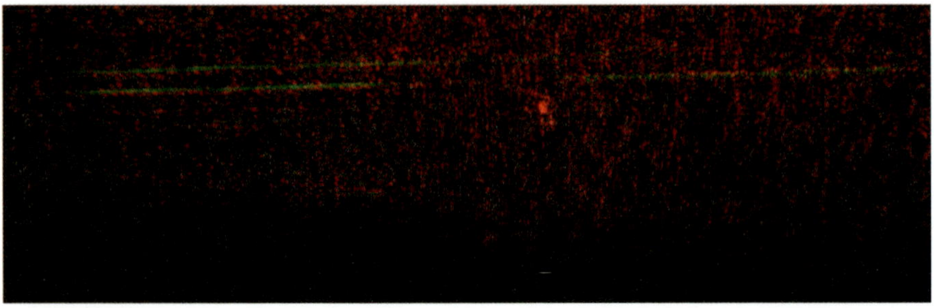

**Color Plate 17** Covariance separation of surface and subsurface responses in the above image. Essentially all traces of the surface response have been removed from the subsurface response—the road side turf is purely green, proving efficient suppression in the subsurface channel.

centers. The mover will appear in the same SAR image position in the two images and be represented by the respective complex numbers

$$C_{\pm} = C\left(1 - e^{\pm i\frac{4\pi}{\lambda_c}\frac{D\dot{\rho}}{v}}\right) \tag{3.105}$$

Forming the quotient $C_+/C_-$, it is possible to determine $\dot{\rho}$

$$\frac{C_+}{C_-} = \frac{1 - e^{i\frac{4\pi}{\lambda_c}\frac{D\dot{\rho}}{v}}}{1 - e^{-i\frac{4\pi}{\lambda_c}\frac{D\dot{\rho}}{v}}} = -e^{i\frac{4\pi}{\lambda_c}\frac{D\dot{\rho}}{v}} \Rightarrow \dot{\rho} = \frac{v\lambda_c}{4\pi D}\text{phase}\left(-\frac{C_+}{C_-}\right) \tag{3.106}$$

Note that for the determination of $\dot{\rho}$ to be unique, it must be asserted a priori that $|\dot{\rho}| \leq V_{\max}$ for some value $V_{\max} << v$ representing the maximum ground mover speed, which is expected. Selecting the displacement $D$ according to

$$D = \frac{v\lambda_c}{4V_{\max}} \Leftrightarrow V_{\max} = \frac{v\lambda_c}{4D} \Rightarrow \left|\text{phase}\left(-\frac{C_+}{C_-}\right)\right| < \pi \tag{3.107}$$

Thus, uniqueness of (3.105) is asserted. A typical separation choice is $D \approx \lambda_c$ whereupon $V_{\max} = v/4$. From (3.93), the angular accuracy $\delta\phi$ with which the mover can be positioned follows as:

$$\delta\phi = \delta\left(\frac{\dot{\rho}}{v}\right) = \frac{\lambda_c}{4\pi D}\delta\ \text{phase}\left(\frac{C_+}{C_-}\right) \approx \frac{\lambda_c}{4\pi D\sqrt{\frac{\Gamma_{TC}}{2}}} \tag{3.108}$$

For instance, assuming $\Gamma_{TC}$ =13 dB (achievable for a moving vehicle but perhaps not a human walker) and $D \approx \lambda_c$, the formula implies that $\delta\phi = 1.7°$.

It is observed that requirements for clutter suppression request a large separation between phase centers, where in fact $D = \lambda_c v/4\dot{\rho}$ is the preferred phase center displacement and also for targets where $\dot{\rho} << V_{\max}$. However, due to the sine function appearing (3.102), this choice makes data insensitive to the mover. It is possible to reconcile the conflicting requirements to some extent, observing that optimal clutter suppression is most acute for targets like human walkers, which are both slow and have a low RCS. For these targets, the angular offset due to motion is small, with a ground mover at 1.5 m/s, the angular offset is $\phi = 3°$ for $v = 25$ m/s. For vehicle target speeds and thus the motion offset are expected larger, making angular compensation important, but the RCS is then sufficiently large not to require optimal clutter suppression. A system, which would handle both situations reasonably well, would rely on one front and one rear antenna, say, separated by 3m. The antennas are arranged to allow bistatic operation, with transmissions with respect to one phase center and reception with respect to the other. The resulting bistatic radar data will have its phase center in between. Also including monostatic operation from the front and rear antennas, three phase centers are obtained, with a separation of 1.5 m. For vehicle targets, a suitable radar platform speed might be 50 m/s, which allows unique mover speed determination at velocities up to about 12 m/s. This mode of operation is suitable for detecting vehicle targets moving in deep hide. For detecting low RCS slow targets like human walkers, the radar mode of operation would be purely monostatic, receiving and transmitting from the either the forward or rear phase center in a ping pong fashion to obtain a high degree of clutter cancellation, while not making any compensation for the mover misalignment in the SAR image. The platform speed might be reduced to 25 m/s to increase the degree of clutter cancellation further.

In conclusion, some remarks will be made on the operational difference between SAR surveillance of stationary and moving targets. Both cases generally call for repeated radar registrations to capture and interpret an ongoing development (or to establish that no development occurs) but at very different revisit frequencies. In the first case, the assumption is that changes develop slowly, and using change detection and returning to scene at sparse intervals (hours and days), the essence of what is happening is being captured. Vast areas could be surveyed in this way. To capture the development of moving targets, the sensor must bear on the scene in a more or less unbroken fashion. A typical requirement is that revisit intervals, within the surveyed area, are sufficiently short to allow target tracking, providing continuous information on the moving target locations. The SAR path in the case of change detection is that of extended straight lines, with the same area covered in renewed flights (the sensor paths need not to be exactly the same). For moving targets, the path must be a loop that the sensor platform goes around in order to continue to bear on a particular scene. For the SAR path, using a fixed wing aircraft as sensor platform is an economic proposition, also simplifying SAR processing by it the capability to fly straight with constant speed. For moving targets, a helicopter is a sensible sensor platform, having the ability to make quick turns in a loop track. Moreover, the helicopter may easily adjust speed to optimize detection performance between stronger more rapid vehicle targets and the slow weak targets such as moving humans. Figure 3.14 shows the principles

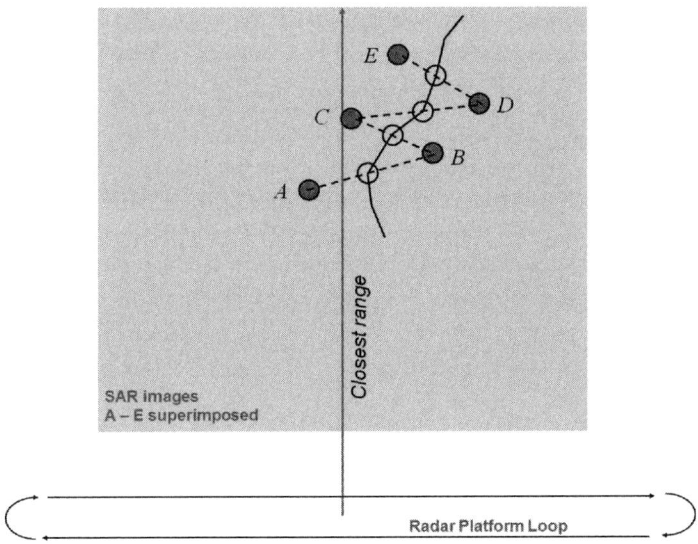


**Figure 3.14** Tracking movers with radar platform in a loop path. Targets can be a human walker. The path length and SAR image size are assumed several kilometers, that is, a leg in the loop will be completed in one to a few minutes. For each return, there is a new image superimposed onto the previous ones. Depending on leg direction mover, SAR image position is offset to one or the other side of the actual mover track (the figure illustrates three legs moving to the left and two legs moving to the right with mover offsets in each case). The actual mover position is estimated by interpolating offset position between the neighboring oppositely directed tracks. For slow targets, the angular offset is relatively small. Using the tracking method, the actual position of the movers remains relatively known throughout the registration time.

for tracking slow targets (e.g., walkers), traversing the scene using such a (typically unmanned helicopter, that is, rotor UAV) loop track.

# References

[1] Carrara, W.G., R.S. Goodman, and R.M. Majewski, *Spotlight Synthetic Aperture Radar: Signal Processing Algorithms*, Norwood, MA: Artech House, 1995.

[2] Hellsten, H., and L.E. Andersson, "An Inverse Method for the Processing of Synthetic Aperture Radar Data," *Inverse Problems*, Vol. 3, No. 1, 1987, pp. 111–124.

[3] Andersson L.E., "On the Determination of a Function From Spherical Averages," *SIAM Journal on Mathematical Analysis*, Vol. 19, No. 1, January 1988, pp. 214–232.

[4] Fawcett, J.A., "Inversion of N-Dimensional Spherical Averages," *SIAM Journal on Applied Mathematics*, Vol. 45, No. 2, 1985, pp. 336–341.

[5] Rocca, F., C. Caffori, and C. Prati, "Synthetic Aperture Radar: A New Application for Wave Equation Techniques," *Geophysical Prospecting*, Vol. 37, No. 7, 1989, pp. 809–830.

[6] Hellsten, H., "Inverse Scattering Analysis of Diffraction Limited SAR," *IEEE Trans. on Antennas and Propagation*, Vol. 38, No.10, October 1990, pp. 1517–1522.

[7] Ulander, L.M.H., and H. Hellsten, "A New Formula for SAR Spatial Resolution," *International Journal of Electronics and Communications*, Vol. 50, No. 2, 1996, pp. 117–121.

[8] Frölind, P.O., A. Gustavsson, M. Lundberg, et al., "Circular-Aperture VHF-Band Synthetic Aperture Radar for Detection of Vehicles in Forest Concealment," *IEEE Trans. on Geoscience and Remote Sensing*, Vol. 50, No. 4, April 2012, pp. 1329–1339.

[9] Hellsten, H., L.M. Ulander, A. Gustavsson, and B. Larsson, "Development of VHF CARABAS II SAR," *Proceedings of SPIE*, Orlando, Florida, Vol. 2747, April 1996, pp. 48–60.

[10] Ulander, L.M.H., P.O. Frölind, and T. Martin, "Processing and Calibration of Ultra-Wideband SAR Data from CARABAS-II," *Proceedings of CEOS SAR Calibration Workshop*, Toulouse, France, October 26–29, 1999.

[11] Martin, T., and L. Pettersson, "Development of VHF CARABAS II SAR," *IEEE Trans. on Antennas and Propagation*, Vol. 4, No.10, April 2000, pp. 494–501.

[12] Hellsten, H., P.O. Frölind, A. Gustavsson, et al., "Ultrawideband VHF SAR Design and Measurements," *Proceedings of SPIE*, Orlando, Florida, Vol. 2217, July 16–22, 1994.

[13] Mirkin, M.I., T.O. Theodore Grosch, T.J. Murphy, et al., "Results of the June 1993 Yuma Ground Penetration Experiment," *Proceedings of SPIE*, Orlando, Florida, Vol. 2217, No. 4, July 16–22, 1994.

[14] Binder, B.T., M.F. Toups, S. Ayasli, et al., "SAR Foliage Penetration Phenomenology of Tropical Rain Forest and Northern U.S. Forest," *Proceedings of IEEE 1995 International Radar Conference*, Alexandria, Virginia, May 8–11, 1995.

[15] Soumekh M., "Moving Target Detection in Foliage Using Along Track Monopulse Synthetic Aperture Radar Imaging," *IEEE Trans. on Image Processing*, Vol. 6, No. 8, August 1997, pp. 1148–1163.

[16] Ulander, L.M.H., and H. Hellsten, "Low-Frequency Ultra-Wideband Array Antenna SAR for Stationary and Moving Target Imaging," *Proceedings of SPIE*, Vol. 3704, No. 149, July 20, 1999.

[17] Hellsten, H., and L.M.H. Ulander, "Airborne Array Aperture UWB UHF Radar—Motivation and System Considerations," 1999 IEEE Radar Conference, April 22–22, 1999, IEEE AES Systems Magazine, Vol. 15, No. 5, May 2000, pp. 35–45.

[18] Jao, J.K., "Theory of Synthetic Aperture Radar Imaging of a Moving Target," *IEEE Trans. on Geoscience and Remote Sensing*, Vol. 39, No. 9, September 2001.

[19] Chapman, R.D., C.M. Hawes, and M.E. Nord, "Target Motion Ambiguities in Single-Aperture Synthetic Aperture Radar," *IEEE Transactions on Aerospace and Electronic Systems*, Vol. 46, No. 1, January 2010.

[20] Jao, J.K., and A. Yegulalp, "Multichannel Synthetic Aperture Radar Signatures and Imaging of a Moving Target," *Inverse Problems*, Vol. 29, No. 5, April 2013.

[21] Moyer, L.R., and M.A. Govoni, "Moving Target Trajectories in Low-Frequency SAR Imagery," *IEEE Transactions on Aerospace and Electronic Systems*, Vol. 50, No. 3, July 2014, pp. 2354–2360.

[22] Dammert, P., and H. Hellsten, "Focusing Ground Movers Using Fast Factorized SAR Backprojection," *Proceedings of EUSAR 2010*, 8th European Conference on Synthetic Aperture Radar, Aachen, Germany, June 7–10, 2010.

[23] Klemm, R., "Applications of Space-Time Adaptive Processing," IET Radar, Sonar & Navigation, MILCOM 97 Proceedings, Vol. 21, 2004.

# 4

# Meter Wavelength SAR System Design

## 4.1 Low-Frequency SAR Design Aspects

Theory has been developed predicting the capability to penetrate vegetation and to isolate radar returns from hidden and buried objects. As was accounted for in the last chapter, radar measurements largely support the predictions made. Using frequencies below 100 MHz and horizontal polarization, the best choice to obtain favorable target to clutter ratios for stationary targets extending in all three dimensions for more than a meter was demonstrated. For smaller objects, even though the cross section is increased by increasing frequency, clutter levels rise even more sharply. Hence, such objects cannot be detected by radar methods when stationary, but the method accounted for regarding moving objects can be applied to humans walking in a forest. In this case, horizontal polarization is again the best option, while favorable frequencies are above 100 MHz, thus providing a good radar cross section (RCS) for human-sized objects. Also for buried objects, the optimal frequency choice is related to object size, selecting the center frequency corresponding to subsurface wavelength equal to object circumference. In this case, a combined use of vertical and horizontal polarizations provides the best possibilities for isolating the target response from clutter. In all applications, detecting stationary objects by synthetic aperture radar (SAR) methods, a close-to-unity fractional bandwidth is essential, according to the discussion in Section 3.2.

Theoretical predictions stating the feasibility of these applications assume an ideally functioning SAR system. Actual radar designs obviously deviate from such ideal performance. This chapter will discuss design principles, with the applications above in mind, and thus seeing to that the system solutions will not significantly impair on the required performance in these cases.

Radar design principles for low-frequency radar have to deviate considerably from microwave practice. This is well exemplified by low-frequency SAR radar antenna design. Requirements on the directional characteristics of the radar signals follow from the general SAR principles in the foregoing chapter and are the required one-sidedness of signal directivity and that the signals must have a wide uniform spread on this one side. The radar signal can be endowed with the appropriate directional properties using the combined antenna beam, formed when transmitting and receiving, that is, it is a matter of designing transmit and receive antenna functions for the radar system. The resulting low-frequency SAR antennas will have low gain (even less than unity gain as will be seen), which is quite contrary to microwave radar design practice. The design aspects on antennas will be analyzed in this chapter. For electrically small antennas, original theory was developed by Wheeler [1] and Chu [2] in the late forties. The approach to antenna theory adopted here is nonhistorical, however, and intended to fit the requisites of meter-wave SAR design.

The discussion so far has concerned the visibility of objects in competing clutter. It is noted that the target to clutter ratio does not depend on transmit power, that is, clutter can be considered as *multiplicative noise* in target detection schemes. Target detection performance is further determined by *additive noise*, present at a certain level in a SAR image, as will be analyzed in this chapter. The SAR image additive noise is the combined effect of all signal sources, which during the SAR integration time provide contributions to the received radar signal, but are statistically uncorrelated to the transmit signal. This statistical independence is the only criterion for additive noise. For instance, additive noise arises in the signal processing chain by digitization rounding off errors. Very significant sources of additive noise are communication signals within the radar band, as is discussed below. Due to its statistical independence to the transmit signal, additive noise will spread in a statistically uniform way across SAR image. It should be obvious from the foregoing chapter that total cross section of the SAR image (reflectivity integrated across the image) increases by the fraction $E_n/E_s$ of the noise energy $E_n$ impinging on the radar antenna compared to the backscattered energy $E_s$ during the SAR registration process (for additive noise entering later in the reception chain, the antenna equivalent noise energy has to be found). Since $E_s$ depends directly on transmitted power, the level of transmitted power will be decisive of the degree to which additive noise is suppressed in the SAR image. It is noted that apart from transmitted energy, the energy of the backscattered signal is determined by several interrelated factors such as surveillance range, platform speed, antenna performance, what receiver filtering is applied, and transmitter power. Their interrelation is expressed by the *radar equation*, to be derived subsequently.

The feasibility of cohabitation between low-frequency SAR and wireless services (typically communication, broadcast and radio navigation) operating within the radar frequency band is a very major concern, in fact determining whether low-frequency SAR will have real practical applicability. A large number of frequency allocations fill the very high frequency (VHF) and ultrahigh frequency (UHF) bands and stretch up into the microwave regime. National and international agreements regulate services to be strictly confined to assigned bands, many of these being quite narrow (tens of kilohertz). The temporal and areal coverage of the services are mostly sporadic and local. Therefore, operating octave order–bandwidth low-frequency SAR, the bandwidth occupied by services ongoing at power levels of significance is invariably just some small fraction of the radar bandwidth (see Figure 4.1 for typical situations).

While the part of the radar spectrum filled with these transmissions (henceforward *radio interference* (RFI)) is moderate, the power with which they impinge on the radar can be significant. A radar transmitter designed so that the power of the backscattered signal is certain to overwhelm the interference would require an impractical amount of transmit power, and will certainly jam those wireless services that cause the interference. However, since the interfering services affect only a small fraction of the radar bandwidth, it is possible to perform receiver filtering, nulling the data in the interfering bands. If only sufficient receiver linearity (dynamic range) can be upheld, the interference can be removed regardless of the power level with which it impinges on radar data. However (as will be discussed), receiver dynamic range must be large, typically 70 dB or better. Obviously, the backscatter radar data across the frequencies of the inferring services is lost by the nulling. The effect of the lost data on SAR image quality is essentially the same as when there are specks of dirt on a camera lens; in both cases, when the percentage of lost data is minor, the degradation in data quality is minor.

As will be exemplified subsequently, radar transmit power will be relatively low for many systems, apart from being spread over a large bandwidth. Even so, if unmitigated, radiated power will still be sufficient to cause interference to nearby in-band services. An evident mitigation step is the ability of the radar to null both transmit and receive signals in the sub-bands of frequency

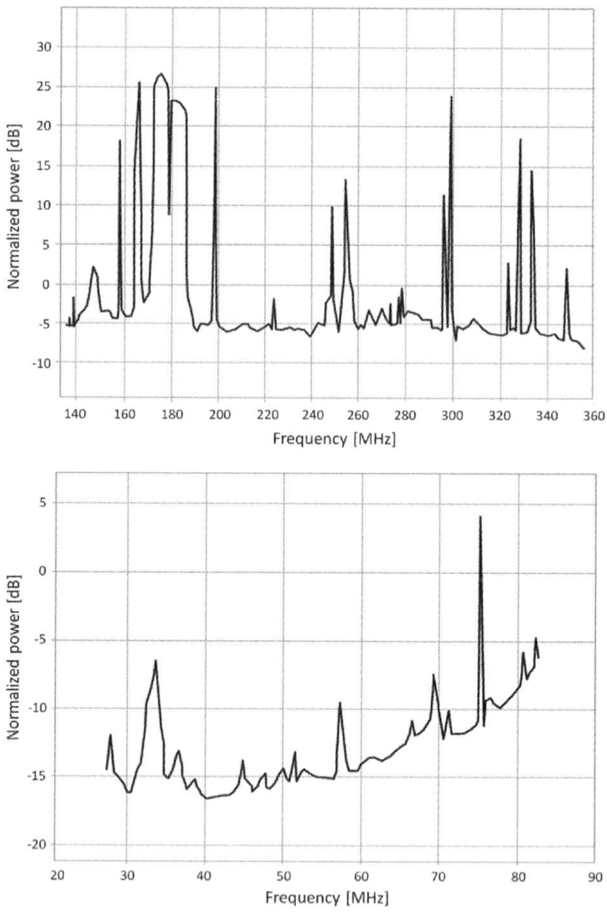


**Figure 4.1** Receiver output for CARBAS III SAR set in passive (nontransmitting) mode in its lower and higher bands. The received spectra illustrate well that exterior noise dominates over interior noise in the lower band, but not so in the higher. In fact, both the high- and low-band antennas are electrically small, so their sensitivity increases with frequency. The 10-dB noise floor slope in the lower band is due to man-made wideband noise, which is abundant below 100 MHz, but attenuated by the frequency-dependent antenna sensitivity. For the high band, the noise floor is flat, which is expected since the exterior noise is much weaker at these frequencies and interior noise expected to dominate. The RFI occupancy is typical, and characteristically, the higher band is more crowded. The 16-MHz-wide signal double peak at 180 MHz is due to two adjacent digital television (in fact DVB-T2) transmitters. In the upper diagram, perhaps 15% of the radar bandwidth is blocked by wireless services. The dynamic range encompassing the signal is seen to be no larger than about 30 dB at the high band and 20 dB at low band. The transmitters causing the interference in the diagrams are all relatively far away, that is, the interference peaks may well be considerably higher.

conflicts. Signals to be nulled on the transmit side would be those used by services, which are active during the radar transmissions with subscribers sufficiently nearby. There are also frequency allocations used for distress signals, which constantly watched for perhaps weak transmissions. These must never be intruded into by radar transmissions.

As it turns out, the nulling of transmit frequencies is more difficult than nulling of receive frequencies. For one SAR radar, transmitters being airborne must be small and lightweight. They are conventionally designed to handle transmit signals of constant amplitude and without any linearity requirement whatsoever. This allows transmitters to operate close to their saturation limit, by which compactness and efficiency are achieved. Spectral nulls directly obtained by a Fourier transform of the nulled spectrum will critically depend on the amplitude modulation. For the spectral nulls to be preserved by the transmitter, the feed signal must be predistorted to compensate for the transmitter amplitude nonlinearity (there is a publication on the subject by Nunn and Moyer [3]). The transmitter itself has to be equipped with appropriate feedback circuitry for keeping its nonlinear characteristics precisely constant and independent of temperature. An alternative approach to spectral nulling, not calling for specialized transmitter equipment, will be developed in Section 4.5. It exploits phase modulation of a transmit signal of constant amplitude, for which algorithms will be provided. The phase modulation survives the nonlinearity of a saturated transmitter.

A second issue with transmit spectral nulling, also to be addressed at the end of this chapter, is its mathematical conflict with the pulsed operation of radars. Indeed, according to Fourier theory, any truly pulsed signal (i.e., having zero amplitude for finite time intervals) cannot be zero for finite spectral intervals. In practice (as the preceding chapters have provided several examples of the truly pulsed signal), truncation of bandwidth/time duration causes an infinite tail of declining sidelobes in the temporal/spectral domain. For pulsed radar, the sidelobes prohibit spectral notches of finite bandwidth to be created. The possible remedy is radar operating in a nonpulsed (i.e., continuous) fashion. The combination of such radar and phase modulated signals have been demonstrated (as per 2016 only in the laboratory) to provide the required transmit notching capabilities. Incidentally, the continuous operation is also beneficial for nulling received narrow band interference, since also the radar pulse creates sidelobes to the received interference, which thus absorbs a wider than necessary part of the radar band.

Apart from minimizing the impact on surrounding radio services, also data rate considerations put requirements on SAR transmit waveforms. The high speed with which modern digital hardware runs opens for new and more uncompromising waveform schemes, with continuous operation notching being a promising venue for low-frequency SAR frequency sharing. Reduced data rates do, however, make systems simpler and cheaper to realize, and consequently, the waveforms yielding these simplifications are important. The minimum data speed required for SAR operation depends on swath width and is set by the rate of resolution cells bypassed by the radar platform. A suitable choice of waveform can reduce the rate burden to approach this minimum. Assume as an example of 100 m/s platform speed, SAR resolution 1 $m^2$, and a 5-km surveillance *swath* (the width of the surveyed area). The reflectivity density of each such resolution cell is represented by a complex number, given with a binary accuracy corresponding to the target to noise ratio of the strongest targets in the image. For low-frequency SAR, this ratio is seldom larger than 30 dB, which corresponds to 10-bit complex amplitudes. The resulting data rate is 5 Mbps. On the other hand, for a range resolution of 1m, a bandwidth of 150 MHz is called for. Performing Nyquist sampling at this bandwidth and requiring a 70-dB dynamic range result in the digitization output of 3.6 Gbps. This higher figure gives an idea of front-end data speeds when no reduction is attempted.

For microwave high-resolution SAR, *stretch* waveforms are generally used to achieve the data rate reduction. The stretch waveform utilizes a linearly frequency modulated signal, which when transmitted covers a range interval considerably wider than the intended swath illuminated by the radar antenna. The bandwidth of the signal backscattered from the swath is thus much smaller than the radar bandwidth, and after receiver front-end bandpass filtering, the digitization rate can be correspondingly reduced. While excellent for standoff imaging at ranges much larger than aircraft altitude, applying the stretch method to low-frequency SAR is difficult due to the relatively steep depression angles that are required. At low frequency, the ground below the radar platform may well be flat compared to wavelength (it is the likely situation when wavelengths are of meter order, and is particularly pronounced when the radar platform traverses flatland or water tables). In this case, any component of the transmit signal hitting in the direction of the ground normal undergoes specular refection, that is, the radar antenna receives the scattering echo (attenuated by the Fresnel reflection coefficient) from its mirror copy at a range twice the platform altitude; this return is referred to as the *nadir echo*. Compared to the diffuse scattering from the intended SAR imaging range interval, the nadir echo can be exceedingly strong. The well-proven method to eliminate its influence is by having the radar receiver turned off during the nadir echo. While this is not possible for stretch waveforms, a useful waveform that achieves the double purpose of reducing data rates and avoiding the nadir echo is based on the so-called *stepped frequency*. This method splits the full radar cycle into a steps of transmission/reception for one sub-band at a time, with transmission and reception periods set so that the backscatter from nadir has gone when the returns from the intended radar range impinges on the radar. Formulas for stepped frequency waveforms will be provided in the chapter, as will methods to mitigate nadir echo influence also for waveforms in which the nadir echo is actually received. Mitigation is then performed by signal processing.

## 4.2 Characterization of Additive Noise

Noise, in the sense used here, is just another term for measurement error. There is an ultimate limitation to achievable measurement accuracy, which is set by temperature of the measurement device. Temperature manifests itself by causing unavoidable and uncontrollable random excitations to all the degrees of freedom of a system, thus adding energy. The larger the number of degrees of freedom, the larger the energy caused by temperature will be. If there is no dependence between the degrees of freedom, the mean thermal energy per degree of freedom systems is $\kappa K$, where $\kappa = 1.38 \times 10^{-23}$ J/K is Boltzmann's constant and $K$ is the temperature in degrees kelvin (the letter $K$ is used to avoid confusing with time $T$). The ideal gas law is an example—this follows from assuming three degrees of freedom per gas particle and, thus, that $3\kappa K$ is the particle mean potential energy, with respect to the force excised by the particle randomly hitting the gas volume walls. Also, electromagnetic fields are subject to the effects of temperature. Not unlike an ideal gas, at low and microwave frequencies, thermal energy is manifest in attributing the mean energy $\kappa K$ to each particular "state" in which the field is present. Here, the application is that of a field impinging on the radar antenna, across a frequency band of bandwidth $B$ and during the SAR integration time $2T$. The sampling theorem states that the signal is completely characterized by $4BT$ real samples thus constituting its states. The thermal energy carried by the signal is then $4BT\kappa K$. Hence, $2\kappa K$ is the spectral density of the incident thermal noise.

As stated above, any field impinging on the radar antenna that is uncorrelated to the transmit signal spreads uniformly across the SAR image and therefore behaves just as the truly thermal noise. If the field impinging on the radar antenna has a power $P$, it thus acts to enhance noise temperature to a value $K' = K + P/2B\kappa$ above the physical temperature $K$. Also, any distortion or error independent of the transmit signal, which may arise inside the radar receiver, may be represented as contributing to an antenna temperature, artificially raised above the environmental temperature.

For microwave frequencies, noise arises above the thermal noise floor mainly due to imperfections in the signal handling by the reception and digitization stages. At UHF and VHF frequencies, major noise contributions are due to the abundance of radio services impacting wideband radar reception, but there are also wideband electromagnetic noise sources adding to the thermal noise. The major noise contributions have man-made atmospheric or galactic origin. They all decline with frequency and have little significance above 1 GHz. A standard reference for the characteristics of this interference environment is the recommendations of the International Telecommunication Union [4]. The following is a summary of main characteristics picked from this source and adapted to the situation for meter-wave SAR.

Noise can be categorized as galactic noise emanating from the sky and partly (but not entirely) from the Sun, as well as atmospheric stemming from lightning, and man-made noise originating from spark plug emissions, rectifiers, digital hardware, and wideband emissions from wireless equipment. A low-frequency SAR antenna is essentially isotropic, so the anisotropy of these different noise sources is of no concern. The different noise sources are further characterized by their power density, suitably understood as median values translated into noise temperatures. For different forms of noise, the temperature will vary differently with frequency. Atmospheric noise is the dominating noise form below 10 MHz, but drops below the 300K level of environmental noise at the frequencies $\lesssim$20 MHz. Galactic noise drops below the environmental noise temperature at frequencies $\lesssim$100 MHz. Also at frequencies below 100 MHz, it can be neglected, since man-made noise almost certainly will have a stronger presence. An empirical formula for the man-made noise level compared to $K = 300\text{K}$ is

$$K_{\text{mm}} = F_{\text{mm}} K;\ \left[F_{\text{mm}}\right]_{\text{dB}} = c - d\,{}^{10}\log[f]_{\text{MHz}} \tag{4.1}$$

Here $f$ denotes frequency, $c$ is in the interval $67.2 < c < 76.8$, and the parameter $d$ is equal to 27.7. This leads to a noise figure $F_{\text{mm}}$ between 35 and 15 dB for a VHF SAR operating across a frequency interval 30 to 80 MHz. For a UHF SAR operating in the interval 150 to 350 MHz, the noise figure $F_{\text{mm}}$ varies between 17 and 0 dB. In the following, the exterior noise figure is assumed to be 26 dB below 100 MHz, that is, a noise temperature of around 120,000K will be assumed. In the low UHF band, a noise figure 6 dB corresponding to a noise temperature of around 1,200K will be assumed. The impact of man-made noise, in particular below 100 MHz, is well illustrated in Figure 4.1.

First, consider the dynamic range required to perform the previously discussed cancellation of narrow-band RFI. The accuracy of modern receivers is typically set by their analog-to-digital converter (ADC) performance. The high-level amplitudes of interference must on no account saturate analog-to-digital (AD) conversion, since this leads to broadband errors at the amplitude level of the inference, which would be disastrous for reception performance. RFI power calculations can be based on "Free space loss" and effective radiated power, since the radar receiver are placed high. The antenna temperature is

$$2\kappa K_{\text{RFI}} B = \frac{\lambda^2}{4\pi d^2} P_{\text{ER}} \tag{4.2}$$

The receiver dynamic range should be selected so that receiver noise is not dominated by internal noise but the unavoidable external noise, with the reception gain set so that received RFI is not saturating the ADC. The dynamic range therefore has to be better than the ratio of antenna temperature due to man-made noise and RFI. Dominant RFI sources are commercial broadcast (typically television) transmitters with ERP (short form for effective radiated power) today being in the regime 80 dBm (old analog TV transmitters operate at up to 90-dBm ERP). Extensive antenna arrangements are used on such transmitters to narrow the radiated beam to nearly horizontal propagation. With the SAR platform operating at normal altitudes, say above 1 km, the radar will only be hit by the broadcast beam at a fairly significant distance from the transmitter, say more than 1 km. Setting $\lambda \approx 5$m for the VHF band and $\lambda \approx 1$m for the UHF band, the antenna temperatures to the broadcast interference become $1.4 \times 10^{12}$K and $1.4 \times 10^{10}$K, respectively. Given that man-made noise sets the antenna temperatures to 120,000K and 1,200K, as discussed above, the minimum required dynamic range is found to be 70 dB in both cases. Ideally, a receiver should provide an ample margin on top of this minimum requirement, 85 dB can be a suitable dynamic range requirement, basically provided by a 14-bit ADC.

The radar equation is the fundamental relation for determining the transmitter power required to achieve a certain RCS sensitivity at a certain range. It will now be derived based on the additive noise assumptions made. The noise spreads uniformly across the SAR image causing an error $\sigma_0^N$ defining the accuracy when determining SAR image reflectivity. The accuracy for determining the RCS from the SAR image becomes $\sigma^N = \sigma_0^N \Delta A$ with $\Delta A$ denoting resolution. For RCS values significantly below $\sigma^N$, the signature of a point-like target will be totally hidden.

Basic to the radar equation is the radar antenna performance. For low frequency, SAR antenna design is a challenge, since the antenna cannot be much larger than the wavelength and must also function for large fractional bandwidths. There are also strict requirements on the antenna radiation pattern. To analyze these aspects, a relevant understanding of the physical principles behind antennas must be acquired. A fundamental performance parameter for the antenna turns out to be what will here be termed *transmissivity*, defined as the ratio $\Gamma$ between the power fed into the antenna terminal and the power transmitted. There is the related concept of *directivity* $G(\boldsymbol{n})$, which is the ratio between the transmitted power flux in any particular direction $\boldsymbol{n}$ and the mean transmitted power flux over all directions. Expressed differently, directivity is the ratio between directional power flux and $1/4\pi$ times the total transmitted power. Antennas smaller than or at most of the size equal to half the wavelength will have smoothly varying directivity, never strong in any particular direction. For antenna significantly smaller than half the wavelength, transmissivity will be much less than unity. The term *antenna gain* is used for antenna transmissivity and directivity combined, that is, $\Gamma G(\boldsymbol{n})$. Semantically, it understands an entity, larger than unity, which is the situation for microwave radar. Presently, the term is avoided.

According to the basic antenna transmission/reception reciprocity, the same transmissivity and directivity properties also hold when the antenna is used for receiving. While transmitted fields are basically spherical, received fields normally occur and are analyzed as plane waves. Even though plane waves have infinite power, it is still possible to apply the transmissivity (the term receptivity is superfluous and will not be used) concept, defining it by means of the ratio $\vartheta$ of power received at antenna terminal compared to the power reflected by the antenna as an outgoing spherical wave. The relation $\vartheta = \Gamma/1 - \Gamma$ applies to lossless antennas (assumed here), so $\Gamma$ follows according to $\Gamma = \vartheta/1 + \vartheta$. Though $\vartheta$ does not depend on the incidence direction $\boldsymbol{n}$, the antenna directivity determines the dependence of the terminal power (or equivalently the reflected power) on $\boldsymbol{n}$.

The high gain of microwave radar antennas is obtained using antenna apertures many wavelengths across. The generic microwave aperture antenna is characterized by an essentially flat antenna surface with area $A_{\text{ant}}$, with a field impinging in a direction close to the normal of the surface from a source at the distance $r$. The impinging angular power flux $P_{\text{in}}$ is transformed into received power $P_{\text{RX}} = p_{\text{in}}A_{\text{ant}}/4\pi r^2$. Similarly, on transmit, a spherical wave emerges from the antenna in the form of a narrow beam centered on the surface normal. The angular power flux within the transmitted beam is then $p_{\text{out}} = P_{\text{TX}}/\Delta\theta^2$, which means that the areal power flux hitting the target is $\hat{p}_{\text{out}} = P_{\text{TX}}/(r\Delta\theta)^2$. Expressed in directivity $G(\boldsymbol{n}) = 4\pi/\Delta\theta^2$ (with $\boldsymbol{n}$ pointing toward the target), the relation $\hat{p}_{out} = P_{TX}G(\boldsymbol{n})/4\pi r^2$ is found. The backscattered angular power flux follows, knowing the RCS $\sigma(-\boldsymbol{n})$, and is $p_{in} = \sigma(-\boldsymbol{n})\hat{p}_{out}/4\pi = \sigma(-\boldsymbol{n})P_{TX}G(\boldsymbol{n})/(4\pi r)^2$ from which the radar equation in the form $P_{RX} = \sigma(-\boldsymbol{n})P_{TX}G(\boldsymbol{n})A_{ant}/(4\pi r)^2r^4$ follows.

To introduce directivity also for reception, consider that the target may be located in a direction $\boldsymbol{n}$ well to the side of the antenna disk normal. Expressions can set up to incorporate this case, writing $P_{RX} = p_{in}G(\boldsymbol{n})A_{ant}/r^2G_{max}$ instead of $P_{RX} = p_{in}A_{ant}/r^2$. Since disk beam width is $\Delta\theta = \lambda/\sqrt{A_{ant}}$, it follows that $G_{max} = 4\pi/\Delta\theta^2 = 4\pi A_{ant}/\lambda^2$ so $P_{RX} = \lambda^2G(\boldsymbol{n})p_{in}/4\pi r^2$. The power received becomes (which is the standard expression for the radar equation)

$$P_{RX} = G(\boldsymbol{n})^2\lambda^2\sigma(-\boldsymbol{n})\frac{P_{TX}}{(4\pi)^3r^4} \tag{4.3}$$

Consider next a body of an arbitrary shape and size, used as an antenna. In cases of arbitrary transmissivity, the above expression $P_{RX} = p_{in}G(\boldsymbol{n})A_{ant}/r^2G_{max}$ generalizes to $P_{RX} = p_{in}\Gamma G(\boldsymbol{n})A_{ant}/r^2$ in which $A_{ant}$ may no longer be the antenna physical area. For the disk antenna, it is the disk area divided by $G_{max}$. The antennas presently of relevance have $G_{max} \approx 1$, however. The power flux impinging on the target is noted to be $\Gamma G(\boldsymbol{n})P_{TX}/4\pi r^2$, so the power backscattered to the antenna and thus the radar equation obtains a form useful for any type of antenna and any wavelength

$$P_{RX} = \Gamma^2G(\boldsymbol{n})^2A_{ant}\sigma(-\boldsymbol{n})\frac{P_{TX}}{(4\pi)^2r^4} \tag{4.4}$$

While the aggregate $\Gamma^2G(\boldsymbol{n})^2A_{ant}$ may be difficult to assess in the general case, it is noted that for unity transmissivity $G(\boldsymbol{n})^2A_{ant}$ can be computed from the RCS of the body. Imagine two experimental arrangements: one in which a standard aperture antenna is connected to the radar transceiver and the transmitted power is back scattered with the arbitrary antenna being the target and another arrangement where the transceiver is connected to the dipole and the aperture antenna is the target. In both cases, the antenna used as a target has its antenna terminal short circuited to reflect the received field. In the two experiments, the received power will be the same, namely, denoting by $\sigma_{ant}$ the RCS of the scattering structures, alternatively used as an antenna

$$P_{RX} = \sigma(-\boldsymbol{n})\sigma_{ant}(\boldsymbol{n})\frac{P_{TX}}{(4\pi)^2r^4} \tag{4.5}$$

The only property of the aperture antenna in this relation is its RCS, so evidently, the same relation is valid for any other target that has the same RCS. In other words, knowing the cross section of a structure used as an antenna, the radar equation can be set up.

Equation (4.5) is not valid for small antennas, that is, if $\Gamma < 1$. In this case, the signal impinging on the antenna will not be totally backscattered by the reflective short circuit of the antenna terminal, since the antenna partially only conveys a signal at the terminal into an emitted field. By reciprocity, the impinging field is only partially turned it into a received signal at the terminal. The reduced transmissivity of electrically small antennas will be analyzed as a combined effect of that the antenna capability to radiate reduces, whereas its capacitive impedance increases, as size diminishes. Expressed differently, for a small antenna, the voltage drop due to radiation power loss will be a small fraction $\hat{U} = \mu U$ of the overall voltage $U$, to which the antenna mainly responds as a capacitive reactance ($\mu$ is a constant characteristic for the antenna). The reciprocal case is that an impinging field provides a voltage $U$ across the antenna, while the power received is determined by the voltage $\hat{U} = \mu U$. Since power is proportional to voltage squared, the relation $\sigma(\boldsymbol{n}) = \mu^4 G(\boldsymbol{n})^2 A_{\text{ant}}$ can be set up. If the dipole is fed through a transmission line, the reflections back into the transmissions line and thus transmissivity can be similarly expressed in $\mu$, as will be investigated further. The radar equation then follows from (4.4). The rule $\sigma(\boldsymbol{n}) = \mu^4 G(\boldsymbol{n})^2 A_{\boldsymbol{n}}$ is still useful, since the RCS results of Chapter 2 can be exploited to find expressions for $\mu$.

Size restrictions forces low frequency SAR antennas to be basically electric dipoles (though several modifications of the basic electric dipole are conceivable and may in fact be necessary, as will be discussed). The radar equation will now be worked out in detail in the case where dipoles are utilized for the radar antenna.

First, consider such dipoles with length from half the wavelength and up. The slender cylinder RCS expression in (2.213) can be used for determining $G(\boldsymbol{n})^2 A_{\boldsymbol{n}}$. Doing so, it is observed that (2.213) is derived on the assumption that the backscattered field at the surface of a finite cylinder is the same as if the cylinder was infinitely long. The formula does not apply to cylinders short compared to wavelength or even in the case of metallic, that is infinitely conducting, cylinders. In the latter case, narrow band scattering resonances are added to the solution for antenna lengths that are multiples of half the wavelength. For a wideband system, the resonances will be of no influence and (2.213) applies. It follows that for the RCS to be significant, the cylinder radius must be sufficiently large; here it will be assumed that $|\ln(0.890535\, ak)| << \pi/2$ so that the denominator can be assumed constant. The logarithmic term is a factor of ten smaller than the constant term when the radius to wavelength ratio is larger than ≈ 10%. Then, $\sigma \approx 4\ell^2/\pi$. Since (2.213) applies to directions orthogonal to the antenna axis, thus providing maximum directivity, (4.6) becomes

$$\sigma(\boldsymbol{n}) = G(\boldsymbol{n})^2 A_{\boldsymbol{n}} = \frac{4\ell^2}{\pi}\left[\frac{\sin\left(\frac{2\pi\ell\cos\theta}{\lambda}\right)}{\frac{2\pi\ell\cos\theta}{\lambda}}\right]^2$$
$$\Rightarrow P_{\mathrm{RX}} = 4\lambda^2\sigma(-\boldsymbol{n})\left[\frac{\sin\left(\frac{2\pi\ell\cos\theta}{\lambda}\right)}{\frac{2\pi\ell\cos\theta}{\lambda}}\frac{\frac{\ell}{\lambda}}{2}\right]^2\frac{P_{\mathrm{TX}}}{(4\pi)^3 r^4} \tag{4.6}$$

with $\theta$ being the angle between direction to target and dipole axis.

A properly designed half wavelength antenna has negligible transmissivity losses. Just based on the $\sin(\pi\cos\theta)/\pi\cos\theta$ antenna radiation pattern, the maximum dipole directivity is

$$G(\boldsymbol{n})_{\max} = \frac{1}{\oint \frac{\sin(\pi\cos\theta)}{\pi\cos\theta}d\Omega} = 1.62 \tag{4.7}$$

Comparing (4.6) and (4.4), the directivity estimate becomes $G_{\max} = 2$, that is, (4.6) remains a reasonable approximation down to dipoles of half wavelength size.

Antenna beam width $\Delta\theta$ can be defined in analogous to resolution formula (3.7). Presently, applied to (4.6) and restricted just to the polar angle, two important cases are

$$\Delta\theta = \frac{1}{G_{\max}{}^2}\int_{-\pi/2}^{\pi/2} G(\boldsymbol{n})^2 d\theta = \int_{-\pi/2}^{\pi/2}\left(\frac{\sin\left(\frac{2\pi\ell\cos\theta}{\lambda}\right)}{\frac{2\pi\ell\cos\theta}{\lambda}}\right)^2 d\theta = \begin{cases} 56^\circ \Leftarrow \lambda = 2\ell \\ 28^\circ \Leftarrow \lambda = \ell \end{cases} \tag{4.8}$$

When wavelength increases and surpasses twice dipole length, the radiation patter remains essentially unchanged (though transmissivity goes down), that is, $\Delta\theta \approx 60^\circ$ remains valid.

From (4.8), it follows that half wavelength and smaller dipoles are of particular importance for low frequency SAR, since, as was discussed in Chapter 3, for along track SAR resolution to become of half wavelength order, it is required that an integration angle of the order 60° is kept across the radar frequency band. A steerable antenna beam (so-called *spotlight SAR*) makes this type of resolution achievable also with the directive antennas used in microwave SAR. When wavelengths becomes of meter order, size restrictions make such solutions impractical, while a fixed wide angle beam is an obvious approach. A half-minimum-wavelength (or indeed an even shorter) dipole antenna, oriented along the platform path, offers an essentially ideal radiation pattern. The arrangement yields horizontal polarization, while for vertical polarization, dipoles can be stacked transversely into an end-fire antenna array to obtain a similar radiation pattern. In the vertically polarized case, the antenna might be extended beyond half wavelength, thus improving wideband transmissivity. The restriction lies in this case more in the difficulties in integrating a long vertical antenna in an aircraft. An end fire array is the obvious and integrationwise practical way of obtaining both the required antenna beam steering to just one side of the radar aircraft (this will be further addressed in the next section) and reducing radiation outside the 60° sector, compared to what is the case for the single horizontal dipole.

The drawback of the half-minimum-wavelength dipole antenna is of course its inefficiency for the longer radar wavelengths. As will be seen, this is handled well by small systems, while for large systems, with a requirement for operation at large ranges, improved efficiency is desirable. There are several options. One is to compromise with the requirement of 60° beam width for the shortest wavelength by allowing dipoles to be longer than half the shortest wavelength. There will be degradation in resolution attained. However, even a slight size length increase will make the antenna dipoles significantly more efficient, while the resulting resolution degradation will be minor.

Another approach is to abandon the pure electric dipole for antennas geometries providing a dipole-like pattern with near unity transmissivity over a larger band. A log periodic antenna is an example. This arrangement must involve compensation for that the wideband radar signal does not originate from a single point, which is the assumption in the SAR theory of Chapter 3. The main drawback of the log periodic arrangement is that it becomes large in two rather than one dimension, and can be difficult to integrate on an arbitrary, perhaps small, aircraft. Magnetic dipoles (i.e., loop antennas) offer still another approach. Orienting the plane of the loop to include the SAR path, radiation becomes adapted for essentially horizontally polarized

SAR imaging, but may not be confined to a sector set by antenna size. The so-called Pi antenna (which will be analyzed in the next section) constitutes an adaption of this type of antenna to the requirements of low-frequency SAR imaging, and will be analyzed in more detail.

Initially, it will however be assumed that entire wideband antenna system is based on electric dipoles with length equal to or smaller than half the shortest wavelength $\lambda_{min}$, that is, $\lambda_{min} = 2\ell/\alpha$; $\alpha \leq 1$. For the previously considered low-frequency SAR frequency bands, dipole lengths will typically be (depending on the particular band selected) of the order 0.5m to 3m.

According to the previous discussion on electrically small antennas, the radar equation in its form (4.4) will be determined from the RCS of electrically small cylinders. As discussed at several instances in Chapter 2, for objects smaller than the wavelength, shape is not decisive for RCS. Rather, RCS is determined by the body volume $V$. This is well illustrated by comparing the small sphere RCS (2.190) with the thin cylinder transverse electric (TE) expression (2.213). Both expressions are noted to be of the form $CV^2/\lambda^2$ for some constant $C$. Even though the thin cylinder might be thought a shape, which would be exceedingly ineffective in providing TE backscattering, the fact is that it is only about half the RCS of a sphere with the same volume. Only by allowing the object dimension to grow in the $\boldsymbol{E}$ field direction will the cross section grow significantly. When the length grows to $\lambda/2$ antennas becomes fully efficient, the radar equation assumes the form (4.6). As has been discussed, cross section may be understood as the aggregate $\mu^4 G(\boldsymbol{n})^2 A_{\boldsymbol{n}}$ (i.e., $\mu^4 A = CV/\lambda^4$) in the limit of large wavelength and for broadside radiation directions. With the requirement $\mu = 1$, if $\lambda = 2\ell$, it becomes reasonable to assume $\mu = 2\ell/\lambda$ for $\lambda \leq 2\ell$, which is sufficiently but not fully correct, as will be shown in the next section.

For a dipole antenna, fed through a transmission line, transmissivity will depend on the impedance of the transmission line. For the slightly nonoptimal, though practical feed impedance selection, the relation $\Gamma = 2\mu^3/3$ (with $\mu = 2\ell/\lambda$ assumed) will be a good approximation, as is demonstrated in the next section. As seen, when $\lambda = 2\ell$, the equation (4.6) becomes valid up to a received power attenuation by 4/9, which is due to the feed mismatch For $\lambda > 2\ell$, the antenna diagram remains the same as for $\lambda = 2\ell$. The following expressions are arrived at for radar operating within a small bandwidth within this wavelength regime:

$$P_{\mathrm{RX}} = 16\ell^2 \sigma(-\boldsymbol{n}) \left[ \frac{\sin(\pi \cos\theta)}{\pi \cos\theta} \right]^2 \frac{4}{9} \left( \frac{\ell}{\lambda/2} \right)^6 \frac{P_{\mathrm{TX}}}{(4\pi)^3 r^4} \tag{4.9}$$

For later reference, (4.9) is restated in the following form, superficially similar to the standard form (4.5) of the radar equation:

$$
\begin{aligned}
P_{\mathrm{RX}} &= \Gamma^2 \lambda_{\min}{}^2 \sigma(-\boldsymbol{n}) \left[ \frac{\sin(\pi\cos\theta)}{\pi\cos\theta} \right]^2 \frac{P_{\mathrm{TX}}}{(4\pi)^3 r^4}; \\
\Gamma &= \frac{8}{3} \frac{\ell}{\lambda_{\min}} \left( \frac{\ell}{\lambda/2} \right)^3 = \frac{4}{3} \alpha^4 \left( \frac{\lambda_{\min}}{\lambda} \right)^3
\end{aligned}
\tag{4.10}
$$

The subsequent analysis generalizes (4.10) to the wideband situation of low-frequency SAR. Basic to this analysis is that the receiver noise energy, during the registration time $2T$, spreads uniformly across the SAR image. Given a certain level of transmit power, this noise floor will define a lower limit to the SAR reflectivity called the *noise equivalent reflectivity.*

Denote $\omega_{\max} = 2\pi c/\lambda_{\min}$, where $\lambda_{\min} = \ell/2$. The radar receiver is assumed equipped with the appropriate analog and/or digital filters cutting off any noise outside the transmitted radar band $\omega_{\min} \le \omega \le \omega_{\max}$ and the Doppler band $-\pi/v\tau \le \Delta\omega \le \pi/v\tau$ (with $v$ radar platform ground speed). Denote by $\Omega_{\mathrm{BOX}}$ the corresponding region in the $\omega,\Delta\omega$-plane. The noise energy $4\kappa KBT$ received under registration time $2T$ is accordingly zero outside $\Omega_{\mathrm{BOX}}$, while inside the noise impinging on the antenna and the internal noise is assumed uniformly distributed. The noise distribution of the received signal may not be uniform if there are frequency-dependent transmission losses (which indeed is the case) or frequency-dependent gain variations. Recall from Chapter 3 that the SAR image spectral support (the nonzero part of the SAR image Fourier transform) assumes the shape of an annulus segment in the $k_\rho,k_y$-plane. The received signal spectral amplitude $f_\omega^{\mathrm{RX}}$ maps to a point in the $k_y,k_\rho$-plane, according to $k_y = \Delta\omega/v$, where $\Delta\omega = \omega - 2\pi n(\omega)/\tau$, and $k_\rho = \sqrt{(2\omega/c)^2 - (\Delta\omega/v)^2}$ . The outer and inner radii of the annulus segment are $k_{\max} = 2\omega_{\max}/c$ and $k_{\min} = 2\omega_{\min}/c$, while the angle of the segment is the SAR integration angle $2\Phi$. In the $\omega,\Delta\omega$-plane, the SAR image corresponds to a trapezoid shaped region $\Omega_{\mathrm{TRAP}}$. Since (approximating trigonometric relationships to linear) $2\omega_{\max} = 2v\Phi\omega_{\max}/c$ within the trapezoid, the angle of the trapezoid is $2\Phi' = 4\Phi v/c$ in the $\omega,\Delta\omega$-plane (as is further illustrated in Figure 4.2). The noise outside $\Omega_{\mathrm{TRAP}}$ is irrelevant to the SAR image and will be filtered away by the SAR image formation process. Hence, if the received noise energy is uniformly distributed within $\Omega_{\mathrm{BOX}}$, the noise energy of the SAR image will be the received noise energy reduced by the area ratio $A_{\mathrm{TRAP}}/A_{\mathrm{BOX}}$.

Since the reflectivity density determined from (3.21) is to represent RCS, which is a ratio between radiation power fluxes, it must remove the effects

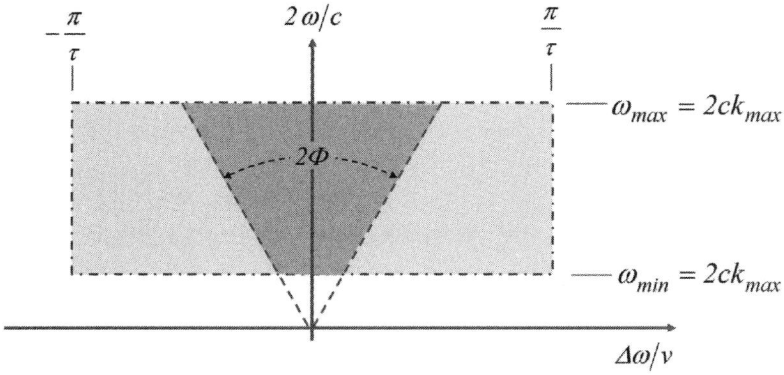


**Figure 4.2** Receiver filter window mapping into SAR image data. The receiver frequency $\omega$ and Doppler shift D$\omega$ have been rescaled as $\omega \to 2\omega/c$ and $\Delta\omega \to \Delta\omega/v$, thus corresponding to SAR image wave numbers $k$ and $k_y$. While the outer light gray rectangular box signifies $\Omega_{BOX}$, the darker gray trapezoid area is $\Omega_{TRAP}$ corresponding to the sector annulus in the $k_\rho, k_y$-plane, constituting the SAR image spectral support. The SAR integration angle $2\Phi$ gives the angle of the trapezoid. Linearizing trigonometric relations, $\Delta\omega_{max} = 2v\Phi\omega_{max}/c$, that is, the angle of the trapezoid in the $\omega$,D$\omega$-plane is $2\Phi' = 2\Delta\omega_{max}/\omega_{max} = 4v\Phi/c$.

of antenna transmissivity. Antenna gain, characterizing microwave radar antennas, will seldom change significantly across the radar bandwidth, and the antenna gain will be one of several gain parameters collectively settled by calibration of the finished SAR image. Presently, however, antenna transmissivity vary strongly with frequency. Leveling of the transmissivity effects is therefore a step, which must be conducted prior to SAR processing. With the antennas illuminating the SAR scene in an essentially broadside fashion, the angular dependence factor in (4.6) can be assumed unity. Transmissivity compensation is carried out so that (4.10) agrees with (4.5). Accordingly, the required reflectivity density adjustment is $\gamma_{\boldsymbol{k}} \to \gamma_{\boldsymbol{k}}/\Gamma = 3(k_{max}/k)^3\gamma_{\boldsymbol{k}}/4\alpha^4$. Rescaling is conveniently carried out for the entire received signal as $f_\omega^{RX} \to 3(\omega_{max}/\omega)^3 f_\omega^{RX}/4\alpha^4$ since the noise outside $\Omega_{TRAP}$ is anyway filtered out in the SAR image formation.

The amplitude compensation affects the power quantities by the amplitude compensation squared. Before amplitude compensation, the receiver noise energy increment $d^2\varepsilon$ at frequency $\omega$ and within an area cell $d\omega d\Delta\omega$ inside $\Omega_{BOX}$ is

$$d^2\varepsilon = 4\kappa KBT\frac{d\omega}{\left(\omega_{max} - \omega_{min}\right)}\frac{d\Delta\omega}{\frac{2\pi}{\tau}} = \frac{\kappa KT\tau}{\pi^2}d\omega d\Delta\omega \tag{4.11}$$

Here, contributions to temperature $K$ are from receiver internal and external sources. The latter contribution is reduced by the receptivity losses, which are equal to the transmissivity losses, so the overall temperature can be stated as $\Gamma K_{\text{ext}} + K_{\text{int}}$. It follows that with the amplitude adjustment imposed the noise energy of the SAR image is:

$$\begin{aligned} \varepsilon_{\text{TRAP}} &= \frac{\kappa T\tau}{\pi^2} \int_{\Omega_{\text{TRAP}}} \frac{\Gamma K_{\text{ext}} + K_{\text{int}}}{\Gamma^2} d\omega d\Delta\omega \\ &= \frac{\kappa T\tau}{\pi^2} \int_{\Omega_{\text{TRAP}}} \frac{3}{4\alpha^4}\left(\frac{\omega_{\max}}{\omega}\right)^3 \left(K_{\text{ext}} + \frac{3}{4\alpha^4}\left(\frac{\omega_{\max}}{\omega}\right)^3 K_{\text{int}}\right) d\omega d\Delta\omega \end{aligned} \tag{4.12}$$

Carrying out the integration

$$\begin{aligned} \varepsilon_{\text{TRAP}} &= 3\frac{\kappa T\tau}{\pi^2}\frac{v\omega_{\max}{}^2\Phi}{c\alpha^4}\left[\left(\frac{\omega_{\max}}{\omega_{\min}} - 1\right)K_{\text{ext}} + \frac{3}{16\alpha^4}\left(\frac{\omega_{\max}{}^4}{\omega_{\min}{}^4} - 1\right)K_{\text{int}}\right] \\ &\approx \kappa T\tau\Phi \frac{vc}{\lambda_{\min}{}^2\alpha^4}\left(12\frac{\beta}{1-\beta}K_{\text{ext}} + \frac{9}{4\alpha^4}\frac{1}{(1-\beta)^4}K_{\text{int}}\right) \end{aligned} \tag{4.13}$$

The result is here expressed in fractional bandwidth $\beta = 2\pi B/\omega_{\max} = 1 - \omega_{\min}/\omega_{\max}$. The approximation assumes $(1 - \beta)^4 = << 1$, that is, fractional bandwidth must not be too small.

Observe that in (4.13), it has been implicitly assumed that the receiver is continuously open and always across the entire radar band (the situation can realized by continuous operation with simultaneous transmission and reception or, hypothetically, by infinitely short transmit pulses). To be accounted for here is a fundamental relation, determining the maximum ground area coverage rate of a SAR set up to operate in this fashion. The noise energy (4.14) captured during the registration time will be uniformly spread over the ground area thus covered during the registration time. Equating this noise energy per unit area with the back scattered energy, obtained by integration (4.6) over registration time $2T$, provides the estimate for the noise equivalent cross section, which is obtained for a certain transmit power level.

However, in the normal situation of pulsed radar, the radar receiver is operating intermittently. Moreover, transmit waveforms may be chosen to relax the requirement of full bandwidth reception. The aforementioned method of stepped frequency waveforms is an example, in which the radar band is covered by a suite of frequency staged narrow-band pulses. Then the receiver may be tuned to the more narrow bandwidths of these pulses and received noise power may thereby be reduced. In the first case, the receiver noise will be limited in proportion to relative time of the full repetition interval during which the receiver is open. In the stepped frequency case, the receiver noise is reduced by the number of subpulses required to span the full radar bandwidth. While these waveforms reduce noise, the subsequent treatment on waveforms will show that the reduction of receiver opening time or receiver bandwidth in each of the cases is proportional to limitation in the range interval covered and thus the ground area covered by the radar. The particulars concerning waveform therefore do not affect the fundamental issue of transmit mean power, while the continuous full bandwidth assumption enables the argument to be kept on a principal level.

For continuous receiver operation, a signal repetition period $\tau$ implies that the radar covers ranges up to $r_{\max} = c\tau/2$. For a SAR swath parallel to the SAR platform path, the swath outer edge is limited to $\rho_{\max} = r_{\max}\cos\Phi$. For the swath width $\Delta\rho = \rho_{\max} - \rho_{\min}$, the relation $\Delta\rho \le c\tau/2$ follows. The relation between receiver data and the SAR image spectrum is defined by $\Delta\omega = \omega - 2\pi n(\omega)/\tau$ and $k_\rho = =\sqrt{(2\omega/c)^2 - (\Delta\omega/v)^2}$. The first expression requires $|\Delta\omega| \le \pi/\tau$ for the determination of $\Delta\omega$ to be unique. As was discussed in Chapter 3, $\Delta\omega$ has the meaning of Doppler shift for each of a set of discrete transmitted frequencies. In other words, the Doppler measurement frequency $1/\tau$ must obey the sampling criterion for the Doppler bandwidth. The second relation states that because transmit and Doppler frequencies maps into SAR image wave vectors, there is a relation $|\Delta\omega| \le 2v\omega/c$. This relation sets Doppler bandwidth to $2v\omega/c$. In conclusion, aircraft velocity and radar repetition rate must obey the relation $2v\omega/c \le \pi/\tau$ (i.e., $\tau \le \lambda_{\min}/4v$). In combination with $\Delta\rho \le c\tau/2$, the overall relation $\Delta\rho \le \lambda_{\min}c/8v$ is implied. The rule

$$\dot{A} = v\Delta\rho \le \frac{c\lambda_{\min}}{4} \tag{4.14}$$

follows for the area increase rate along the SAR swath, recalling that both sides of the SAR path are covered.

According to (4.16), the upper limit to surveillance rate is independent of platform speed; a slower speed allows larger swath width without Doppler bandwidth being compromised. Actual numbers are illustrative. With the platform speed $v$ = 100 m/s and $\lambda_{\min}$ = 1m, it follows that $\dot{A} \le 75$ km$^2$/s. For a stepped frequency system involving, say, 16 frequency steps, the area coverage rate becomes reduced 16-fold, that is, the area coverage rate is less than 4.6 km$^2$/s. Moreover, the receiver will not be constantly open, implying a further reduction in surveillance rate. Typically, the maximum slant range may be close to 20 km (providing suitable incidence angles for a platform operating at 10 km), while the surveillance area rate may be of the order 1 $km^2/s$. In any case, the receiver noise power reduction will agree with the reduction in area surveillance area compared to the continuous full bandwidth case.

From (4.16) and assuming a registration time $2T$, the noise energy per unit area of the SAR image is $e_0^N = \varepsilon_{TRAP}/T\dot{A}$. Considering the case of equality in (4.14) and replacing $\tau \to \lambda_{min}/4v$ in (4.13), the following expression is obtained:

$$e_0^N = \kappa\Phi\frac{1}{\lambda_{min}{}^2\alpha^4}\left(6K_{ext}\frac{\beta}{1-\beta}+\frac{9}{8\alpha^4}K_{int}\frac{1}{(1-\beta)^4}\right) \tag{4.15}$$

With receiver amplitude rescaling, the radar equation in the form (4.10) is valid, but with $\Gamma = 1$. The noise equivalent reflectivity at some SAR image position $\rho, y$ follows. $P_{TX}$ is understood as mean transmit power and $r \approx \rho/\cos[(y - vt)/\rho]$. Integrating over time $2T$, the backscattered energy per unit area is obtained. Doing so, the range variation can be substituted with an effective range according to

$$\frac{1}{r_{eff}{}^4} = \frac{1}{2T}\int_{-T}^{T}\frac{dt}{r^4} \tag{4.16}$$

For the typical integration angle $2\Phi = 60°$, the effective range is $r_{eff} \approx 1.04\rho$, so the effective range may well be approximated with slant range. Restrict the attention to backscattered energy per unit resolution area (denoted $e_0$) by replacing $\sigma$ with $\sigma_0$

$$e_0 = \sigma_0\lambda_{min}{}^2\frac{P_{TX}T}{(4\pi)^3\rho^4} \tag{4.17}$$

Equation (4.17) applies to an actual target and the backscattered energy from this. In the SAR image, without any ability of right/left discrimination, the target will be represented by an equal amplitude at mirror positions on either side of the SAR path, the amplitude being just half the actual target backscattering amplitude. The reflectivity values $\hat{\sigma}_0$ obtained from the SAR image will therefore be just one-fourth of the actual target reflectivity $\sigma_0$. The visibility of a target in the SAR image will depend on to what degree this reduced reflectivity may dominate the noise that is present. The noise level per unit area is provided by (4.15). The noise equivalent reflectivity is therefore obtained by equating (4.15) with (4.17) divided by four, that is

$$\kappa\Phi\frac{1}{\lambda_{\min}{}^{2}\alpha^{4}}\left(6K_{\text{ext}}\frac{\beta}{1-\beta}+\frac{9}{8\alpha^{4}}K_{\text{int}}\frac{1}{(1-\beta)^{4}}\right)=\frac{\sigma_{0}^{N}}{4}\lambda_{\min}{}^{2}\frac{P_{\text{TX}}T}{(4\pi)^{3}\rho^{4}}$$
$$\Rightarrow\sigma_{0}^{N}=(4\pi)^{3}\frac{\kappa\Phi}{P_{\text{TX}}T}\frac{\rho^{4}}{\lambda_{\min}{}^{4}\alpha^{4}}\left(24K_{\text{ext}}\frac{\beta}{1-\beta}+\frac{9}{2\alpha^{4}}K_{\text{int}}\frac{1}{(1-\beta)^{4}}\right) \tag{4.18}$$

Noting that the recording time $2T$ required for attaining the SAR resolution corresponding to an integration angle $2\Phi$ is determined by $T = \rho\Phi/v$

$$\sigma_{0}^{N}=(4\pi)^{3}\frac{\kappa}{P_{\text{TX}}}\frac{v\rho^{3}}{\lambda_{\min}{}^{4}\alpha^{4}}\left(24\frac{\beta}{1-\beta}K_{\text{ext}}+\frac{9}{2\alpha^{4}}\frac{1}{(1-\beta)^{4}}K_{\text{int}}\right) \tag{4.19}$$

The noise equivalent reflectivity is noticed to be independent of resolution (i.e., registration time or integration angle $2\Phi$).

As numerical examples of (4.19), consider the CARABAS III system mentioned in Chapter 3. It operates with the ground speed $v$ = 25 m/s at a maximum frequency of 82.5 MHz (i.e., $\lambda_{\min}$ = 3.75m). The fractional bandwidth is $\beta$ = 0.67. The interior and exterior noise temperatures are set to 1,200° and 120,000° respectively, as per the discussion above. The antenna dipole length is 1.5m, implying $\alpha$ = 0.825. Assume the operating range to be 5 km. With the mean transmit power of 2W, the noise equivalent reflectivity becomes −25 dB. The conclusion is preliminary however, since resolution of the right/left ambiguity (addressed in the next section) may further affect the power budget.

For the existing CARABAS III system, the mean transmit power is of the order 1W, while the acquired SAR imagery extends a little less than 5 km (curtailed by fading due to the about 1 km altitude limit for the small helicopter used). In this imagery, limitations due to additive noise are not evident. When the right/left separation has been included, the experienced and predicted transmit power requirements are in agreement.

For a large system in the same frequency band, assume the ground speed $v$ = 150 m/s and operating range 20 km (the platform altitude should be around 10 km). Interior and exterior noise temperatures are as before, while the antenna dipole length is set to 1.88m (i.e., $\alpha$ = 1). The noise equivalent reflectivity becomes −25 dB with 300W mean transmit power, which is an achievable power requirement and allows also large systems to be considered (from the results of the next section, the requirement will actually be halved by right–left separation).

The transmit power required critically depends on wavelength as is evident when this type of low-band configuration is scaled to the CARABAS III high band 137.5 to 357.5 MHz. For the high band, the interior and exterior noise temperatures are set to 1,200° in view of the reduced exterior noise. The antenna dipole length for CARABAS III high band is 0.4m, implying $\alpha = 1$. Fractional bandwidth is $\beta = 0.62$, while available transmit power is the same as for the low band. In this case, (4.19) implies the noise equivalent reflectivity of −25 dB for ranges no more than 1 km. For the large system example, the power requirements become forbidding and rule out the possibility of using electrically small antennas for this application.

Consider the human walker moving target application within the high CARABAS III band, which was discussed in the preceding chapter. In this case, the bandwidth was 55 MHz, whereas wavelength, (considered to be minimum wavelength) was 1.2m. Hence, the fractional bandwidth was just $\beta = 0.22$. The range was 5 km, and with the currently used high-band antennas, $\alpha = 0.7$. A noise equivalent reflectivity of −26 dB was required. The transmit power required becomes 1W. The crucial circumstance resulting in this moderate power request is the modest fractional bandwidth.

Evidently, the use of electrically small antennas is quite suitable for radar frequencies below 100 MHz but less so for still low frequencies in the band 100 to 500 MHz. Not to compromise resolution, the concept of nondirectional antennas must be retained also at these frequencies, but near unity transmissivity becomes a strong need. At the same time, as wavelength becomes smaller, such solutions become size wise manageable and concepts like Vivaldi elements might be considered to satisfy this requirement, apart from the afore-mentioned log-periodic antennas. Note that if antenna arrays are utilized, the antenna elements must employ independent waveforms to retain nondirectionality. The saving in the required transmit power gain is then at best equal to the number of elements used. The elements can be set to operate in the multiple-in multiple-out mode, but this does not affect the power requirements.

According to the discussion, a further question to be resolved here is how (4.19) is modified in the case of unity transmissivity across the radar band. Again, transmissivity compensation is carried out to take the radar equation into the form (4.17). For the type of unity transmissivity, isotropic antenna considered the received power scales as wavelength squared, so the required rescaling of the received signal is $f_\omega^{RX} \to f_\omega^{RX}/\omega_{max}$. Thereby, rescaling will suppress noise, rather than increasing it as the case above. In analogous to (4.13), performing integration of noise spectral density in the region $\Omega_{TRAP}$ yields

$$\varepsilon_{\text{TRAP}} = \int_{\Omega_{\text{TRAP}}} \frac{\kappa T \tau}{\pi^2} \frac{\omega}{\omega_{\max}} \left[ K_{\text{ext}} + \frac{\omega}{\omega_{\max}} K_{\text{int}} \right] d\omega\, d\Delta\omega$$

$$= 16\kappa T \tau \Phi \frac{vc}{\lambda_{\min}{}^2} \left\{ K_{\text{ext}} \frac{1}{3}\left[1-(1-\beta)^3\right] + K_{\text{int}} \frac{1}{4}\left[1-(1-\beta)^4\right]_{\omega_{\min}}^{\omega_{\max}} \right\} \quad (4.20)$$

$$\approx \kappa T \tau \Phi \frac{vc}{\lambda_{\min}{}^2}\left(5K_{ext} + 4K_{\text{int}}\right)$$

where again it assumed that $\beta$ is large. By full analogy with the derivation of (4.19)

$$\sigma_0^N = (4\pi)^3 \frac{\kappa}{P_{\text{TX}}} \frac{v\rho^3}{\lambda_{\min}{}^4} \left( \frac{5}{2} K_{\text{ext}} + 2K_{\text{int}} \right) \quad (4.21)$$

For the previous example of a large system with the ground speed of $v$ = 150 m/s and the operating range of 20 km, for the CARABAS III high band 137.5 to 357.5 MHz, (4.17) implies that 500W mean transmit power is required for $\sigma_0^N = -25$ dB. At the low band and the same system configuration, (4.21) implies 60W mean transmit power for $\sigma_0^N = -25$ dB. The power penalty of using the electrically small antenna at the low band is thus 300/60 = 7 dB, which seems quite tolerable in view of the significant reduction in antenna size which is achieved. The crucial factor in the low penalty is the high level of exterior noise, at frequencies below 100 MHz. The penalty is significantly higher above 100 MHz, where exterior and interior noise levels can be considered similar. Again, the power requirements become halved by right/left separation.

## 4.3 Antenna System Basics

The following section provides an outline treatment of antennas for low-frequency SAR, starting with electric dipoles shorter than or equal to half wavelength. Ther impedance and transmissivity properties of such dipoles will be treated. Arrays of such dipoles for the required beam steering to one side of the radar aircraft will be analyzed. Mounting arrangements for handling coupling and reflection from the radar aircraft will be touched upon.

For antennas not larger than half the wavelength, directional aspects of radiation will be of minor importance and can be neglected for radiation relatively close to the broadside, which is the situation in the antenna arrangements to be considered. When fed through a transmission line, the unavoidable impedance mismatch of small antennas causes reflections back into the feed line, reducing transmissivity. The relation between the radiated field and the field inside the antenna, a conducting cylinder for dipole antennas, is represented by energy continuity equation (2.8). Since energy fluctuations inside the cylinder are averaged out in the course of time, the equation can be stated as

$$c^2\varepsilon_0 \oint_{\text{cylinder}} \langle \boldsymbol{E}\times\boldsymbol{H}\rangle \bullet d\boldsymbol{A}_a = \int_{\text{cylinder}} \langle \boldsymbol{E}\bullet \boldsymbol{j}\rangle dV_a = \langle UI\rangle \tag{4.22}$$

Here, $\langle\rangle$ represents time averages. On the right hand side appears voltage across the dipole $U$ and current $I$, where the voltage may be split into $U = \hat{U} + U_{\text{react}}$ with $\hat{U}$ causing radiation and $U_{\text{react}}$ being a reactance voltage drop. Since $\langle U_{\text{react}} I\rangle = 0$

$$c^2\varepsilon_0 \oint_{\text{cylinder}} \langle \boldsymbol{E}\times\boldsymbol{H}\rangle \bullet d\boldsymbol{A}_a = \frac{\hat{U}^2}{2R} = \mu^2\frac{U^2}{2R} \tag{4.23}$$

The resistance quantity $R$ is known as *radiation resistance*. Integration over the cylinder surface can be substituted for an integral over the transmitted power flux over any larger region encompassing the cylinder. At the surface of such a larger region, the radiated field will soon settle to an outgoing spherical wave with the $\boldsymbol{H}$ field determined by the $\boldsymbol{E}$ field yielding (2.25) for the power flux. Select the region to be integrated over large enough for (2.25) to apply and small enough for field attenuation inside it not to be noticeable. Hence

$$\begin{gathered} \mu^2\frac{U^2}{R} = c\varepsilon_0 E^2 \oint_{\text{antenna}} dA_a = \frac{A_{\text{ant}}}{\ell^2} c\varepsilon_0 U^2 \\ \mu = \frac{2\ell}{\lambda} \Rightarrow R = \frac{\ell^2}{2A_{\text{ant}} c\varepsilon_0}\left(\frac{2\ell}{\lambda}\right)^2 \end{gathered} \tag{4.24}$$

Consider the cylinder impedance in detail. The ohmic resistance will here be assumed negligible so the real part is the radiation resistance. The imaginary part consists of an inductive reactance $Z_L$ and a capacitive reactance $Z_C$. As is well known, $Z_L = \omega L = 2\pi cL/\lambda$ and $Z_C = 1/\omega C = \lambda/2\pi cC$, where $C$ and $L$ are the cylinder capacitance and inductance, respectively. With $\mu$ meaning the fraction of the dipole voltage that is left for radiation, the expression $\mu = 2\ell/\lambda$ is only an approximation. However, for the sake of notational compactness $\mu$

will henceforward be used as an abbreviation for $2\ell/\lambda$, making the fractional voltage interpretation of $\mu$ an approximate one only. With this understanding, the impedance can be stated as

$$Z = \frac{\ell^2}{A_{\mathrm{ant}} c\varepsilon_0}\mu^2 + i\left(\frac{\pi c L}{\ell}\mu - \frac{\ell}{\pi c C}\frac{1}{\mu}\right) = Z_0\mu^2 + i\left(Z_C\frac{1}{\mu} + Z_L\mu\right) \tag{4.25}$$

Reliable estimates for the constants $Z_0$, $Z_C$, and $Z_L$ are best found by matching to an electromagnetic solver for electrically small dipoles. A comparison with simulations is illustrated in Figure 4.3, with the reactive and resistive antenna behavior got from MATLAB Antenna Toolbox. A reasonable agreement between (4.25) and the simulated graphs are obtained assuming $Z_0$ = 30Ω and $Z_C = Z_L$ = 95Ω in (4.25). Model and simulations are seen to agree very closely at low frequencies. For frequencies approaching unity (the wavelength corresponding to twice the antenna length), (4.25) is not expected to be fully valid and indeed implies $Z_{\mathrm{re}}$ = 30Ω, whereas the simulation gives $Z_{\mathrm{re}}$ = 144Ω. This disagreement between the simulation and model at frequencies

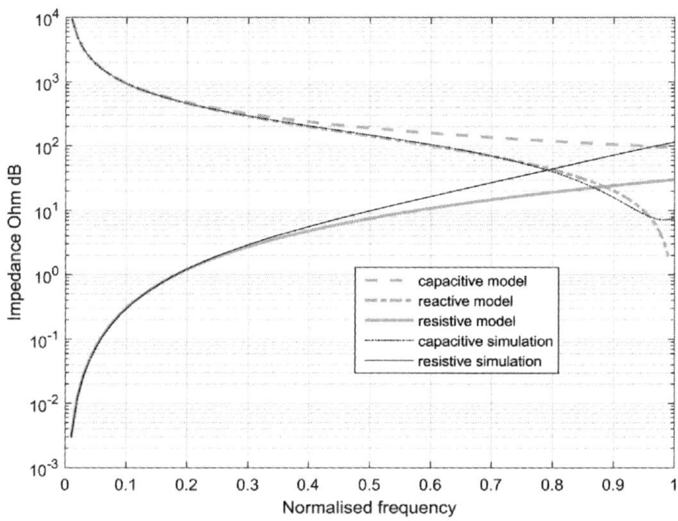


**Figure 4.3** Impedance of cylinder dipole using the MATLAB Antenna Toolbox dipole antenna model with radius 5% of dipole length. The simulation and parametric models developed in the text are matched to establish radiation resistance, capacitance, and inductance constants. Frequency is normalized to the ratio of wavelength to twice the cylinder length. For the parametric model, only the capacitive part is of relevance for the lower frequencies and will be retained; the plot indicates the approximation in doing so.

close to unity has a negligible impact on the transmit power calculations of the preceding section, since the required power is largely determined by the equalization required for the low-frequency part of the band. In fact, (4.25) will be further simplified by neglecting the inductive reactance contribution in (4.25) (i.e., $Z_L = 0$ henceforward), since it has little importance at lower frequencies and thus for transmit power.

With the values $\varepsilon_0 = 8.854187187 \times 10^{-12}$ F/m and $c = 299{,}792{,}458$ m/s, it is found that $1/c\varepsilon_0 = 377\Omega$. If the antenna volume is considered a sphere of diameter $\ell'$, its surface area is $A_{\text{ant}} = \pi\ell'^2$, so $\ell'/\ell = \sqrt{377/\pi \times 30} = 2$. The self-capacitance of a conducting sphere of diameter $\ell''$ is $C = \pi\varepsilon_0$. Comparing with the capacitive impedance at unity frequency $\ell/\pi cC = 95\Omega$, it is found that $\ell/\ell'' = \pi^2 c\varepsilon_0 \times 95 = 2.5$. A number of that order is what is intuitively expected.

Adopting this impedance model (4.25), the dependence of $\Gamma$ on $\mu$ follows for the electrically small cylinder, being fed via a transmission line and performing as an antenna. With the feed impedance being $Z_F$, the reflection coefficient is $(Z - Z_F)/(Z + Z_F)$. The power not reflected is transmitted, so

$$\Gamma = 1 - \left|\frac{Z - Z_F}{Z + Z_F}\right|^2 \tag{4.26}$$

Assuming the transmission line is lossless $Z_F$ is real, whereupon

$$\Gamma = \frac{4Z_{\text{re}}Z_F}{\left(Z_{\text{re}} + Z_F\right)^2 + Z_{\text{im}}^2} \tag{4.27}$$

Evidently, transmissivity will depend on $Z_F$. The value of $Z_F$ can be adjusted by a transformer and the quest is which value of $Z_F$ maximizes $\Gamma$. This optimum is found as

$$\begin{aligned} \frac{1}{4Z_{\text{re}}}\frac{\partial\Gamma}{\partial Z_F} &= \frac{Z_{\text{re}}^2 - Z_F^2 + Z_{\text{im}}^2}{\left[\left(Z_{\text{re}} + Z_F\right)^2 + Z_{\text{im}}^2\right]^2} = 0 \\ \Rightarrow Z_F &= \sqrt{Z_{\text{re}}^2 + Z_{\text{im}}^2} = |Z| \end{aligned} \tag{4.28}$$

Consequently, transmissivity is at best

$$\Gamma = \frac{2Z_{\text{re}}}{|Z| + Z_{\text{re}}} \tag{4.29}$$

It is observed that since $Z$ depends on wavelength, so does transmissivity. To investigate, insert model (4.25) (where as agreed, $Z_L = 0$) into (4.27)

$$\Gamma = \frac{4Z_0\mu^2 Z_F}{\left(Z_0\mu^2 + Z_F\right)^2 + Z_C^2\dfrac{1}{\mu^2}} \approx \mu^4\frac{4Z_0 Z_F}{\mu^2 Z_F^2 + Z_C^2} \tag{4.30}$$

The first term in the denominator can be neglected unless $Z_F$ à $Z_C/\mu$. Then transmissivity depends on wavelength as $\mu^4$. However, selecting $Z_F$ optimally for each frequency

$$Z_F = \sqrt{Z_0^2\mu^4 + Z_C^2\frac{1}{\mu^2}} \approx \frac{Z_C}{\mu} \tag{4.31}$$

In this case

$$\Gamma = \frac{2Z_{\text{re}}}{|Z| + Z_{\text{re}}} = \frac{2Z_0\mu^2}{\sqrt{Z_0^2\mu^4 + \dfrac{Z_C^2}{\mu^2}} + Z_0\mu^2} \approx \frac{2Z_0\mu^3}{Z_C} \approx \frac{2}{3}\mu^3 \tag{4.32}$$

Here, $Z_C/Z_0 = 95/30 \approx 1/3$. The result is of theoretical importance only, since impedance frequency adaptivity required according to (4.31) is difficult to realize technically. An obvious compromise, adopted in this text, is to select a fixed feed impedance, providing optimal transmissivity at the lower edge of the radar band (where antenna performance is weakest), that is, to select $Z_F = Z_C/\mu_{\text{min}}$. Figure 4.4 demonstrates that the cubic approximation in (4.32) underestimates the optimal transmissivity, but quite well approximates this nonoptimal choice. A fuller comparison with simulations shows that the cubic approximation relatively well approximates the nonoptimal transmissivity in the wide regime $\lambda_{\max} \leq 10\,\ell$ for the lower band edge. It thus well encompasses the relevant situations for low-frequency SAR.

The required steering of the antenna beam to just one side of the radar aircraft will now be treated. It is recalled that the reason for this requirement is the fundamental inability of the SAR process to separate returns from either side of the radar path, that is, the SAR image obtained with a single omnidirectional antenna (like a dipole element) is a linear combination of the right-hand side and left-hand side images. The apparent method of separation is to acquire two independent linear combinations of the two images and solve the corresponding linear equation for each of them. The independent

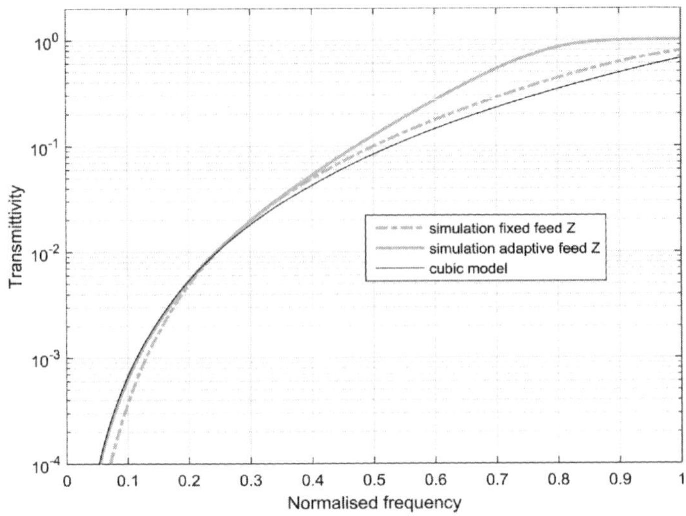


**Figure 4.4** Comparison of transmissivity estimates obtained by simulation and cubic model. The fixed feed impedance is 307Ω and is set for minimum $\mu = 0.3$. It is observed that cubic transmissivity constitutes a conservative estimate of actual transmissivity while being in perfect agreement with adaptive (optimal) feed impedance in the low-frequency limit. For frequencies approaching unity, it is seen to agree well with transmissivity for fixed feed impedance.

combinations are suitably obtained by obtaining independent SAR imagery from each of two transversally separated antenna elements.

The initial SAR system configuration is depicted as in Figure 4.5(a), where the reciprocal transfer function compensation and the associated noise growth are observed. Including the right/left separation, the arrangement becomes as in Figure 4.5(b). Note, however, that the equation system solution involves subtraction between the two acquired ambiguous SAR images. Unless an efficient antenna element set up is found, the solution may impose a significant signal loss or equivalently a noise increase, which will be added to the penalty of using small antennas. Suitable arrangements for narrow-band beam steering are obvious and well known. In the present wideband case, to obtain the required high efficiency one-sidedness, a 90° phase shift, imposed on the transmit signals fed into each antenna, turns out to be crucial.

To keep the analysis on a principal level, the two elements are characterized by a one-dimensional model of right and left transmissions with respect to the SAR flight path. Radiation in other directions follows a smooth transition between right and left radiation and will be a minor impact on the power budget. The antenna elements are assumed to be set up symmetrically on either

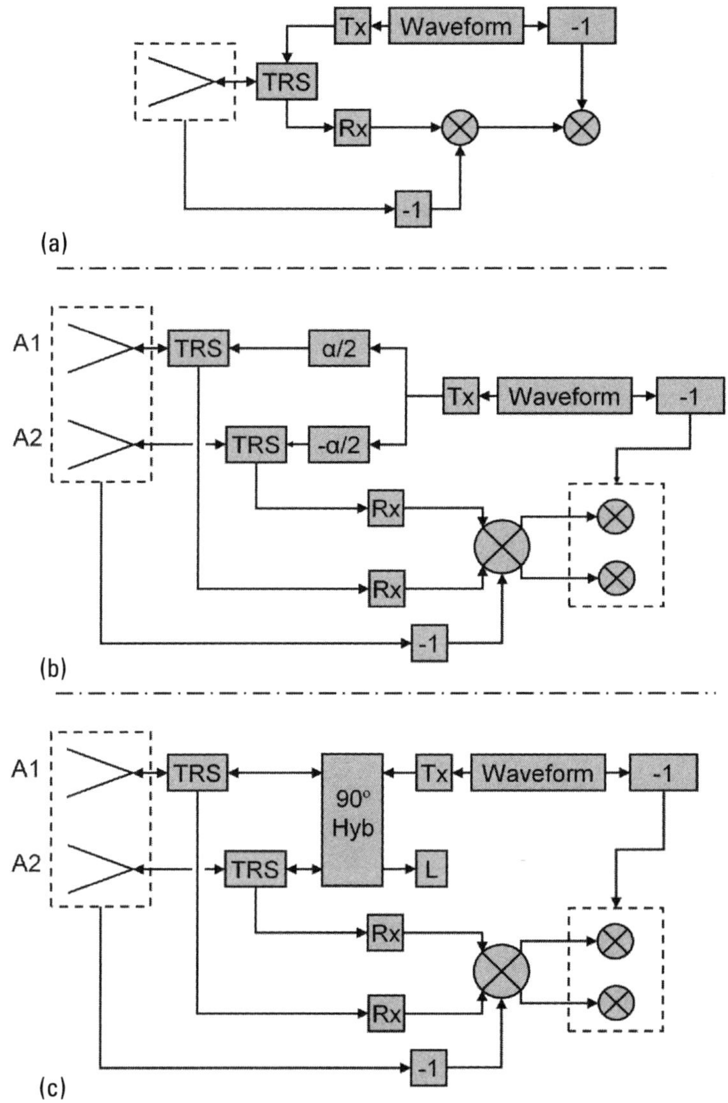


**Figure 4.5** Three stages of ultrawideband low-frequency radar system development. (a) Acquiring spectral flatness. (b) Right/left beam shaping. (c) Small antenna standing wave removal. RX: receiver. TX: transmitter. "–1" symbolizes inverse filtering by the transfer function reciprocal. As indicated in the figure and which was explained in Chapter 3, the filtering removes waveform dependence in the reflectivity density reconstruction. Moreover, filtering can be extended to compensate for antenna effects and in particular right/left separation, as in (b). The final stage of development is combining steering of the transmit beam with cancellation of the strong feed reflections in the case of electrically small antennas, as in (c).

side of the path so that the signal can propagate to and fro from either element to the ground. Transmissivity between each antenna element and the ground at the same side will be unity after equalization, whereas to the other side, it assumes some value $\Gamma_0 < 1$ due to the shadowing of the other element. Antenna transfer characteristics are modeled by a $2 \times 2$ linear transform between the required right and left SAR image slant range reflectivity densities $\gamma^R_{2k}$ and $\gamma^L_{2k}$ and the received frequency response $z_R$,$z_L$ in either antenna. Consider initially that the transmit radar beam is shaped for one-sidedness by imposing a (possibly frequency varying) phase shift $2\alpha$ between the two antennas. Then for unit magnitude transmitted signal in either element

$$\begin{pmatrix} z_R \\ z_L \end{pmatrix} = \begin{pmatrix} e^{-\frac{ikd}{2}} & \Gamma_0 e^{\frac{ikd}{2}} \\ \Gamma_0 e^{\frac{ikd}{2}} & e^{-\frac{ikd}{2}} \end{pmatrix} \begin{pmatrix} \gamma^R_{2k} & 0 \\ 0 & \gamma^L_{2k} \end{pmatrix} \begin{pmatrix} e^{-\frac{ikd}{2}} & \Gamma_0 e^{\frac{ikd}{2}} \\ \Gamma_0 e^{\frac{ikd}{2}} & e^{-\frac{ikd}{2}} \end{pmatrix} \begin{pmatrix} e^{i\alpha} \\ e^{-i\alpha} \end{pmatrix} \tag{4.33}$$

The structure of the equation follows from Figure 4.6 (left). When inverted, (4.33) yields

$$\begin{aligned} \gamma^R_{2k} &= \frac{1}{e^{-ikd} - \Gamma_0^2 e^{ikd}} \frac{e^{\frac{-ikd}{2}} z_R - \Gamma_0 e^{\frac{ikd}{2}} z_L}{e^{\frac{-ikd}{2}} e^{i\alpha} + \Gamma_0 e^{\frac{ikd}{2}} e^{-i\alpha}} \\ \gamma^L_{2k} &= \frac{1}{e^{-ikd} - \Gamma_0^2 e^{ikd}} \frac{e^{\frac{-ikd}{2}} z_L - \Gamma_0 e^{\frac{ikd}{2}} z_R}{\Gamma_0 e^{\frac{ikd}{2}} e^{i\alpha} + e^{\frac{-ikd}{2}} e^{-i\alpha}} \end{aligned} \tag{4.34}$$

Ideally, $\Gamma_0$ is effectively zero and no additional step of right–left separation is required. The opposite and most challenging case is when $\Gamma_0$ is effectively unity, which is the typical situation for electrically small antennas and indeed is the situation for the CARABAS III system. In this case, (4.34) assumes the form

$$\begin{aligned} \gamma^R_{2k} &= \frac{i}{4} \frac{i}{\sin(kd)} \frac{e^{-\frac{ikd}{2}} z_R - e^{\frac{ikd}{2}} z_L}{\cos\left(\frac{\alpha - kd}{2}\right)} \\ \gamma^L_{2k} &= \frac{i}{4} \frac{1}{\sin(kd)} \frac{e^{-\frac{ikd}{2}} z_L - e^{\frac{ikd}{2}} z_R}{\cos\left(\frac{\alpha + kd}{2}\right)} \end{aligned} \tag{4.35}$$

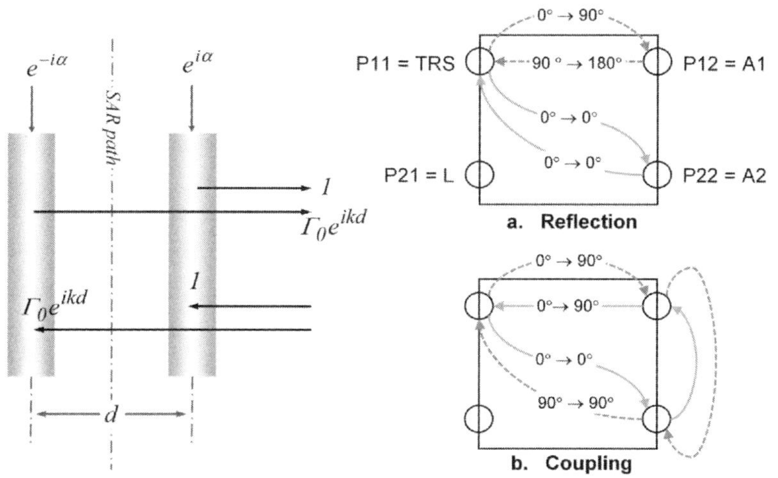


**Figure 4.6** Left: two dipoles set up for wideband separation of right and left returns. The separation is obtained by solving a linear equation, where coefficients are provided by the transmit and receive antenna transfer characteristics. The antenna arrangement is right–left symmetric, so the coefficients are also right/left symmetric with only the right-hand coefficients shown. A transmit phase shift $2\alpha$ between antennas steers the transmit beam to one side and is inverted to steer to the opposite side. Right: the behavior of a 90° hybrid during transmission and reception of radar signals in the configuration of Figure 4.5(c).

Exterior noise is modeled as having the left- and right-hand Fourier components $n_{\text{ext},R}$ and $n_{\text{ext},L}$ impingent on the antennas from either side. The components themselves are incoherent but with equal variance. Also assume that there are incoherent left and right internal noise components $n_{\text{int},R}$ and $n_{\text{int},L}$ of equal variance. In the case of no transmitted radar signal, the received signal in the spectral domain is pure noise becomes

$$\begin{aligned} z_R &= e^{-\frac{ikd}{2}} n_{\text{ext},R} + e^{\frac{ikd}{2}} n_{\text{ext},L} + n_{\text{int},R} \\ z_L &= e^{-\frac{ikd}{2}} n_{\text{ext},L} + e^{\frac{ikd}{2}} n_{\text{ext},R} + n_{\text{int},L} \end{aligned} \tag{4.36}$$

The noise is expected to vary randomly. Considering the right-hand image, by insertion into (4.35) and averaging over time

$$\left\langle \left| \gamma_{2k}^{R-\text{noise only}} \right|^2 \right\rangle = \frac{2\sin^2(kd)\left\langle \left| n_{\text{ext},R} \right|^2 \right\rangle + \left\langle \left| n_{\text{int}} \right|^2 \right\rangle}{8\sin^2(kd)\cos^2\left( \dfrac{\alpha - kd}{2} \right)} \tag{4.37}$$

Here $\langle|n_{\text{int},R}|^2\rangle = \langle|n_{\text{int},L}|^2\rangle = \langle|n_{\text{int}}|^2\rangle$. The beamforming noise increase should be as small as possible, which is the case if

$$\sin^2(kd) = 1;\ \cos^2\left(\frac{\alpha - kd}{2}\right) = 1 \Rightarrow kd = \frac{\pi}{2};\ \alpha = \frac{\pi}{4} \tag{4.38}$$

It follows that $d$ should be selected so that d = $\lambda_c/4$ with $\lambda_c$ wavelength at center frequency. For instance, the CARABAS III low-band center frequency is 55 MHz, so $d$ = 1.35m in this case. With $\alpha = \pi/4$, the unit power fed into the twin elements will correspond to amplitudes $\sqrt{2}e^{i\pi/4}$ and $\sqrt{2}e^{-ii\pi/4}$ fed into either element. Modifying the right column vector in (4.33) accordingly, the expressions in (4.35) become rescaled as $\gamma_{2k}^{R/L} \to \gamma_{2k}^{R/L}/\sqrt{2}$. The beamforming noise factor, that is, the noise change due to the beamforming compared to transmit power, thus becomes twice the quantity $\langle|\gamma_{2k}^{R\text{-noise only}}|^2\rangle$ in (4.37). Splitting the noise factor into internal and external contributions and noting $kd = \pi f/2f_c$

$$\begin{aligned} C_{\text{ext},bf} &= \frac{1}{2}\frac{1}{\cos^2\left[\frac{\pi}{4}\left(1 - \frac{f}{f_c}\right)\right]} \\ C_{\text{int},bf} &= \frac{1}{4}\frac{1}{\sin^2\left(\frac{\pi}{2}\frac{f}{f_c}\right)\cos^2\left[\frac{\pi}{4}\left(1 - \frac{f}{f_c}\right)\right]} \end{aligned} \tag{4.39}$$

The situation is illustrated in Figure 4.7 for CARABAS III low band. Since exterior noise is totally dominating at the low band, there is an implied 50% reduction in the transmit power figures found in the preceding section for radar in this band. Above 100 MHz, exterior noise levels reduce, but a suitable receiver design will not let internal noise dominate over exterior noise, which is still significant. Thus, performance will be set by exterior noise. The methods for right–left separation therefore entail a 3-dB reduction of the required transmit power in the examples discussed in connection with (4.19) and (4.21).

A final step of low-frequency SAR development refers to Figure 4.5(c) and Figure 4.6 (right). Clearly, the low transmissivity of the small antennas discussed corresponds to exceedingly large standing waves, that is, reflections at the antenna port. It turns out that that the insertion of a 90° hybrid in the radar schematic layout precisely satisfies the double role of creating the required 90° phase shift between the antennas and effectively handling the antenna reflections.

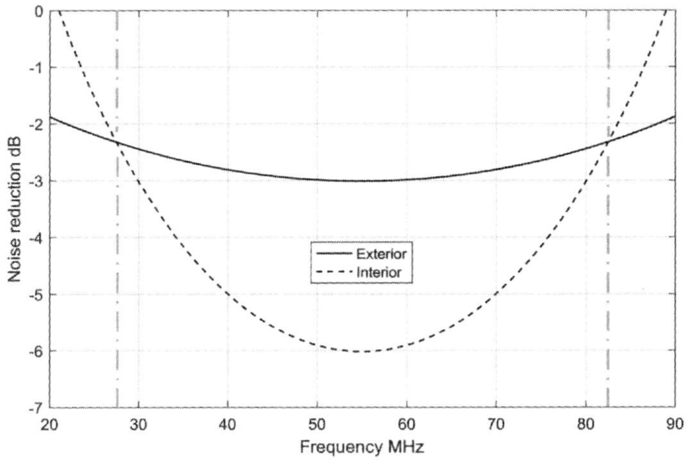


**Figure 4.7** Noise reduction obtained by right–left separation for the CARABAS III low band (band limits indicated by the vertical dashed lines). The dipole separation is 1.35m, providing a symmetric degradation of the noise reduction at the band edges. The noise reduction with respect to the dominating exterior noise is seen to remain flat within a single decibel across the utilized frequency band.

Referring to Figure 4.6 (right), the 90° hybrid yields for an input signal in port P11 phase-shifted output signals in ports P12 and P22, with a relative 90° phase shift between them. The 90° hybrid is a reciprocal four-port device in the sense it has the same performance if input and output ports are substituted according to P11 ↔ P12 and P21 ↔ P22. In the present application, antennas are connected to P12 and P22 and the transmit/receive switch (TRS) to P11. Hence, as the components of the transmit signal reflected at the ports P12 and P22 are subjected to net phase shifts which differ by 180° and thus cancel at P11 (there is a 180° phase difference between the red and blue signal paths in Figure 3(a)). At P21, the components will however add constructively. In the application, this combined reflected signal is sunk by a load L connected to P21. Thus, the transmit signal components that are reflected by the antenna due to its low radiation efficiency will not cause any reflected signal impingent on the TRS or the transmitter itself.

While reflected signals cancel, signals coupling without delay or phase shift between P12 and P22 add constructively at port P11 (see the blue and red signal paths in Figure 3(b)). Thus, the coupling between the antennas gives rise to a signal impingent on the TRS output and the transmitter. At low frequencies, the low radiation efficiency makes the coupling effects small.

In all, the proposed design solves the otherwise major issue of connecting the antennas to a standard transmit power generation design.

The location of the radar twin dipole array with respect to the airframe is an important design issue. Unless mounted in front or to the rear of the airframe (putting the airframe in the dipole radiation null), overall radiation performance will be affected by interaction with airframe. Front and rear mountings were used in the original CARABAS II and I systems, respectively. Drawbacks were partly mechanical integration issues. More fundamentally, the by necessity unsymmetrical nature of a coaxial feed for antennas so mounted disrupted the desired broad band 60° wide radiation diagram. If the antenna diagram was perfectly known, it might still be compensated for by equalization, similar to what is practiced for the small antenna frequency dependence. In practice, knowledge proved too imprecise to render these attempts successful, and optimal resolution was never attained.

Whatever the location of the antenna pair on the airframe is, there is an important advantage in a right/left symmetric placement. With a symmetric arrangement, toggling between transmissions to the right and left redundant data is obtained. Not to be explicitly accounted for here, the redundant data is sufficient to determine the transfer characteristics of the aircraft to the radar signal. For a wideband system, these characteristics are quite complex and can be very difficult to determine with any accuracy by electromagnetic modeling, so such a method is of great benefit. Figure 4.8 shows three examples of symmetric antenna arrangements. Another common characteristic of these is the significant distance to which the antennas extend from the platform belly.

The vicinity to the belly has the same influence on antenna characteristics as has the vicinity to the ground surface for the scattering from shallow ground objects, which was analyzed in Chapter 2. Coarsely, the belly can be modeled as an infinitely extended conductive plane, whereby it is straightforward to estimate its effect on a field radiated. The horizontal polarization fading calculation mimics the argument leading up to (2.5.85), though here just one-way propagation will be considered. Consulting Figure 2.6, it is seen that with the dipoles located horizontally at a distance $D$ below the fuselage, 2 $D = L/2$, while $\Phi(\theta,\theta) = 1$ and $R = -1$ with the notations of Chapter 2. Then

$$\frac{\left|\boldsymbol{E}_{\text{tot}}(\theta)\right|}{\left|\boldsymbol{E}_{\text{dip}}\right|} = \left| e^{-\frac{ikD}{\cos\theta}} - e^{-\frac{ikD\cos(\pi-2\theta)}{\cos\theta}} \right| = \left|\sin(kD\cos\theta)\right| \tag{4.40}$$

Here is the angle of radiation with respect to the nadir direction. The case of interest is that of relatively shallow propagation angles (typical for SAR

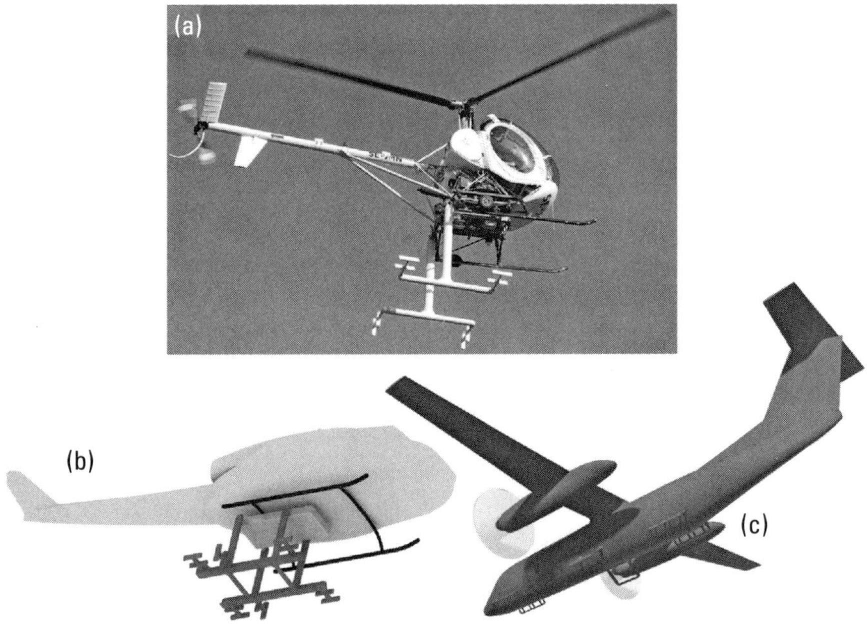


**Figure 4.8** Low-frequency SAR antenna concepts. (a) The CRABAS III system featuring electrically small dipoles for the bands 27.5 to 82.5 MHz (low band) and 137.5 to 357.5 MHz (high band). The upper band antennas are duplicated for DPCA moving target purposes, as described in Chapter 3. As seen, antennas on the right and left sides provide horizontal or vertical polarization, respectively. All the antennas are paired to be excited with a 90° phase shift according to principles described in the text. The low-band dipoles are 1.5m long and the high band dipoles are 0.5m long. The vertical pillar holds the antenna arrangement at the necessary distance of 1m below the landing skids, which effectively acts as a ground plane. Antennas are attached to a hydraulic actuator, enabling the antennas to swing back to a ground parking position, beside the tail boom. The complete system including actuator is held tight to the landing gear by means of the helicopter cargo hook, and can be emergency dropped if required. (b) A Pi antenna-based system for a larger system on a Bell Huey helicopter. The arrangement features both GMTI capability and subsurface capability by exploiting full polarimetric capability at both the high and low bands. The system folds by swinging to the front and is held by the cargo hook as before. (c) A U-antenna-based system for a Dash 8, again allowing low-band polarimetric SAR and high-band polarimetric SAR/GMTI. The antennas are fixed but provide sufficient ground clearance. In this case, antennas have good right/left separability by themselves and the 90° phase shift feed is not used. The right-hand antennas are used for horizontal polarization to the right and vertical polarization to the left and vice versa.

imaging), that is, $\theta = \pi/2 - \Delta\theta$, where $\Delta\theta$ is relatively small (down to 20°), and when $D$ is short compared to wavelength. Then

$$\frac{\left|\boldsymbol{E}_{\text{tot}}(\theta)\right|}{\left|\boldsymbol{E}_{\text{dip}}\right|} \approx \left|kD\Delta\theta\right| \tag{4.41}$$

For instance, $D$ = 1m for the CARABAS III system with where maximum wavelength is $\lambda_{\text{max}} \approx$ 10m. Then $kD \approx 0.6$, so for $\Delta\theta = 0.5$, it follows that $|\boldsymbol{E}_{\text{tot}}(\theta)| \approx 0.3|\boldsymbol{E}_{\text{dip}}|$. It should be stressed that this model is overly pessimistic. An aircraft belly has finite extension and is rounded or carry other structures, which are all circumstances preventing any deep destructive cancellation between the antenna and its belly mirror image. Thus, in practice, as indeed is the case for the CARABAS III system (the CARABAS III landing skids have a function similar to the reflecting plane), attenuation is less pronounced than what is indicated by the model. To predict the antenna function accurately and in a specific case, a 3D model of the entire platform has to be incorporated in the antenna performance simulation. Due to large wavelengths, the model is not required to be accurate in any detail though the salient reflecting structures (like the helicopter landing skids) must all be included.

While the electrically small dipole antennas discussed so far fulfill the requirements for high-quality SAR imaging, they are only suitable for small systems for which transmit power requirements are modest, and where the smallness of the antenna system and not transmit power is the overriding concern. In contrast, a large system may require kilowatts of transmit power even with a fully efficient antenna. Obviously, overall concerns like keeping emitted power to minimum, and not unnecessarily taxing the platform power supply, will take precedence over small overall dimensions for the antenna system. This said for wavelengths of several meters, antenna integration issues will remain a challenge and antenna designs must certainly be practical in this sense. The main requirements on antennas for large systems are thus:

1. Near unity transmissivity across the radar band.
2. Along track beam width of the order 60° to either side of the radar aircraft.
3. Beam confined vertically to steer to intended SAR swath on either side of aircraft.
4. Antenna should allow both horizontal and vertical polarizations.
5. Antenna only imposing small drag on aircraft.
6. Design must allow mechanical rigidity, being lightweight at the same time.

Good solutions to all these requirements have been found in loop monopoles (here referred to as U-antennas) and their enhancements into *Pi antennas*. The latter will now be discussed. The discussion will be restricted to a generic example. The Pi antenna is characterized by more geometry parameters than the dipole, and there is a multitude of variations possible, all affecting the antenna characteristics. No attempt to provide their optimization will be done, and indeed such an optimization will depend on the detailed circumstances in the particular application considered. The U antenna is such a special case and will not be analyzed explicitly.

The Pi antenna is a two port device, standing erect (the $y$-direction) on a ground plane (in practice represented by the aircraft body in the vicinity of the antenna location), with reflection and coupling given by the S-parameters $S_{11}$ and $S_{12}$. In the analysis, the entire loop of the antenna structure and its mirror image in the ground plane will be considered. The voltage distribution for this loop is shown in Figure 4.9, where the selectability of polarization by means of

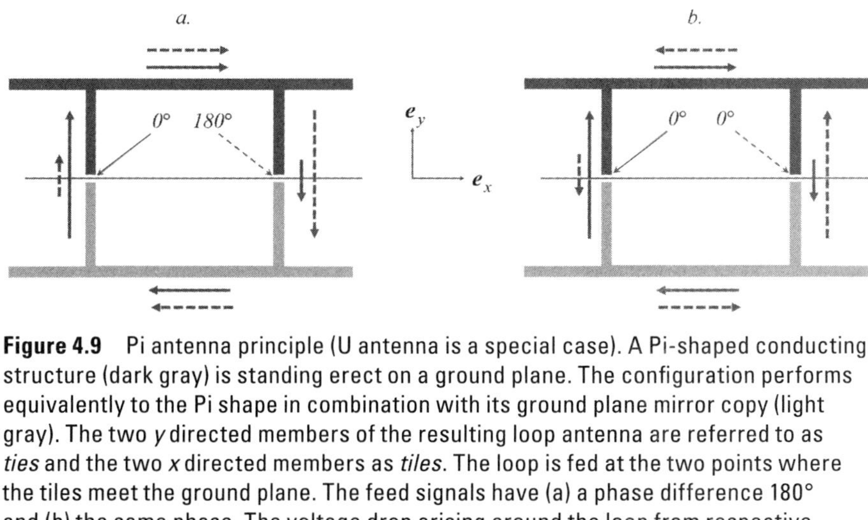


**Figure 4.9** Pi antenna principle (U antenna is a special case). A Pi-shaped conducting structure (dark gray) is standing erect on a ground plane. The configuration performs equivalently to the Pi shape in combination with its ground plane mirror copy (light gray). The two $y$ directed members of the resulting loop antenna are referred to as *ties* and the two $x$ directed members as *tiles*. The loop is fed at the two points where the tiles meet the ground plane. The feed signals have (a) a phase difference 180° and (b) the same phase. The voltage drop arising around the loop from respective feed is depicted as the continuous and dotted vectors. The largest drop occurs at the tie containing the feed, while (due to the power lost by radiation) it is smallest at the opposite tie. In (a), there is a strong excitation for the tiles, while in (b), there is only excitation for the ties. In particular, in the direction of the viewer, there is no net radiation in (a) and only $y$-polarized radiation in (b). Indeed, the arrangement in (a) has fading characteristics reminiscent of the belly-mounted electrical dipole described. The difference from the dipole arrangement is the phase diversified feeds of the tiles, resulting in the required widening of the tile radiated beam. Similarly, the performance in (b) is that of the ties behaving as two-phase diversified parallel dipoles, providing a suitably wide beam in the $z$ direction (that of the viewer).

phase difference is noted. The antenna structure is assumed two-dimensional with a tile length ($x$-extension) of 3.25m, a tie height ($y$-extension) of 2m, a 1.25m tie separation, and an overall 0.125m width of all antenna members. The frequency band considered is 27.5 to 82.5 MHz. The numerical assessment of performance is carried out by means of MATLAB Antenna Toolbox. Evaluating $S_{11}$ and $S_{12}$, the transmissivity follows as $\gamma = 1 - |S_{11}|^2 - |S_{12}|^2$, that is, transmissivity combines reflection and coupling losses. From the results in Figure 4.10, it is concluded that $Z_F \approx 200\Omega$ is a suitable compromise for best overall transmissivity across the frequency band.

The radiation pattern is a consequence of the voltage distributions in Figure 4.1. The result depends to some extent on the coupling between the ports and thus the feed impedance. The overall pattern can be obtained as the superposition of the radiation pattern from the radiated field for each feed with the other feed terminated by the selected 200Ω impedance. From Figure 4.9, the two cases are noted to differ by the $\boldsymbol{E}$ field in the $x$ direction

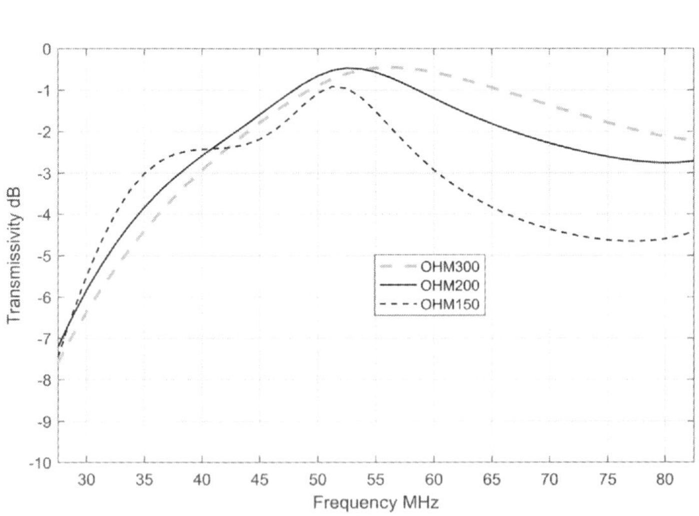


**Figure 4.10** Transmissivity in the studied case of a 3.25 × 1m Pi antenna, used for the 27.5 to 82.5 MHz band, and for three choices of feed impedance. The transmissivity takes into account both reflection and coupling. As expected, the transmissivity is relatively good and frequency dependence is relatively flat up to wavelengths around twice the antenna length, beyond which it starts to drop as for electrically small antennas. However, the drop is small compared to what was the case for the 1.5m dipoles, and the requirement for transmit power increase will therefore not be very significant. The antenna does, however, require a transmitter design accepting nearly total reflection of the transmitted signal at the lower end of the radar band. A suitable compromise for feed impedance seems to be 200Ω.

being reversed/kept and the $\boldsymbol{E}$ field in the $y$ direction being kept/reversed for the same/opposite phase in the terminals.

The voltage distributions in Figure 4.9 and the radiation pattern are connected through Kirchhoff's integral theorem. With the theorem stipulated in the form (2.112), the resulting field distribution is the sum of the net voltage contributions along each of the antenna tile and tie constituents. Several general conclusions can be drawn. Consider the situation in the two principal two-dimensional cases of the $yz$ and $xz$ planes. From Figure 4.9, for in-phase feed, there is no field contribution from the tiles, so the resulting polarization will be "vertical," that is, orthogonal to the ground plane. In particular, the radiated $\boldsymbol{E}$ field in the $yz$ plane will have its direction within that plane. Since the $\boldsymbol{E}$ field is orthogonal to the direction of radiation, the in-phase radiation pattern will be (up to a tie voltage dependent normalization $|\boldsymbol{E}_{\text{tie}}|$, determined by the transmissivity)

$$\frac{\left|\boldsymbol{E}_{\text{in phase}}(\theta)\right|}{\left|\boldsymbol{E}_{\text{tie}}\right|} = |\sin\theta| \tag{4.42}$$

Similarly, for opposite feed phases, the tile voltage contributions will cancel in the $yz$ plane, that is, radiation will be "horizontal" with respect to the ground plane. The situation is then analogous to (4.40), so

$$\frac{\left|\boldsymbol{E}_{\text{opp phase}}(\theta)\right|}{\left|\boldsymbol{E}_{\text{tile}}\right|} = |\sin(kH\cos\theta)| \tag{4.43}$$

where $H$ is half tie length (i.e., $H$ is the distance to ground plane or the aircraft body).

Now analyze radiation in the $xz$ plane. For in-phase feed, that is, vertical polarization, the pattern again follows in analogous to (4.40):

$$\frac{\left|\boldsymbol{E}_{\text{opp phase}}(\theta)\right|}{\left|\boldsymbol{E}_{\text{tile}}\right|} = |\cos(kD\cos\theta)| \tag{4.44}$$

where and $D$ is half the tie separation. For opposite phase feed, the situation is more complicated since the radiation pattern in this case is determined by the relative strength between $\boldsymbol{E}_{\text{tile}}$ and $\boldsymbol{E}_{\text{tie}}$. From MATLAB Antenna Toolbox, the field pattern is obtained for each feed separately. Adding these patterns according to the symmetry of Figure 4.9 provides the result, which is illustrated in Figure 4.11. As expected, there is widening of the antenna beam, even though transmissivity is that of the physically larger and high efficiency antenna structure adopted. The resulting wideband pattern is in fact similar to that of the small dipole, with beam widths for the different frequencies (calculated according to (4.3)) provided in Figure 4.11. Comparing Figures 4.4 and 4.10, it is concluded that the 3.25m long Pi antenna has more than 10 dB higher transmissivity than 1.5 dipole (adopted for CARABAS III) across the band 27.5 to 82.5 MHz.

A suitable mounting of the Pi antenna is to have tiles horizontal and ties vertical, as for the helicopter in Figure 4.8(b). Then the opposite phase feed produces a horizontally polarized radiation pattern similar to the dipole arrangement of the electrically small antenna in Figure 4.8(a), while in-phase feed provides vertical polarization in the normal sense of the word. An alternative arrangement is mounting Pi antennas on the aircraft body sides as in Figure 4.8(c). Preferably, locations should be chosen so that by mounting the antennas normal to the surface, they become tilted downward corresponding to the depression angle of the SAR image registration, that is, at some angle 30° to 45°. In this case, the horizontally polarized radiation beam from each antenna has the appropriate direction to the same side of the aircraft, while the "opposite side" transmissivity $\Gamma_0$ of (4.34) will be a relatively small number. For vertical polarization, the antennas have transmission nodes to

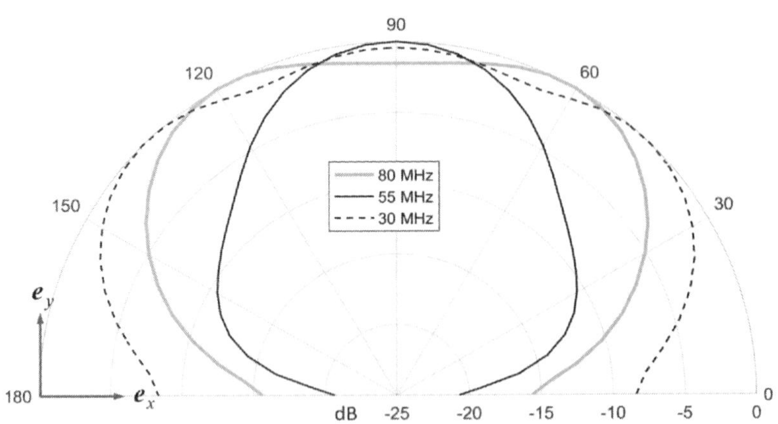


**Figure 4.11** Radiation pattern for the Pi antenna example in the case of polarization in the $y,z$-plane and 180° fed phase difference. The feed impedance is 200Ω. With the antenna tile horizontal, the radiation pattern is effectively that of horizontal polarization in the slant range plane. The design thus achieves combining high wideband transmissivity with the required radiation beam width ≳60°. Presently, the beam width goes from 113° at 30 MHz to a minimum of 55° at 55 MHz and up to 97° at 80 MHz.

the same side, but diffract around the aircraft to illuminate the opposite side. Again, $\Gamma_0$ will be small.

As will be discussed in the next section, the extra degrees of freedom of using a right–left symmetric antenna arrangement, with separate receiver channels for the right- and left-hand sides, nevertheless remains a requirement. However, transmissions to just one side are possible. In certain cases, transmission to both sides is required and a suitable arrangement is then that these to occur in ping-pong fashion alternating the transmit pulses to go to either of the right- and left-hand side antennas. Selecting polarization is suitably achieved by inserting a 180° wideband hybrid between the antenna and the transceiver. The hybrid feed port on the transceiver side is made selectable by a switch.

Side mounted Pi antennas is the preferred choice for larger aircraft, where ground clearance is sufficient for such antennas not to interfere with ground. Dash 8 and Bombardier Global Express have been considered candidate aircrafts. For a smaller aircraft, there is generally not sufficient space available for side mounting (doors, wing, and landing gear being the typical obstructions). The belly mounted antenna is then the suitable alternative. Beechcraft King Air, Twin Otter or Dornier 228, and several helicopter types are good candidates for this type of installation. Its drawback is that it must be made collapsible (in Figure 4.8(b), the antenna folds forward to be smooth with the underside of the movement actuator). For UAVs, the body is generally too narrow for a belly mount and the side mounted antennas are preferable (fixed or flexible depending on ground clearance). For a larger aircraft, high efficiency and thus physically large antennas are needed to minimize the significant transmit power required. For small systems (which UAV systems generally are), transmit power is of single watt order and size rather than antenna efficiency is the priority. Pi and U antennas, which are electrically small, seem a good choice, since their attachment to the vehicle body by the two ties makes them a mechanically and aerodynamically more sturdy alternative than using a single pillar containing the feed of the dipole as in Figure 4.8(a). The fragility of this particular design limits possible platform speed and thus restricts application to, for instance, that of the light Schweitzer 300C helicopter in Figure 4.8(a).

## 4.4 Waveforms

Radar in its simplest form is pulsed, with the impulse response resolution $\Delta\rho$ determined by transmit pulse duration $\Delta t$ according to $\Delta\rho = c\Delta t/2$. In this case, since radar range is determined by mean transmit power, high resolution forces high peak power. High peak power requires high voltage, implying large costly transmitter equipment, apart from causing electromagnetic interference. As was demonstrated in Chapter 3 (see (3.3)), high peak power can be avoided, since by precise knowledge of the transmit signal, the target impulse response can be retrieved also when $\Delta t >> 2\Delta\rho/c$. The transmitted signal can take any form, while resolution is entirely determined by its bandwidth according to $B = c/2\Delta\rho$. However, this freedom in selecting waveform only applies in principle, since there are a number of harsh practical constraints, being quite decisive when determining the transmit waveform.

The first such requirement is that of constant transmit amplitude, which is a consequence of keeping peak transmit power as low as possible. Actually, radar power amplifiers can be kept very simple and compact if operating in a saturated mode in which case there are no options but constant output amplitude.

A second constraint is that in order to obtain maximum power efficiency, the transmit signal spectral density should be as constant as possible. Indeed, the solution of (2.8) can be viewed as including an equalization process for spectral inhomogeneities. Consider a case of modifying a constant spectral density transmit signal into that of inhomogeneous spectral density $f_\omega^{TX} \rightarrow \alpha_\omega f_\omega^{TX}$, but without modifying mean transmit power, that is, $\langle \alpha_\omega^2 \rangle = 1$. As a consequence, the received signal will be modified as $f_\omega^{RX} \rightarrow \alpha_\omega f_\omega^{RX}$. Since additive noise is unchanged, the mean noise power added to the reflectivity becomes changed by a factor $\langle 1/\alpha_\omega^2 \rangle$. Since $\langle \alpha_\omega^2 \rangle \langle 1/\alpha_\omega^2 \rangle \geq 1$ is a well-known relation between harmonic and arithmetic means, it can be concluded that the spectrally flat transmit waveform is optimal for minimizing noise impact on reflectivity.

A further set of restrictions on waveform follows from the requirements of a nonambiguous ordering of data samples into range and slow time dependence. As was analyzed in Chapter 3, the use of a periodic transmit signal is critical in that it has a discrete transmit spectrum, allowing this sorting and thus the construction of the reflectivity density $\gamma(r,t)$. The condition is that $|\Delta\omega| < |\pi/\tau|$, where $\tau$ is the signal repetition rate and where $\gamma_{\omega,\Delta\omega}$ is the Fourier transform of $\gamma(r,t)$, that is, $\Delta\omega$ is Doppler shift. Thus, signal repetition rate must be selected with respect to radar platform speed, that is, Doppler bandwidth.

Given $\tau$, there is a range interval $\Delta r = c/2\tau$ in which the reflectivity density can be uniquely determined. Increasing the signal repetition rate, the Doppler nonambiguity is still maintained, but doing so reduces $\Delta r$. Clearly $\Delta r$ must be selected with a sufficient safety margin so that extraordinary strong but far off targets are not folded in, becoming erroneously imaged as weaker targets within the intended SAR swath. It was noted that condition (4.14) is quite generous regarding the possible areal coverage rate thus implied. Thus, the risk of range ambiguities is small as long as $\Delta r$ is only governed by platform speed. However, in order to minimize the required pulse transmit power, pulses should be launched at a rate making the radar only receive data for ranges up to the farthest range required for imaging the intended SAR swath. This rate is significantly higher than what is needed for Doppler nonambiguity. Achieving this while still avoiding range ambiguities is possible by the stepped frequency waveforms to be described.

As mentioned in Section 4.1, a stepped frequency waveform has the additional advantage of reducing the data sample rates, which are significant and tended to be beyond what could practically achieved until just the very last few years (at the time of writing, 2016). The stretch waveform, which was mentioned to be routinely used for microwave SAR, also decouples swath and repetition rate and implies a data reduction similar to what is obtained by stepped frequency waveforms. It is ideal for long-range radar operation and but less useful in the low-frequency case.

The freedom in selecting waveforms is significantly enhanced by allowing simultaneous transmission and reception. For an advanced radar, not using short pulses as the means to attain resolution, pulsing is merely a way of filtering to achieve the required very high degree of isolation between the transmit and receive signals. Thus, pulsed operation is not a strict requirement for radar operation, but has for low-frequency SAR the important disadvantage of not allowing designs, which are optimal for frequency sharing with in-band wireless services. In particular, there is a requirement on the transmit waveform to contain sharp notches for any reserved sub-band within the radar band, which require continuous waveforms.

One continuous waveform technique is *bistatic radar*, in which transmitter and receiver antennas are separated by distance and beam shaping not to interfere with each other. For low-frequency radar, bistatic operation is of course difficult due to the diffuse antenna pattern. Another method is what is called *frequency modulated continuous wave radar* (FMCW) in which frequency modulation of the transmit signal allows separation between the currently transmitted signal and the range response occurring at a different previous frequency of the transmit signal. Still another method, *continuous arbitrary waveform radar* (CAW), relies on cancellation, that is, just by knowing the transmit signal with sufficient accuracy it can be subtracted out of the received signal to attain transmit/ receive isolation. The FMCW technique is reminiscent of the stretch method, relying on range filtering by means of a frequency swept waveform. Just as bistatic radar, also FMCW is less useful for low-frequency SAR due to the diffuse radiation patterns. More fundamentally frequency swept waveforms are inconsistent with sharp notches. As will be demonstrated in Section 4.5, sharp notches can be obtained with phase modulated waveforms, though these will not be swept frequency waveforms. The CAW method is very well suited for low-frequency radar, and is in fact the method required for notch waveforms.

Among the various waveforms mentioned, stepped frequency has a particular advantage in enabling range gating to suppress the nadir echo, as mentioned in Section 4.1. Though using stepped frequency waveforms in this way is a reliable and well-proven technique, it is not the only option of nadir echo suppression. An alternative technique will be developed exploiting the degrees of freedom of the two-element antenna array for subspace filtering. The method yields a high degree of suppression and can be applied to independently of the waveform selected.

The nadir echo will now be analyzed. As mentioned in Section 4.1, it arises by the essentially omnidirectional low-frequency antenna radiation pattern being specularly reflected by the ground. For undulating ground, there may be several ground patches where the ground instantaneously happen to exhibit slopes so inclined that the ground normal points to the radar platform and specular reflections of the transmitted signal impinges on the antenna. When ground is rough on the scale of wavelength, the reflections are temporal occurrences. Although the backscattering amplitude can be instantaneously high, these will contribute relatively little to the overall backscattered energy. However, if ground is flat compared to wavelength, which very often is the case for low-frequency SAR, there will be constant specular reflection from right below the aircraft. The specular reflection might in this case be considered as the signal impinging from an antenna right below the radar platform at a distance of twice its altitude and reduced by twice the ground reflection coefficient. It can thus be modeled as a phenomenon arising by one-way propagation, whereas backscattering from oblique incidence angles arises as backscattering from spherical waves and thus two-way propagation. The nadir echo will thereby be very pronounced. There is another approach to demonstrate the important difference between these two scattering cases, based on Kirchoff's integral theorem (2.112). According to this approach, the reflection occurs from the circular region of the so-called first Fresnel zone below the aircraft, for which the radius increases with altitude, thus again reducing the scattering range attenuation to become effectively that of one-way propagation.

Assume a generic situation of perfectly flat ground, in which case the nadir echo is the only signal received. Then, the following relation prevails among mean radiated power $P_{\mathrm{TX}}$, antenna transmissivity $\Gamma$ (as defined in connection with (4.10)), the ground index of refraction $\eta$, platform altitude $h$, and the received power:

$$P_{\mathrm{RX}} = \left(\frac{1-\eta}{1+\eta}\right)^2 \Gamma^2 \frac{\lambda_{\min}{}^2}{4\pi(2h)^2} P_{\mathrm{TX}} \tag{4.45}$$

In the sequel, it is assumed that receive amplitude has been frequency equalized, so $\Gamma = 1$. In terms of field amplitudes and reflectivity density, (4.45) can be restated by means of (3.15) and (3.16)

$$\begin{aligned} f_{\mathrm{RX}}(t_2) &= \frac{1-\eta}{1+\eta}\frac{\lambda_{\min}}{4\sqrt{\pi}}\frac{c}{2}\int_{-\infty}^{t_2} \delta(r-h) f^{\mathrm{TX}}(t_1)\frac{dt_1}{r} \\ \Rightarrow \gamma^{\mathrm{nadir}}(r,t) &= \frac{1-\eta}{1+\eta}\frac{\lambda_{\min}}{4\sqrt{\pi}}\delta(r-h) \end{aligned} \tag{4.46}$$

The expression is inserted in SAR inversion formula (3.31) in order to analyze the impact on the SAR image. Also invoking the slow time Fourier transform of $\gamma_{\Delta\omega}^{\mathrm{nadir}}(r,t) = \gamma^{\mathrm{nadir}}(r,t)\delta(\Delta\omega)/2\pi$ and noting the time independence of $\gamma^{\mathrm{nadir}}$

$$\begin{aligned} \hat{\gamma}_{k_\rho,\Delta\omega/v}^{\mathrm{nadir}} &= \frac{v|k_\rho|}{2\pi}\int_0^\infty\int_{-\infty}^\infty J_0\left(r\sqrt{k_\rho{}^2 + \left(\frac{\Delta w}{v}\right)^2}\right) e^{-i\Delta\omega t}\gamma^{\mathrm{nadir}}(r,t)\,r\,dr\,dt \\ &= \frac{1-\eta}{1+\eta}\frac{h\lambda_{\min}}{4\sqrt{\pi}}\frac{|k_\rho|}{2\pi}J_0\left(h\sqrt{k_\rho^2 + \left(\frac{\Delta w}{v}\right)^2}\right)\int_{-\infty}^\infty e^{-\frac{i\Delta\omega y}{v}}dy \end{aligned} \tag{4.47}$$

It follows that:

$$\hat{\gamma}_{k_\rho}^{\mathrm{nadir}}(y) = \frac{1-\eta}{1+\eta}\frac{h\lambda_{\min}}{4\sqrt{\pi}}|k_\rho| J_0(h|k_\rho|) \tag{4.48}$$

Represent reflectivity density as a complex quantity with $\hat{\gamma}_{k_\rho} = 0$ if $k_\rho < 0$. Use the asymptotic Bessel function approximation $J_0(z) \approx \sqrt{2/\pi z}\cos(z - \pi/4)$. Then

$$\begin{aligned} \hat{\gamma}_{k_\rho}^{\mathrm{nadir}}(y) &= \frac{1-\eta}{1+\eta}\frac{\lambda_{\min}}{4\pi}\sqrt{2hk_\rho}\cos\left(hk_\rho - \frac{\pi}{4}\right) \Rightarrow \\ \hat{\gamma}^{\mathrm{nadir}}(\rho,y) &\approx e^{-i\frac{\pi}{4}}\frac{1-\eta}{1+\eta}\frac{\lambda_{\min}}{8\pi}\int_0^\infty \sqrt{2hk_\rho}\,e^{i(\rho-h)k_\rho}dk_\rho \end{aligned} \tag{4.49}$$

Here, only the contributing exponential term of the cosine function has been kept. With the relation $\omega = ck_\rho/2$, the expression is restricted to a finite bandwidth. The square root within the integrand will vary slowly so approximately that

$$\hat{\gamma}^{\text{nadir}}(\rho, y) = e^{-i\pi/4}\left(\frac{1-\eta}{1+\eta}\right)^2 \frac{\lambda_{\min}}{2\pi c}\sqrt{\frac{h\omega_c}{c}} \int_{\omega_c-\pi B}^{\omega_c+\pi B} e^{i2(\rho-h)\frac{\omega}{c}} d\omega$$
$$\Rightarrow \left|\hat{\gamma}^{\text{nadir}}(\rho, y)\right|^2 = \left(\frac{1-\eta}{1+\eta}\right)^2 \left(\frac{\lambda_{\min}}{2\pi}\right)^2 \frac{h\omega_c}{c} \frac{\sin^2\left[2\pi(\rho-h)\frac{B}{c}\right]}{(\rho-h)^2} \tag{4.50}$$

Reexpress (4.6) in fractional bandwidth $\beta = 2\pi B/\omega_{\max}$, noting that $\omega_c = \omega_{\max}(1 - \beta/2)$. The final reflectivity expression for the nadir response becomes

$$\left|\hat{\gamma}^{\text{nadir}}(\rho, y)\right|^2 = \left(\frac{1-\eta}{1+\eta}\right)^2 \frac{h\lambda_{\min}}{2\pi}\left(1-\frac{\beta}{2}\right) \frac{\sin^2\left[2\pi(\rho-h)\frac{\beta}{\lambda_{\min}}\right]}{(\rho-h)^2} \tag{4.51}$$

Considering the limit $\rho \to h$, the reflectivity the nadir echo is obtained as

$$\left|\hat{\gamma}^{\text{nadir}}(h, y)\right|^2 = 2\pi\left(\frac{1-\eta}{1+\eta}\right)^2 \frac{h}{\lambda_{\min}} \beta^2\left(1-\frac{\beta}{2}\right) \tag{4.52}$$

As examples, regard a CARABAS III system operating at $h$ = 3,000m and for typical ground index of refraction $\eta$ = 5. For the low band (27.5 to 82.5 MHz), $\lambda_{\min}$ = 4m and $\beta$ = 0.67 yielding $\hat{\gamma}^{\text{nadir}}(h, y)$ = 28 dB. For the high band (137.5 to 357.5 MHz), $\lambda_{\min}$ = 0.6m and $\beta$ = 0.62 yielding $\hat{\gamma}^{\text{nadir}}(h, y)$ = 34 dB. The useful part of the SAR image typically has its closest range at 45° incidence (i.e., $\rho = \sqrt{2}h$). The sidelobe residues must be on par with the noise equivalent reflectivity, that is, of the order −30 dB for typical foliage penetration applications and even less for certain subsurface applications. The roll off rate therefore must be not less than providing a 60-dB peak value reduction at $\rho \geq \sqrt{2}h$. Using (4.7) for the roll-off rate, this condition is seemingly fulfilled. However, most importantly, (4.7) does not take into account calibration accuracy and sources of data degradation, for example, RFI spectral notches. From the existing evidence, a 60-dB sidelobe suppression within a SAR image is possible at the low band for a point reflector, where sidelobes spread in two dimensions. For a linearly extended reflector, like the nadir echo, the spread is just one-dimensional, and the roll off rate will be significantly less. For the high band, the roll off rate is faster due to the shorter wavelength. Moreover, in contrast to the low band, ground is more seldom flat compared to wavelength at the shorter wavelengths, making average nadir return values smaller than (4.8). Hence, nadir echo interference is not an as critical issue for the high band.

The mentioned possibilities to mitigate the nadir echo impact in the SAR image are either to select a waveform, which avoids the nadir echo by restricting time samples by frequency stepping or by filtering, exploiting the antenna degrees of freedom. In the latter case, the nadir echo appears in the SAR image, but at a subdued level. Both approaches will be treated with advantages/disadvantages analyzed.

Since a frequency sweep is the suitable waveform for keeping constant transmit amplitude, the range response cannot be reconstructed to the full bandwidth unless the entire transmit signal is received. Thus, if the radar is set to operate with receiver opening in such a way that only the last part of the transmit signal is received from the nadir, the reconstruction of the nadir occurs at a bandwidth less than the full radar bandwidth and thus with a resulting reduced roll-off rate. For instance, in a stepped frequency scheme, the spectrum of the nadir response would consist of a set of narrow band peaks, each with bandwidth smaller than the full radar bandwidth. It follows that there are two alternatives, namely, either that the transmit signal is received in its entire also for the nadir echo or that the nadir echo is avoided entirely. Note that while avoiding the nadir echo, some part of the ground in between the intended SAR swath and the nadir echo still becomes partially represented in the received spectrum, with resulting leakage of its response into the SAR image. This is normally not a problem since the response is expected nonspecular and thus weak.

Figure 4.12 represents the geometry of the two situations of receiving and avoiding the nadir echo. Assume that the diagrams represent an interval of radar operation corresponding the transmission of radar frequency sweep. For a stepped frequency scheme, the transmission covers just a fraction $B_{\text{step}}$ of the full radar bandwidth $B$ and the entire process of transmitting, receiving, and switching the system between transmission and reception last just for an interval of time $t_{\text{dwell}}$ (called dwell time) being the same fraction of the full radar period $t_{\text{rec}}$ (to be referred to as recurrence time). The transmit time $t_{\text{TX}}$ determines the transmit duty cycle $\chi = t_{\text{TX}}/t_{\text{dwell}}$, that is, the ratio between mean and peak transmit power. The dwell time spans the four processes of 1) transmission, 2) switching from transmission to reception, 3) reception, and 4) switching back to reception, whereupon $t_{\text{dwell}} = t_{\text{TX}} + t_{\text{TX}\to\text{RX}} + t_{\text{RX}} + t_{\text{RX}\to\text{TX}}$. According to (4.12), the maximum range $r_{\max}$ is determined by $t_{\text{RX}}$ and $t_{\text{TX}\to\text{RX}}$

$$r_{\max} = \frac{c}{2}\left(t_{\mathrm{TX}\to\mathrm{RX}} + t_{\mathrm{RX}}\right) = \frac{c}{2}\left[t_{\mathrm{dwell}}(1-\chi) - t_{\mathrm{RX}\to\mathrm{TX}}\right] \tag{4.53}$$

Including or avoiding nadir minimum range yields the relations

$$\begin{aligned} &\text{incl. nadir: } h = \frac{c}{2}\left(t_{\mathrm{TX}} + t_{\mathrm{TX}\to\mathrm{RX}}\right) = \frac{c}{2}\left(t_{\mathrm{dwell}}\chi + t_{\mathrm{TX}\to\mathrm{RX}}\right) \\ &\text{excl. nadir: } r_{\min} = \frac{c}{2}\left(t_{\mathrm{TX}} + t_{\mathrm{TX}\to\mathrm{RX}}\right) = \frac{c}{2}t_{\mathrm{dwell}}\chi + h \end{aligned} \tag{4.54}$$

Note that the minimum range equals the minimum slant range $r_{\min} = \rho_{\min}$, whereas with a 60° integration angle $r_{\max} = 2\rho_{\max}/\sqrt{3}$. In the previous chapters, it has been analyzed that for efficient foliage penetration, SAR imaging minimum and maximum slant ranges must be selected to keep incidence angle in an interval 60° to 45° across the SAR image. As will be discussed in Chapter

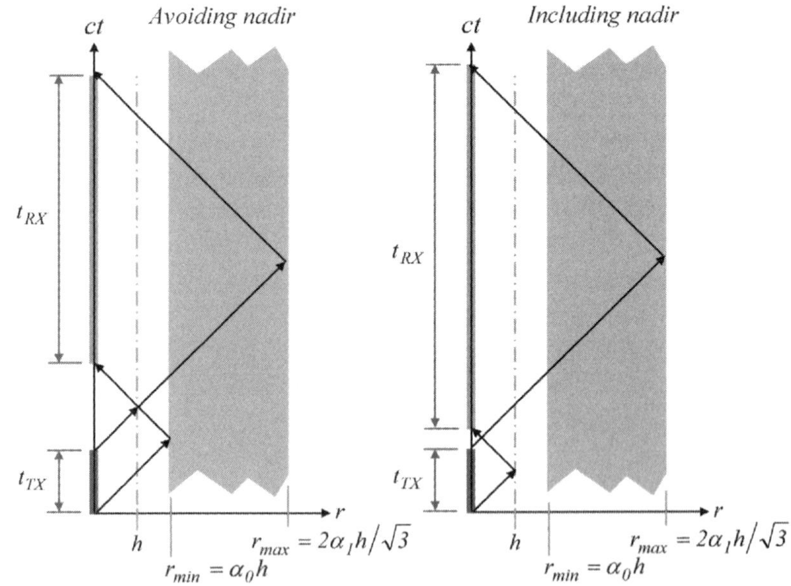


**Figure 4.12** Waveforms for avoiding and including nadir. According to the diagrams, maximum range is selected to correspond to transmit to receive switch time plus receive time. Thereby, the full transmitted signal will be received from the farthest range. When including nadir, the return of the full transmit signal from the nadir also falls into the receive interval. Avoiding nadir, the return of the transmit signal stays outside the receive interval. In this case, the minimum range must be separated from nadir by the length of the transmit signal. Thereby, the transmit signal fully lands inside the receive interval for the minimum range.

6, very shallow incidence, with the SAR image in a span 70° to 80°, is useful in certain subsurface applications. Generally, denote $\rho_{max} = \alpha_1 h$ and $\rho_{min} = \alpha_0 h$, where $\alpha_0 = \sqrt{2}$ and $\alpha_1 = 2$ are the typical foliage penetration choices. The two waveform types impose their respective limitations on achievable swath width and duty cycle. Excluding or including nadir, (4.9) and (4.10) imply

$$\begin{aligned} \chi_{excl} &= \frac{1}{\dfrac{2\alpha_1}{\sqrt{3}(\alpha_0 - 1)} + 1 + \dfrac{c}{2h(\alpha_0 - 1)} t_{RX \to TX}} \\ \chi_{incl} &= \frac{1 - \dfrac{ct_{TX \to RX}}{2h}}{\dfrac{2\alpha_1}{\sqrt{3}} + 1 + \dfrac{c(t_{RX \to TX} - t_{TX \to RX})}{2h}} \end{aligned} \tag{4.55}$$

If switch times are negligible and $\alpha_0 > 2$ (i.e., incidence is shallow), excluding waveforms have the larger duty cycle. In most cases, however, $\alpha_0 < 2$. In these situations and again neglecting switch times, the excluding waveform gives the smaller duty cycle. With the constants $\alpha_0 = \sqrt{2}$ and $\alpha_1 = 2$, that is, the typical foliage penetration geometry, it is found that $\chi_{excl} \approx 15\%$ and $\chi_{incl} \approx 30\%$.

Not to impair significantly on duty cycle, the switch times $t_{TX \to RX}$ and $t_{RX \to TX}$ must preferably be smaller and certainly not much larger than $2h/c$. Depending on platform and application, altitude may vary from very low altitudes like 30m for small low-power systems (for subsurface imaging using shallow incidence) to 15,000 m or even more for large- and high-power foliage penetration of stationary and moving target systems. In the first case, switch times should preferably be smaller than 1 $\mu$s, whereas in the latter case, switch times of a few microseconds are acceptable. With the advent of GaAs and GaN semiconductor materials, the required switch performance has become achievable. Traditionally, PIN diode switches offered the best solution, but could not mange switch times faster than a 10-$\mu$s forcing system to use high peak power with poor duty cycle.

The analysis now turns to the composition of frequency steps into a suite covering the full radar bandwidth. A fundamental requirement for stepped frequency waveforms is that calibration and system stability have been achieved to such a level that in the synthesis of the step data, any discontinuities or even spectral variations arising due to stepping are mitigated to the level of system noise. The challenge of an accurate reconstruction of the range spectrum of course becomes even more accentuated when the nadir return is present in the data.

In previous days, when analog bandpass filtering of the received frequency step data was a necessity, it proved very difficult to apply precision inverse filtering to regain a flat spectrum. However, at around 2015, ADC and digital-to-analog (DAC) technology reached the level of being able to handle bandwidths of a few hundred megahertz, with 14 bit or better dynamic range. Thus, today the bandwidth and dynamic range requirements discussed for low-frequency radar can be coped with by entirely digital means. A level of control of the filtering, quite beyond what can be achieved by analog means, is possible. An obvious architecture is shown in Figure 4.13. The complete transmit waveform is split into $N$ sub-band waveforms representing the frequency steps, all with equal bandwidth. These are obtained as neighboring frequency points, of the spectrum of the full transmit signal, defined at the dwell time spectral density. Being inverse Fourier transformed by a fast Fourier transform (FFT) and fed through a DAC, they become time signals with bandwidth limited to the sub-band of the frequency points representing them. Since the waveform of the radar remains fixed, it is not necessary to perform the FFT for each frequency in real time. For each sub-band, the waveform can be represented by a stored set of time samples to be fed directly to the DAC. A sequencing unit selects and feeds the sub-band waveform to the transmitter and antenna. A TRS then sets the system to reception mode, in which signals received through the antenna are bandpass filtered for the full radar band, digitized by an ADC, whereupon an FFT is applied on the received dwell period data. Those frequency points covered by the transmitted sub-band are used to retrieve the range spectrum, and the other are just noise data and not used. Repeating the process while sequencing through all steps, data for the full bandwidth range response is thus acquired. As pointed out above, the recurrence time $\tau$ available for completing the cycle is set by Doppler bandwidth, according to $|\Delta\omega_{\max}| = 2v\omega_{\max}/c < \pi/t$ or $v\tau < \lambda_{\min}/4$. That is to say, repeating the cycle with the transmit waveform fixed (and thus transmit signal periodic in slow time), there is a time variation in the frequency point data corresponding to a time variation in the ground reflectivity response due to the platform movement. Performing a Fourier transform of this time variation for each frequency point, the result can be represented as a function $f^{\mathrm{RX}}_{2\pi n/\tau+\Delta\omega}$. With $C_{n(\omega)}$ being the amplitudes of the transmit signal frequency points, the ground reflectivity function $\bar{\gamma}(r,t)$ can thereby be determined from (3.21).

The scheme must be analyzed for conflicts between the contradictory requirements of signals being band limited, as well as having finite time duration and also constant amplitude. The latter condition is due to the nonlinearity of radar transmitters as was mentioned. The basic waveform implied by the constant amplitude requirement is that is that of a frequency sweep. It must

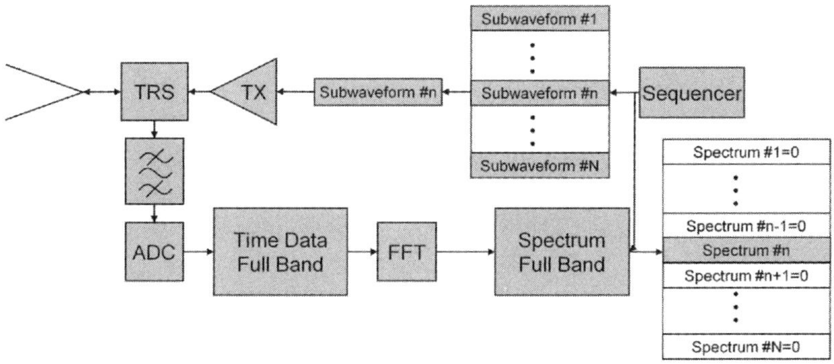


**Figure 4.13** The principle of a fully digital stepped frequency radar. Sub-bands formed by subsets of the frequency points of the full transmit waveform are used for the step waveforms. The hardware realization of the scheme (not detailed in the illustration) suitably reads the step waveforms as time samples from a memory, with these loaded into the memory when setting up the radar and obtained by performing an inverse FFT of the selected waveform. The step waveforms are fed through the power amplifier TX and the TRS to the antenna unit. The received signal fed from the antenna and TRS is bandpass filtered for the full radar bandwidth, whereupon it is digitized by an ADC and Fourier transformed. Only the frequency points corresponding to the current step waveform are saved, representing the range response for each sub-band. By executing all frequency steps, the full range response spectrum is obtained.

be set up to cover the full radar bandwidth $B$ during a total signal time $Nt_{\text{TX}} = Nt_{\text{dwell}}\chi$. The subwaveforms are obtained as sequential segments of the sweep of length $t_{\text{TX}} = t_{\text{dwell}}\chi$, each covering the frequencies of a step. The subwaveforms are zero padded to become of length $t_{\text{dwell}}$. The zero padding modifies the frequency content of the waveforms by a sinc function convolution. The convolution must not lead to any significant widening of the step bandwidth, so it is required that step bandwidth is significantly larger than the reciprocal of $t_{\text{TX}}$, that is, $Bt_{\text{dwell}}\chi/N >> 1$. If this requirement is not respected, folding errors will follow in the range response reconstruction. In consequence, there is an upper limit to the number of possible frequency steps and actual waveforms must be chosen accordingly.

Examples of possible waveforms, including and excluding nadir, serve to further illustrate the arguments. Consider the CARABAS III low band of 27.5 to 82.5 MHz for a platform operating with ground speed $v$ = 150 m/s at $h$ = 7,000m. Interior and exterior noise temperatures are as before 1,200° and 120,000° with a noise equivalent reflectivity of –25 dB aimed at. Transmit/receive switching is assumed to be possible within times $t_{\text{TX}\to\text{RX}} = t_{\text{TX}\to\text{RX}} = 2\ \mu\text{s}$.

Transmit power is determined by the maximum surveillance range, which for foliage penetration applications will be about twice the platform altitude, that is, presently of the order 15 km. Ambiguous range responses are caused by the periodicity of the transmit signal. Because of the cubic range reduction as well as shallow angle ground fading of the ground response, these ambiguous returns become exceedingly weak and thus negligible if sufficiently far away. The recurrence time $\tau$ is selected to create a sufficient safety margin. Presently, $1/\tau$ = 950 Hz is chosen, implying a 158-km range for the closest received range ambiguity. With this recurrence rate, eight frequency steps provide a dwell time, matching the required maximum surveillance range. The implied dwell rate is $1/t_{\text{dwell}}$ = 7.6 kHz. Avoiding nadir in each frequency step and with reception time resulting in a swath covering the required 45° to 70° incidence span, a possible duty cycle is $\chi$ = 10%. With the 55-MHz full bandwidth, eight frequency steps imply a step bandwidth of $B_{\text{step}} = 6.88$ MHz. Since transmit time becomes $t_{\text{TX}} = 13\ \mu$s, it is clear that $B_{\text{step}} >> 1/t_{\text{TX}}$, that is, selected parameters satisfy the condition for accurate range response reconstruction. With the selected recurrence rate $(\lambda_{\min}/4)/v\tau = 5$, that is, the sampling rate is five times higher than required by Doppler bandwidth. Decreasing recurrence rate allows even larger ambiguity range, while a large number of frequency steps are required to match the useful surveillance range. The penalty in this increased number will be shorter transmit times, whereas the step bandwidth is reduced. Hence, the requirement $B_{\text{step}} >> 1/t_{\text{TX}}$ becomes challenged, which must be avoided.

Considering the required transmit power, it is recalled that $\varepsilon_{\text{TRAP}}$ in (4.13) is the noise energy captured by the receiver in the case of a waveform that does not restrict coverage rate from expression (4.14). Presently, these relations are modified, both in that noise in the SAR image becomes suppressed to the ratio of the oversampling factor $(\lambda_{\min}/4)/v\tau = 5$, and since the receiver filters noise to the step bandwidth, the noise is correspondingly reduced compared to a receiver open to the full bandwidth. By the stepped frequency principle, the noise reduction so achieved precisely corresponds to a reduction of area coverage rate. In all, the radar equation, either in the form (4.19) for electrically small antennas or in the form (4.21) for unity transmissivity, applies. For unity transmissivity, the implied mean power is 14W, that is, the implied peak power is 140W. Attaining one-sidedenness with a two-element end fire array, the required transmit power is halved, as discussed in Section 4.3. Adapting to a realistic antenna (e.g., a Pi antenna), transmissivity will be less than unity. To what degree depends on antenna design details, but for a typical Pi antenna, transmissivity is in accordance with Figure 4.10. Just by visual inspection on the figure, mean transmissivity across the band (and after equalization of the received spectrum) is seen to be roughly −3 dB. Hence, the required peak power in the example will be of the order 140W.

Waveforms including nadir are also possible for the same imaging geometry. These permits the transmit duty cycle to be increased leading to a reduction of required peak transmit power. For instance, $\chi$ = 25% is permissible when including nadir. A decrease in dwell rate is required when extending the transmit duty cycle. A value $1/t_{\text{dwell}}$ = 5.6 kHz is arrived at by setting recurrence rate $1/\tau$ = 1,400 Hz and adopting four frequency steps (note that the FFT base radar architecture dictates the number of frequency steps to be powers of 2). The ambiguity range thereby decreases to 108 km. With this waveform, $B_{\text{step}}$ = 13.8 MHz, while $t_{\text{TX}}$ = 45 $\mu$s, so $B_{\text{step}} \gg 1/t_{\text{TX}}$ is well satisfied. Radar equation (4.21) implies 6W so by the same arguments as when the excluding nadir peak power becomes 24W.

The 55-MHz bandwidth is suitably handled by a 220-MHz clock rate for the digital parts of the radar architecture in Figure 4.13. With receive time determined by maximum range and $t_{\text{TX}\rightarrow\text{RX}}$ according to (4.9), the number of received sample becomes 15,346, avoiding nadir and 28,579 including nadir. To perform the FFTs the records are zero-padded to 16,000 and 32,000 samples, respectively. The architecture thus requires that FFTs of that size are performed at the rate of 7.6 and 5.6 kHz, respectively. Processing capabilities of this magnitude started to become readily available with the field programmable gated array development taking place up to and around 2010. Today, it is well below state of the art limits for high speed FFT processing. The data output rate is noted to be 0.65 and 1.27 Msample/s for excluding and including nadir, respectively.

For cases of higher center frequency and larger bandwidths, the CARABAS III high band serves as an example. The radar band is 137.5 to 357.5 MHz, so the bandwidth is the 220 MHz. Being designed around 2010, ADC and AC technologies were not available for coping directly with the full bandwidth, according to the schematics of Figure 4.13. Instead, the radar exploits an intermediate downconversion step, which divides the full bandwidth in four 55-MHz segments, each of which is downconverted to the CARABAS III low band 27.5 to 82.5 MHz, which is thus used as baseband for the higher band. Today, this intermediate step is unnecessary; state-of-the-art ADCs and DACs with a 14- to 16-bit dynamic range and a 1-GHz sampling rate are available off the shelf.

The actual use of low-frequency SAR often ends up in requirements for waveforms including nadir. One reason is that these more efficiently use available transmit power. Another reason is that excluding nadir puts more demanding requirements on fast switch times (by a factor 2.5, assuming $\alpha_0$

= $\sqrt{2}$ in according to (4.11)). Above all, when considering continuous waveform radar architectures, as is the subject of the next section, waveforms including nadir are the only alternative. The subject now will be an efficient method to impose digital beam forming, removing the nadir in SAR image for these waveforms.

The method to be described is applicable to a multichannel linear receiving system recording in $N$ channels onto which less than $N$ interference sources are impinging and energywise overwhelming the impinging useful data. The method is reminiscent of the so-called *music algorithm* in finding dominant signal components as eigenvectors in a signal space (as opposed to finding peaks in the signal spectrum), but is presently adapted to the context of nadir suppression and right/left separation. The method presumes reception conditions to be stationary, that is, statistically constant during registration time. The method then applies statistical arguments to find the linear combination of data, which minimizes the interference.

By the method provided in Section 4.2, right–left separation is achieved by means of two radar receive channels. To be also able to reject nadir, at least one further channel is required. The one available method staying with just two antenna elements is to actively use transmit beam steering, which was described in Section 4.2. In fact, right–left separation together with nadir removal is possible by toggling the transmit beam between right and left within the signal repetition period. By the method, both sides of the aircraft are imaged in parallel. For radar modes requesting maximum coverage rate, such a functionality is anyway requested, while for more focused search, one side imaging would be preferred, since it reduces transmit power and radio frequency emissions.

Assume a two element type of antenna arrays, where transmit beam direction is steered by reversing feed phase, as discussed in the previous section. To simplify geometrical understanding, it is suitable to combine the receive signals from the two elements with phase shifts, so that both the transmit and receive signals represent beams steered either to the left or the right. As such, they possess directivity, which is maximum in the direction to which they are steered, but is generally nonzero in the opposite as well as the nadir direction. These directivity values are represented as coefficients $x_{RR}$, $x_{LR}$, $x_{RL}$, $x_{LL}$, $x_{RN}$, and $x_{LN}$, with the first index representing the direction of transmit beam steering and the second index the direction of the transmitted signal. Note that due to the right/left symmetric antenna arrangement, $x_{RR} = x_{LL}$, $x_{LR} = x_{RL}$, and $x_{RN} = x_{LN}$. The coefficients are valid for either transmission or reception. Let $z_1$ and $z_2$ denote the time-dependent data received by transmitting to the right and receiving by steering the receive beam to right and left, respectively. Analogously, let $z_3$ and $z_4$ denote the time-dependent data received by transmitting to the left and receiving by steering to right and left, respectively. Then (for the sake of compactness, the coefficients $2k$ are suppressed for reflectivity density; however, all components below are frequency $k$ dependent)

$$\begin{pmatrix} z_1 \\ z_2 \\ z_3 \\ z_4 \end{pmatrix} = \begin{pmatrix} x_{RR}^2 & x_{RL}^2 & x_{RN}^2 \\ x_{LR}x_{RR} & x_{LL}x_{RL} & x_{LN}x_{RN} \\ x_{RR}x_{LR} & x_{RL}x_{LL} & x_{RN}x_{LN} \\ x_{LR}^2 & x_{LL}^2 & x_{LN}^2 \end{pmatrix} \begin{pmatrix} \gamma^R \\ \gamma^L \\ \gamma^N \end{pmatrix} \tag{4.56}$$

From (4.33), assuming $\Gamma_0 = 1$ and inserting $\alpha = \pi/4$ and $kd = \pi f/2f_c$

$$x_{RR} = x_{LL} = \cos\left[\frac{\pi}{4}\left(\frac{f}{f_c} - 1\right)\right]; x_{RL} = x_{LR} = 2\cos\left[\frac{\pi}{4}\left(\frac{f}{f_c} + 1\right)\right] \tag{4.57}$$

From nadir, by the same arguments as for (4.33)

$$x_{RN} = x_{LN} = 2\Gamma_1 \cos\left(\frac{\pi}{4}\right) = \sqrt{2}\Gamma_1 \tag{4.58}$$

Here, $\Gamma_1$ represents the directivity of either antenna element in the nadir direction, which may be affected by the presence of the neighboring antenna element as well as the aircraft structure itself. Because the nadir response is so strong, unavoidable errors in estimating $\Gamma_1$ will make attempts to eliminate the nadir influence by a deterministic solution of equation system (4.12) ineffective. Since imaging condition remains significantly stationary, statistical methods relying on averaging suggest themselves as the preferable alternative.

As in any situation of collecting data in the presence of a prevailing stationary interference, source subspace methods (where music is one in particular) prove highly effective. Presently, such a method can be developed, by considering the four-dimensional complex space $\boldsymbol{C}^4$ to which the data $\boldsymbol{z}$ belong. It is recognized that $\boldsymbol{z}$ is a linear combination $\boldsymbol{z} = \gamma^R\boldsymbol{x}^R + \gamma^L\boldsymbol{x}^N + \gamma^N\boldsymbol{x}^N$ of the three column vectors

$$\begin{aligned} \boldsymbol{x}^R &= \left(x_{RR}^2; x_{LR}x_{RR}; x_{RR}x_{LR}; x_{LR}^2\right) \\ \boldsymbol{x}^L &= \left(x_{RL}^2; x_{LL}x_{RL}; x_{RL}x_{LL}; x_{LL}^2\right) \\ \boldsymbol{x}^N &= \left(x_{RN}^2; x_{LN}x_{RN}; x_{RN}x_{LN}; x_{LN}^2\right) \end{aligned} \tag{4.59}$$

In the $\boldsymbol{z}$ space, consider the direction $\boldsymbol{n}$ along which the RMS value of the data reaches its maximum by forming the scalar product between $\boldsymbol{n}$ and $\boldsymbol{z}$, that is

$$\left\langle \left| \boldsymbol{n}^{\dagger} \boldsymbol{z} \right|^{2} \right\rangle_{\forall t} = \max \tag{4.60}$$

where the dagger denotes transpose—Hermitean transpose whenever the entries are complex.

The nadir response $\gamma^{N}\boldsymbol{x}^{N}$ is the overwhelmingly strongest stable component in $\boldsymbol{C}^{4}$ so that (4.16) finds the direction of the nadir response. As for the practical method of finding $\boldsymbol{n}$, expression (4.16) is noted to actually pertain to a quadratic form for $\boldsymbol{n}$ and can be recast as

$$\left\langle \left| \boldsymbol{n}^{\dagger} \boldsymbol{z} \right|^{2} \right\rangle_{\forall t} = \boldsymbol{n}^{\dagger} \boldsymbol{M} \boldsymbol{n};\ \boldsymbol{M} = \left\langle \boldsymbol{z}\boldsymbol{z}^{\dagger} \right\rangle_{\forall t} \tag{4.61}$$

The $4 \times 4$ matrix $\boldsymbol{M}$ is recognized as a *covariance matrix*. As such $\boldsymbol{M}$ is self-adjoint, that is, $\boldsymbol{M} = \boldsymbol{M}^{\dagger}$. Hence $\boldsymbol{M}$ has four eigenvalues $\lambda_1,\ldots,\lambda_4$ corresponding to four orthogonal eigenvectors if the eigenvalues are distinct. Since $\boldsymbol{M}$ becomes diagonal in the eigenvector base, (4.15) reduces in this base to

$$\begin{aligned}\left\langle \left| \boldsymbol{n}^{\dagger} \boldsymbol{z} \right|^{2} \right\rangle &= \lambda_1^2 \cos^2\theta \cos^2\phi + \lambda_2^2 \cos^2\theta \sin^2\phi \\ &\quad + \lambda_3^2 \sin^2\theta \cos^2\phi + \lambda_4^2 \sin^2\theta \sin^2\phi\end{aligned} \tag{4.62}$$

with $\theta$ and $\phi$ being functions of $\boldsymbol{n}$. It follows that among the eigenvalues are found the maximum value of (4.16). The corresponding eigenvector determines the direction $\boldsymbol{n}$ in (4.15). The required direction $\boldsymbol{n}$ is thus obtained as the eigenvector of this maximum eigenvalue. The three remaining eigenvectors seem to have no general immediate meaning in the context of right-left separation.

Restricting the solution of the right and left responses to the orthogonal complement of the axis along $\boldsymbol{n}$ (which is a three-dimensional complex space) thus eliminates nadir response influence. To retrieve right and left returns in this manner is, once $\boldsymbol{n}$ has been found, to restrict data to its orthogonal complement $z - \boldsymbol{n}(\boldsymbol{n}^{\dagger}\boldsymbol{z})$. From this, there are to approaches to retrieve the right and left responses. Either the three-dimensional vector $\hat{\boldsymbol{z}}$ is projected directly onto either side, that is, into the directions of either vector $\boldsymbol{x}^{R}$ and $\boldsymbol{x}^{L}$. Alternatively, it can be restricted to the two-dimensional vector of the orthogonal complement to the response from either side, which then is projected onto the response from the other side. The former method may be considered a *matched filter* approach to beam forming, the latter an *inverse filtering* approach. Thus, the right and left response estimates are

$$\text{matched: } \hat{\boldsymbol{x}}^{R/L} = \frac{\boldsymbol{x}^{R/L\dagger}\left[\boldsymbol{z} - \boldsymbol{n}\left(\boldsymbol{n}^{\dagger}\boldsymbol{z}\right)\right]}{\left|\boldsymbol{x}^{R/L}\right|\left|\boldsymbol{z} - \boldsymbol{n}\left(\boldsymbol{n}^{\dagger}\boldsymbol{z}\right)\right|}$$

$$\text{inverse: } \hat{\boldsymbol{x}}^{R/L} = \frac{\boldsymbol{x}^{R/L\dagger}\tilde{\mathbf{x}}^{L/R}}{\left|\boldsymbol{x}^{R/L}\right|\left|\tilde{\boldsymbol{x}}^{L/R}\right|} \text{ where } \tilde{\boldsymbol{x}}^{R/L} = \boldsymbol{x}^{R/L} - \frac{\boldsymbol{x}^{R/L\dagger}\left[\boldsymbol{z} - \boldsymbol{n}\left(\boldsymbol{n}^{\dagger}\boldsymbol{z}\right)\right]}{\left|\boldsymbol{x}^{R/L}\right|\left|\boldsymbol{z} - \boldsymbol{n}\left(\boldsymbol{n}^{\dagger}\boldsymbol{z}\right)\right|} \tag{4.63}$$

In these expressions, the vectors in (4.16) based on the model (4.13) are used to steer the data with regard to the response from either side. This approach must be adopted since the responses have about the same magnitude. For the nadir response, being singularly strong, (4.14) is not sufficiently accurate to provide high fidelity cancellation, while the adaptive approach of finding $\boldsymbol{n}$ according to (4.16) guarantees a very high degree of cancellation.

The inverse filtering expression in (4.19) is similar to (4.35) in actually negating the response from the unwanted side. Compared to the matched response, this exclusion comes with the penalty of reduced sensitivity. Independently of the method, there is noise increase caused by nadir removal, which stands in proportion to the magnitude reduction in the projection $|\boldsymbol{x}^{R,L}| \rightarrow |\boldsymbol{x}^{R,L} - n^{\dagger}\boldsymbol{x}^{R,L}|^2$ caused by the projection (4.15). The noise penalty can be computed as

$$C_{\text{nadir remove}} = \frac{1}{1 - \left(\dfrac{\boldsymbol{x}^{R/L^{\dagger}}\boldsymbol{x}^{N}}{\left|\boldsymbol{x}^{R/L}\right|\left|\boldsymbol{x}^{N}\right|}\right)^{2}} \tag{4.64}$$

To this, there is a further noise increase for the inverse filtering method, which can be estimated as

$$C_{\text{other side remove}} = \frac{1}{1 - \left(\dfrac{\boldsymbol{x}^{R/L^{\dagger}}\boldsymbol{x}^{L/R}}{\left|\boldsymbol{x}^{R/L}\right|\left|\boldsymbol{x}^{L/R}\right|}\right)^{2}} \tag{4.65}$$

In Figure 4.14, the factors in (4.18) and (4.19) are shown for the CARABAS III low band. The very small extra noise burden due to the inverse filtering stage is noted. For nadir rejection, the overall noise increase is non-negligible since there is a nearly 6-dB noise increase at the lower band edge. The required transmit power is affected by the mean noise penalty which in the 27.5- to 82.5-MHz band is 2.1 dB for nadir removal and 0.3 dB for right/left inverse filtering. Moreover, exterior noise will to some extent be mitigated in the projection (4.19) (the degree of mitigation has not been computed), so the noise penalty will be smaller when exterior noise is dominating, which is the case below 100 MHz and thus for the CARABAS III low band.

The increased duty cycle possible by allowing the nadir return is of course of particular importance for large systems for which peak power anyway will be large and a power efficient design is a priority. Such systems are likely to exploit near full transmissivity antennas (like the aforementioned Pi and U antennas), side mounded according to the Dash 8 example in Figure 4.8. In (4.13), $\Gamma_0 = 1$ expresses the unity front-to-back lobe ratio. This no isolation radiation pattern is a typical behavior of small antenna arrangements, like the CARABAS III antenna arrangement in Figure 4.8. In contrast, for the larger side mounted antenna arrangements, antenna elements will be well separated and be backed by a reflecting structure of significant size, which leads to a pronounced one-sidedness in their radiation behavior, that is, $\Gamma_0$ will be significantly smaller than unity. Figure 4.14 depicts the result of a full

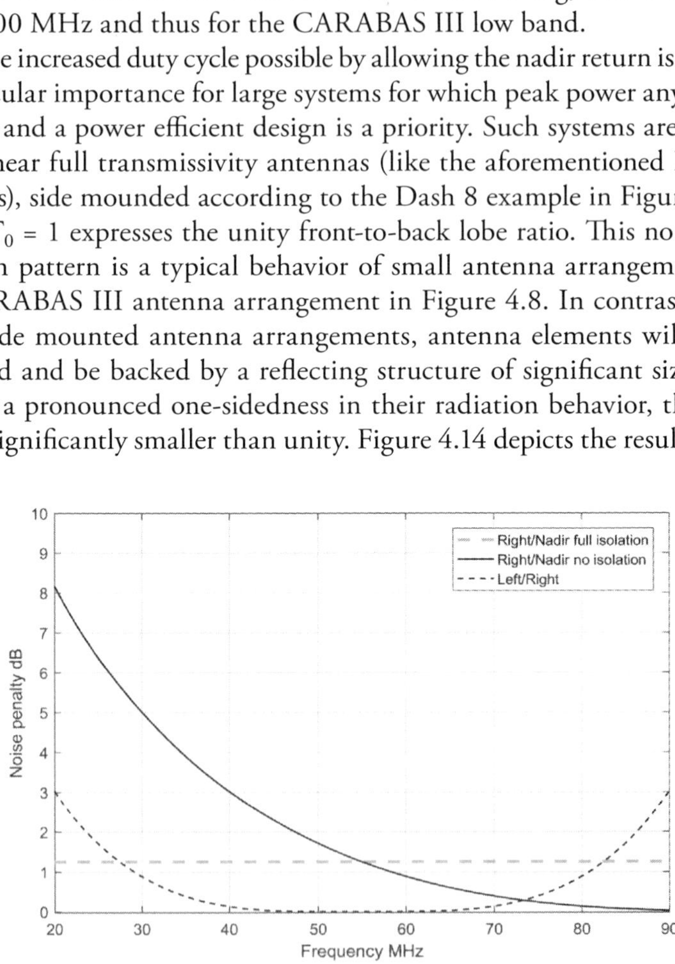


**Figure 4.14** The noise penalty of nadir rejection filtering, transmitting to alternating sides, with a two-element antenna, for the CARABAS III low band (band limits indicated by the vertical dashed lines). By no isolation is understood antenna beam model (4.13) with the same transmit magnitude to either side. By full isolation is understood radiation patterns, which for the right and left elements are entirely directed to the same side. The "left/right" graph is the (notably small) additional penalty of inverse filtering for backlobe removal.

isolation case, assuming $\Gamma_0 = 0$, that is, $x_{LR} = x_{RL} = 0$ in (4.15). As seen, the noise penalty of nadir removal then becomes negligible.

For smaller systems, waveforms including nadir are motivated by other reasons than just a reduction in transmit power, which is not a so important aspect, since peak power anyhow is small. For instance, waveforms avoiding nadir is at times difficult to realize for imaging geometries, which include steep depression angles. Most importantly, the requirement of continuous waveforms, to be discussed in the next section, makes the inclusion of nadir a necessity.

## 4.5 Emission Adaptation

The requirement for notching of parts of the radar spectrum, which are shared by nearby or prioritized wireless services, was mentioned in Section 4.1. While the effect of such services interfering with the radar operation is simply characterized by (4.2), the levels at which radar transmissions interfere with these services are less direct. For instance, they depend on the antenna characteristics and bandwidth of the victim receiver, as well as the waveform characteristics of the radar. The present account is an elaboration of material previously published in [5] and [6].

Consider a ground-to-ground communication setup. These systems sometimes adopt antennas with high horizontal directivity, suppressing signals from the sky. The radar signal may either be captured by such a far off high-gain communication receiver or a receiver with a low directivity but where the radar is more or less straight above the receiver. Some analysis shows that the worst case is the latter. An obvious relation for defining required notch depth $\nu$ can then be stated as

$$\nu = \frac{1}{kK_{\text{ext}}}\left(\frac{\lambda}{4\pi h}\right)^2 \frac{P_{\text{TX}}}{B} \tag{4.66}$$

As before $k$ is the Boltzmann constant, $K_{\text{ext}}$ is the environmental noise temperature, which in the VHF regime is dominated by man-made noise according to (4.45). As an example, consider the two CARABAS III low-band waveforms, excluding and including nadir, which were discussed in the preceding section. With the altitude of $h$ = 7,000m, mean power requirements of 14W and 6W were found, while the dwell rates of $f_{\text{dwell}}$ = 7.6 kHz and $f_{\text{dwell}}$ = 5.6 kHz were selected. Moreover, assuming $K_{\text{ext}}$ = 120,000° and $\lambda$ = 5m, (5.1) implies the notch depth requirements of $\nu_{\text{excl}}$ = 27 dB and $\nu_{\text{incl}}$ = 23 dB, respectively.

The notch depth determined by (5.1) makes the radar signal nondetectable in a victim receiver, if this signal has the character of random noise. However, the signal types discussed so far are modulated by frequency stepping, with further signal shaping by pulsing and frequency sweep. Such modulation increases the instantaneous impact of interference in the victim receiver, to the degree to which the receiver bandwidth is sufficiently large to sense the modulation details. Let ***N*** designate the inference level above the mean value impact in (5.1). If $B_{\text{vict}}$ is smaller than the recurrence rate $1/t_{\text{rec}}$, then $N = 1$. If $B_{\text{vict}} > 1/t_{\text{rec}}$ (as is usually the case), the receiver will sense the radar signal interference as fluctuating. The fluctuation sets the level of instantaneous interference. The amount of fluctuation depends on the instantaneous radar bandwidth $B_{\text{inst}}$. If $B_{\text{vict}} > B_{\text{inst}}$, then the full power of the radar transmission $P_{\text{TX}}/\eta$ will enter into the victim receiver. Since the full power resides within the instantaneous bandwidth, it follows that $N = B/\eta B_{\text{inst}}$ in this case. In an immediate situation $1/t_{\text{rec}} < B_{\text{vict}} < B_{\text{inst}}$, the instantaneous bandwidth is bandpass filtered by the receiver, resulting in a mitigation of its impact, so $N = B/\eta B_{\text{inst}} \times B_{\text{vict}}/B_{\text{inst}}$. In either case, the required notch depth is $\boldsymbol{N\nu}$.

For a linearly frequency swept modulation (LFM) signal $e^{iCt^2}$ (with $C$ the constant FM-rate), let $t_\omega$ denote the time at which the sweep hits a certain instantaneous frequency $\omega$, that is

$$\left[\frac{d}{dt}\left(Ct^2\right)\right]_{t_\omega} = \omega \Rightarrow t_\omega = \frac{\omega}{2C} \tag{4.67}$$

Given the signal bandwidth $B_{\text{step}}$ and $t_{\text{TX}} = \eta t_{\text{step}}$ radar transmit time for a stepped frequency waveform, it follows that $C = \pi B/\eta t_{\text{dwell}}$. From (5.2), for any change in time $\delta t_\omega$, there is a change in frequency according to $\delta\omega = 2\pi B\delta t_\omega/\eta t_{\text{dwell}}$. In this process of changing frequency, according to Fourier theory, there is an unavoidable frequency spread $\Delta\omega$ such that $\Delta\omega\Delta t_\omega \approx 1$, where $\Delta t_\omega$ is the time it will take for the instantaneous frequency to pass through the spread. The spread is the instantaneous bandwidth, that is, $\Delta\omega = 2\pi B_{\text{inst}}$. It follows that:

$$\begin{aligned} B_{\text{inst}} &= \sqrt{\frac{B_{\text{step}}}{\eta t_{\text{dwell}}}} \\ \Delta t_{\omega-\text{inst}} &= \sqrt{\frac{\eta t_{\text{dwell}}}{B_{\text{step}}}} \end{aligned} \tag{4.68}$$

The results may thus be summarized as

$$\mathrm{N}\nu = \begin{cases} \dfrac{1}{kK_{\mathrm{ext}}}\left(\dfrac{\lambda}{4\pi h}\right)^2 \dfrac{P_{TX-\max}}{B_{\mathrm{inst}}} \Leftarrow B_{\mathrm{inst}} \le B_{\mathrm{vict}} \\ t_{\mathrm{dwell}} B_{\mathrm{vict}} \dfrac{1}{kK_{\mathrm{ext}}}\left(\dfrac{\lambda}{4\pi h}\right)^2 \dfrac{P_{TX}}{B_{step}} \Leftarrow \dfrac{1}{t_{\mathrm{rec}}} \le B_{\mathrm{vict}} \le B_{\mathrm{inst}} \\ \dfrac{1}{kK_{\mathrm{ext}}}\left(\dfrac{\lambda}{4\pi h}\right)^2 \dfrac{P_{TX}}{B} \Leftarrow B_{\mathrm{vict}} \le \dfrac{1}{t_{\mathrm{rec}}} \end{cases} \tag{4.69}$$

In the two examples above, for the excluding and including waveforms, the instantaneous bandwidth is $B_{\mathrm{inst}} = 723$ kHz and $B_{\mathrm{inst}} = 556$ kHz, respectively. Consider a typical communication case with $B_{\mathrm{vict}} = 30$ kHz, covered by the middle equation in (4.69). It is found that $N\nu = 42$ dB and $N\nu = 37$ dB, respectively. An important case where the first equation applies is interference to television reception. For instance, for the European DVB-T/T2 standard, the received bandwidth is $B_{\mathrm{vict}} = 8$ MHz. Then $N\nu = 56$ dB and $N\nu = 49$ dB, respectively. This (somewhat surprisingly) high sensitivity of DVB-T/T2 to interference from radar frequency sweeps was corroborated by a study made by the University of Bristol on the issue of the meter wave radar and communication cohabitation. The University of Bristol specializes in propagation simulations for communication and possess for this purpose extensive and validated simulation tools. Within the EDA SIMCLAIRS Program, these were used to test the impact of unnotched CARABAS signals on DVB-T/T2 OFDM. The impact remained noticeable, even when the radar signal power levels were below the threshold for noise suppression by the error correction used in the DVB-T/T2 technique. Indeed, the success of the error correction depends critically on that competing noise must be white. In the case $B_{\mathrm{vict}} > B_{\mathrm{inst}}$ the radar signal impact in the DVB-T channel is nonuniform, thus violating this requirement. Results are published in [7].

As was mentioned in Section 4.1, radar receive notching to remove narrow band interference is relatively unproblematic and can be carried out to a –80 dB level, or even to even higher levels, relying on modern ADC technology. Notching on the transmit side to the depths here established is significantly more challenging. The main hurdles to be overcome are:

1. In contrast to OFDM television transmitters, which support complex amplitude and phase modulation, the large fractional bandwidth of meter-wave radar transmitters as well as size and efficiency constraints makes transmit linearity a difficult task. As stated before,

power amplifiers are best realized (being then also very straightforward) operating in a saturated mode. However, even such transmitters support phase modulation well, whereas they flatten any amplitude modulation of the signal and thus destroy spectral notching requiring amplitude modulation.

2. As will be demonstrated, sharp spectral notches require nonmonotonic frequency change waveforms.
3. The success of spectral notching depends on duty cycle. In particular, as will be exemplified, for notches of finite bandwidth, the pulse sidelobes of a small duty cycle signal will significantly limit achievable notch depths.
4. The notched waveforms must satisfy the various ambiguity and power efficiency criteria previously laid down. A notching technique is preferable in which a suitable unnotched waveform is developed first, whereupon notches are imposed by a modification of this, according to some algorithm. Doing so, the various waveform criteria should remain closely satisfied.
5. A further aspect on that notching should affect the initial waveform to a minimum extent is that new notches can be added to an existing notched waveform with minimum impact on the quality of previous notches. It will thereby be straightforward to find waveforms also in the case of massive notching.

It has been demonstrated that that randomized transmit signals leads to smaller interference impact on in-band wireless services than waveforms that are frequency swept. It is not known to which extent the former type of waveforms can be devised, in a fashion consistent with saturated transmission and also allowing unambiguous Doppler/range reconstruction. Reflecting this ignorance, they will be left out in this book, whereas waveforms consistent with saturated transmitters will be basically frequency swept waveforms or modifications of these.

To see the requirement for nonmonotonic waveforms, denote by $\phi(t)$ the phase function of a phase modulated signal with unity modulus. For the signal to be bandlimited, $\phi(t)$ must vary smoothly in the sense that within the signal band, the rate of change of the instantaneous frequency $\omega_t = d\phi(t)/dt$ is small over the signal periods. Consider the Fourier coefficients

$$c_{\omega}^{F} = \int_{0}^{T} e^{i[\phi(t)-\omega t]}\, dt \tag{4.70}$$

The integral can be approximated with a sum, each term corresponding to a time interval for which the phase of the integrand remains stationary. These instances arise when $\omega$ agrees with instantaneous frequency, that is, when $\omega_t = \omega$. For a monotonic phase sweep, there will just be one such instance for each $\omega$. In the general case, there can be many. Enumerate these $t_\omega^{(n)}$; $n = 1, 2, \ldots$ and the corresponding time intervals $T_n$. Hence

$$c_\omega^F = \sum_n T_n e^{i\left[\phi\left(t_\omega^{(n)}\right) - \omega t_\omega^{(n)}\right]} \tag{4.71}$$

As a consequence, phase modulation by a monotonic frequency sweep (linear or not) leads to a non-zero modulus spectrum, with no ability to effective spectral notching. If the frequency change is nonmonotonic, (5.6) evidently gives room to provide spectral notches by destructive interference among the several terms then making up $c_\omega$.

For individual frequencies, there are many ways for achieving spectral notches by phase modulation (two examples are given in [8] and [9]). These methods leave open if there are better methods. In contrast, the present "phase ripple" method is derived on the basis of causing a minimum modification of an original given waveform. It thereby satisfies requirements 4 and 5 above. As will be plain from the derivation of the method, the solution furnishing the minimum modification is unique for a given unmodified waveform. However, the waveform to be notched can be selected at will, with some waveforms leading to better notching results than others, as will be seen. The examples given start out from an unnotched LFM waveform. Wideband notching is achieved by modifying the initial linear frequency sweep into making a frequency jump across the notch. The coarse notch thus created is further shaped by applying phase ripple phase modulation.

Developing the phase ripple notching method, let $\Phi$ and $\mu_k$ be real-valued time discrete functions with values $\Phi_n$ and $\mu_{k,n}$, given for an even number of moments of time

$$t_n = \frac{n}{f_s};\ n = 0,1,\ldots,N-1 \tag{4.72}$$

Here $f_s$ is the radar clock frequency. To (5.7), there is an associated set of orthogonal frequencies

$$\omega_k = 2\pi \frac{k}{N} f_s;\ k = -\frac{N}{2}, -\frac{N}{2}+1,\ldots,\frac{N}{2}-1 \tag{4.73}$$

Hence, $\mu_{k,n}$ is some time discrete function, particularly for the frequency $\omega_k$. The phase ripple radar waveform is represented as

$$s_n = a_n e^{i\left(\Phi_n - \sum_{k\in K} c_k \cos\mu_{k,n}\right)} \tag{4.74}$$

Here, $a_n$ is the amplitude. The case of a constant amplitude pulse is represented, assuming $a_n = 1$ if $n \le M$ and $a_n = 0$ otherwise for a given $M \le n$.

The notch is constructed by one function $\mu_{k,n}$ for each $k \in K$, where $K$ is the subset of frequencies within (5.9) for which a notch is required. Henceforward, $\mu_{k,n}$ will be referred to as phase ripple functions and $c_k$ as phase ripple coefficients. In order that the functions shall represent a phase ripple without any amplitude modulation, the coefficients $c_k$ are required to be real, which will be the case henceforward.

The number of notches, that is, the set $K$ of frequency numbers, can be large (it is of the order 1000 in several tests made), whereas the coefficient values $c_k$ must be small (as will be seen in the sequel). In the cases of $K$ large, a first-order series expansion of (10) is a less good approximation. However, the approximation

$$s_n \approx a_n e^{i\Phi_n} \prod_{k\in K}\left(1 - ic_k \cos\mu_{k,n}\right) \tag{4.75}$$

remains accurate, since it only depends on the smallness of $c_k$. For $n \le M$, from (5.10)

$$\begin{aligned} a_n e^{i\Phi_n} \prod_{k\in K}\left(1 - ic_k \cos\mu_{k,n}\right) &= a_n e^{i\Phi_n} X_{k'}\left(1 - ic_{k'} \cos\mu_{k',n}\right) \\ \Leftarrow X_{k'} &= \prod_{k\in K, k\neq k'}\left(1 - ic_k \cos\mu_{k,n}\right) \end{aligned} \tag{4.76}$$

For $n > M$, there is no condition on what values $X_{k'}$ takes.

The Fourier transform $\chi^F$ of the time discrete function $\chi$ is

$$\chi_k^F = \sum_{n=0}^{N-1} \chi_n e^{-i2\pi\frac{kn}{N}} \tag{4.77}$$

The notch requirement is

$$s_k^F = 0 \Leftarrow k \in K \tag{4.78}$$

from which it follows that:

$$c_k = -i\frac{\left(ae^{i\Phi_n}X_k\right)_k^F}{\left(ae^{i\Phi_n}X_k\cos\mu_k\right)_k^F} \tag{4.79}$$

Now, $\mu_k$ is going to be found as the function, which determines $c_k$ as the smallest possible real value satisfying (4.11). It is seen that $\mu_k$ is present in both the numerator and denominator. However, since $c_k$ is small $X_K \approx 1$, so the requirement is approximately satisfied by making the denominator modulus maximally large. This is a requirement for stationary phase of the denominator, implying that the modification of the original phase function is actually an additional arbitrary function, defined as a Fourier series with $c_k$ being its Fourier coefficients. Thus, (4.10) becomes

$$s_n = a_n e^{i\left[\Phi_n - i\sum_{k\in K} c_k \cos\left(2\pi\frac{kn}{N} - \Phi_n + \alpha_k\right)\right]} \tag{4.80}$$

Since $X_{k'}$ depends on all coefficients $c_k$; $k \in K$, expression (4.15) constitutes an equation system for $c_k$. The system is heavily nonlinear and cannot be solved by exact methods. An approximate second order perturbation solution can be obtained, however. Assume initially that when determining each particular $c_{k'}$, the remaining coefficients $c_k$ are all assumed zero. A set of first order coefficients $\hat{c}_k$ is determined from $X_K = 1$ exactly. The value of the denominator is the maximum value $A$. In the case $M = N$ in which there is no pulse modulation

$$\left(ae^{i\Phi}\cos\mu_k\right)_k^F = \frac{Ne^{i\alpha_k}}{2} + \frac{e^{-i\alpha_k}}{2}\left(ae^{i2\Phi}\right)_{2k}^F \approx \frac{Ne^{i\alpha_k}}{2} \tag{4.81}$$

In the case that $M < N$

$$\hat{c}_k = -\frac{2ie^{-i\alpha_k}}{M}\left(ae^{i\Phi}\right)_k^F \tag{4.82}$$

Equation (5.13) will be satisfied (in the approximate sense) independently of the value of $\alpha_k$. Since

$$-i\left(ae^{i\Phi}\right)_{k'}^F = \left|-i\left(ae^{i\Phi}\right)_{k'}^F\right| e^{i\text{angle}\left[-i\left(ae^{i\Phi}\right)_{k'}^F\right]} \tag{4.83}$$

choose

$$\hat{\alpha}_k = \text{angle}\left[-i\left(ae^{i\Phi}\right)^F_{k'}\right] \tag{4.84}$$

With the selection $\alpha_k = \hat{\alpha}_k$ in (17)

$$\hat{c}_k = -\frac{2}{M}\left|\left(e^{i\Phi}\right)^F_k\right| \tag{4.85}$$

Hereby $\hat{c}_k$ is real, as required.

Going to second order, the approximation obtained by (5.19) and (5.20) is used for the determination of $X_k$. From (5.11)

$$\hat{X}_{k'} = \frac{\prod_{k\in K}\left[1 - i\hat{c}_k \cos\left(2\pi\frac{kn}{N} - \Phi_n + \hat{\alpha}_k\right)\right]}{1 - i\hat{c}_{k'}\cos\left(2\pi\frac{kn}{N} - \Phi_n + \hat{\alpha}_{k'}\right)} \tag{4.86}$$

The corrected functions $\mu_k$, coefficients $c_k$, and angles $\alpha_k$ are then calculated according to

$$\begin{aligned}
\mu_k &= 2\pi\frac{kn}{N} - \Phi_n + \alpha_k \\
c_k &= \left|\frac{\left(e^{i\Phi_n}\hat{X}_k\right)^F_k}{\left(e^{i\Phi_n}\hat{X}_k \cos\hat{\mu}_k\right)^F_k}\right| \\
\alpha_k &= \hat{\alpha}_k + \text{angle}\left[-i\frac{\left(e^{i\Phi_n}\hat{X}_k\right)^F_k}{\left(e^{i\Phi_n}\hat{X}_k \cos\hat{\mu}_k\right)^F_k}\right]
\end{aligned} \tag{4.87}$$

Applying the method to various LFM waveforms, and in particular applying it to the CARABAS III high and low bands, the outcome can be summarized:

1. Achievable notch depths depend on duty cycle, notch bandwidth, and (to a lesser degree) the total bandwidth of all notches. Figure 4.15 provides an example where the suppression depth of two 40-kHz notches is plotted, for duty cycle varying. Obviously, the requirement for a small transmit duty cycle and a finite notch bandwidth

is contradictory and degrades or even destroys notch performance, depending on how small the duty cycle is. The result can be understood as due to the sidelobes caused by pulsemodulation around individual frequency points, which block the notching of neighboring frequency points. Thereby, there is a deadlock for notching across bandwidths larger than that of the individual frequency points; only zero bandwidth notches are possible for small duty cycles.

2. In hardware tests of notched waveforms with the CARABAS III system, a restriction is that this system is designed for a maximum of 25% duty cycle. For the notch waveforms still allowable, there was a measured 1- to 2-dB degradation of notch depth compared to theoretical prediction. The slight degradation was expected due to the amplifier phase nonlinearity occurring at the onset of the waveform.

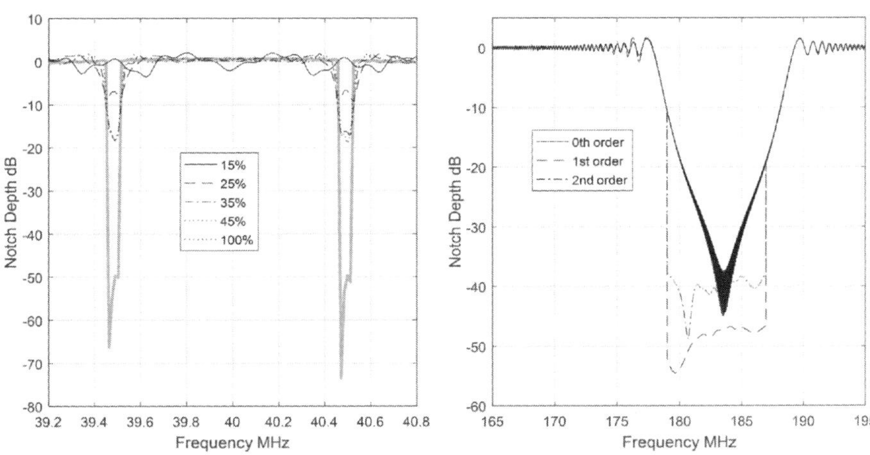


(a) (b)

**Figure 4.15** Phase ripple notch examples. Left: two 40-kHz-wide adjacent second-order phase ripple notches on a CARABAS III low band (27.5 to 82.5 MHz) waveform. Results for different duty cycles are displayed. As seen, even for duty cycle as high as 45%, notch performance is relatively poor and becomes nonexistent for duty cycles less than 15%. Only in the case of continuous operation are deep notches obtained. Right: notching of a typical DVB-T/T2 signal of an 8-MHz bandwidth for a continuous CARABAS III high band (137.5 to 357.5 MHz) waveform. To obtain deep notching using the phase ripple method, the initial frequency swept waveform makes a 10-MHz frequency jump across the notch. As illustrated, this zero-order notched waveform is further shaped by the first- and second-order approximations of the phase ripple coefficients. The varying notch depth results depicted all reflect the approximate character of the notch scheme, which requires that the ripple coefficients are small. Unless the waveform is suitable for implementing the required notch profile, notch depths will be shallow or the notches may possibly not even show up.

3. A 100% duty cycle frequency sweep waveform was used for computing notch depth performance when increasing notch bandwidths and number of notches. For notching 100 randomly dispersed frequencies in the CARABAS III high band, each with a 100-kHz bandwidth, suppression depths of around 28 dB was attained.
4. Results for increasing notch bandwidth indicate high performance for bandwidths ≤ 75 kHz and reasonable performance for bandwidths ≤ 300 kHz. For very large bandwidths like 8 MHz, required for DVB-T/T2, the method requires the initial waveform to be adapted by a frequency jump across the band to be notched. Figure 4.15 shows the result of using this method for one of the DVB-T2 channels in Figure 4.1. Notch depths close to −50 dB are obtained.
5. In general, the second-order correction of the phase ripple coefficients contributes about 10 dB to the suppression depth obtained.

There is a firm implication from the foregoing notching discussion that CW radar is much required for high-performance notching in FOPEN radar. However, the classic FMCW radar principle is not useful, since the bandpass transmit/receive isolation depends on that the transmit signal is instantaneously narrow band. The phase ripple functions creating narrow band notches are active for the entire signal length, that is, 100 $\mu$s in the examples provided. Note that this is not a particular property of this notch method. For mathematical reasons, notch widths ≤ $\Delta f$ require signal durations $\gtrsim 1/\Delta f$. Thus, for notch bandwidths of the order 10 kHz, a signal duration of 100 $\mu$s is required, irrespective of which method is used.

Figure 4.16 shows the basic principle of the previously discussed cancellation radar principle, CAW, designed to enable transmit notching. It has similarities with so-called *full multiplex* communication concepts, in which the transmit signal is canceled in the receiver chain, thus enabling incoming in-band signals, while transmitting. The FOPEN application is the greater challenge, however, in occurring over an octave order bandwidth.

The basic approach for the CAW design is that isolation is achieved in four successive steps (the three cancellation blocks in Figure 4.6 plus pulse compression):

1. An initial isolator, subjected to be highly linear and time stable, and thus accurately characterized by a digital model.
2. A second analog cancellation step, which removes the analog cancellation error of 1 with respect to the digital transmit signal, by making

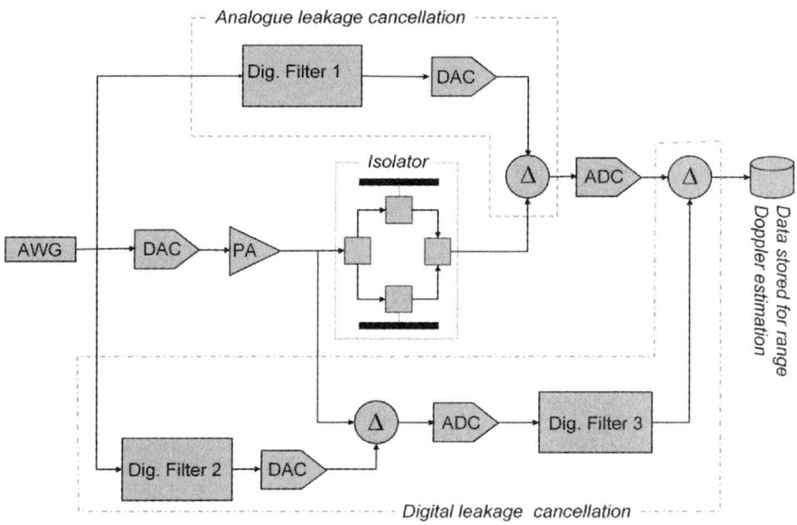


**Figure 4.16** CAW radar principle. In the center is the initial transmit/receive isolation, enabled by a hybrid arrangement for feeding antennas. Initial isolation is followed by analog cancellation, forming a difference signal in a 180° hybrid in the upper loop. The residual error of the analog cancellation is estimated the lower loop by utilizing another 180° hybrid. The error is subtracted digitally whereupon pulse compression (conducted after suitable buffering) removes residual multiplicative errors.

use of that this is accurately known and also that the analog cancellation system characteristics are accurately known.

3. In the output of step 2, the data dynamic range has been reduced to allow AD conversion. Thereupon, a third cancellation stage can be based on the actual transmitted signal being sampled. Errors not accounted for in the transceiver transfer characteristics assumptions in stage 2 are then removed. These errors may be due to additive noise and transmitter nonlinearities.
4. A fourth step of digital cancellation is obtained by pulse compression. Any transmit leakage residuals of multiplicative character stand in proportion to the signal at the moment of transmission. Hence, in the range reconstruction process, they are collected at zero range and are gated out by considering only the relevant radar ranges.

For a narrow-band system, the filters involved would be a mere multiplication of the signals by complex numbers, with phase and magnitude adjusting the signal by the required gain and delay. Presently, filters get more complicated since gain and delay must vary with frequency. The antenna reflections in particular will involve a complicated frequency dependent interaction with the aircraft. However, in the design, this is allowable since the filters are fully digital and the filter coefficients are set by calibration.

Suitably, the isolator in 1 is based on hybrids, being very linear and accurately characterizable. For high performance in stages 2 and 3, the DAC and ADC dynamic range must be large. Since AD and DA conversion is moreover required to occur at least twice the radar bandwidth, the entire design critically relies on high-performance AD and DA conversion.

Laboratory hardware tests of the radar principle have been quite promising; in particular, the anticipated levels of leakage cancellation were reached. The tests emulated CARABAS III high-band operation (137.5 to 357.5 MHz) with 2W continuous transmitted power. ADC sampling rate was 800 MHz and DAC reconstruction rate was 1,600 MHz, while dynamic range was 14 bits. The initial isolator adopted 90° and 180° hybrids to transmit right and left signals, enabling cancellation of the direct transmitter leakage. Resulting initial isolation was better than 30 dB across the band. The isolation prior to pulse compression was measured to be 84 dB. A total isolation of 144 dB was obtained after pulse compression, corresponding to transmit signal noise leakage 23 dB above 300K noise at the receiver input. The receiver internal noise (measured in the absence of transmission) was only 5 dB lower than this value.

The receiver internal noise of the conventionally pulsed transceiver of CARABAS III transceiver is 10 dB lower than the transmit signal noise leakage in the experiment. Memory limitations only allowed for a 10-$\mu$s integration time, whereas 100 $\mu$s is the requirement for a proper system. According to the CAW principles, this should improve transmit receive isolation by a further 10 dB making the CARABAS III and the CAW experimental transceiver performances essentially equal.

Accurate calibration of the different components of the CAW system is required to set the digital filters. Such calibration can be achieved by adding switches and adjustable attenuators to the basic layout in Figure 4.16, allowing that in a calibration mode, the three isolation stages can be characterized individually.

The radar system must be designed to remain temporally stable in between calibrations. Clearly, for low-band operation, the circumstances are favorable for this to be the case. Supposedly, it may be required that for operation in a frequency regime of a few hundred megahertz, some flexing of aircraft and antennas will imply a small spread of the transmit leakage around zero Doppler. To notch a narrow band of radar, data around zero Doppler to remove this leakage is permissible. Below 100 MHz, the flexing amplitudes will be very small compared to wavelength, so such filtering may not be necessary.

## References

[1] Wheeler, H.A., "Fundamental Limitations on Small Antennas," *Proceedings of the IRE*, Vol. 35, No. 12, December 1947, pp. 1479–1484.

[2] Chu, L.J., "Physical Limitations of Omni-Directional Antennas," *Journal of Applied Physics*, Vol. 19, December 1948, pp. 1163–1175.

[3] Nunn, C., and L.R. Moyer, *Spectrally Compliant Waveforms for Wideband Radar*, IEEE Aerospace and Electronics Systems Magazine, Vol. 27, No. 8, August 2012.

[4] International Telecommunication Union, *Radio Noise*, Recommendation ITU-RP.372-11, Geneva, September 2013.

[5] Hellsten, H., and D. Svensson, "Mitigation Methods for FOPEN Radar Impact on Wireless Communications," *2014 Proceedings of the International Telecommunications Symposium*, Sao Paolo, August 2014.

[6] Hellsten, H., et al., "Adaptation of Foliage Penetration Radar to Remove Interference to In-Band Communication," *2015 International IEEE Radar Conference*, Johannesburg, October 2015.

[7] Mellios E., et al., "Low-Frequency Radar Cohabitation with Digital Terrestrial Television," *IEEE Trans. on Broadcasting*, Vol. 59, No. 1, 2013, pp. 84–95.

[8] Selesnick, I. W., and S. Unnikrishna Pillai, "Chirp-Like Transmit Waveforms with Multiple Frequency-Notches," *Proceedings of the IEEE Radar Conference*, May 23–27, 2011.

[9] Gerlach, K., "Spectral Nulling on Transmit via Nonlinear FM Radar Waveforms," *IEEE Trans. on Aerospace and Electronic Systems*, Vol. 47, No. 2, April 2011, pp. 1507–1515.

# 5

# Meter Wavelength SAR Processing

## 5.1 Introduction

In the accounts of low-frequency SAR principles so far given, there still lacks a vital element; namely, a practical method to actually compute the SAR image. Two basic expressions to do so, spectral domain inversion formula (3.32) and time-domain backprojection integral (3.34), have been provided. However, the first expression does not reveal the fact that any actual radar platform will not fly straight and uniformly. The backprojection integral generalizes to expression (3.39), applicable to nonstraight paths, while still approximating the strict inversion formula for straight paths. As some scrutinization of (3.39) shows, in this case, that the difficulty lies in that a direct numerical implementation leads to an overwhelming and impractical computational task. To take a typical example, consider a 5 km × 5 km CARABAS III low-band SAR image. The SAR image path may be 15-km long. To avoid that discretization errors influence radar sensitivity, the image pixel size is selected to be 0.5m, while the radar data is represented with the same density along the SAR path. For the direct implementation of (3.39), the range values from each SAR path sample point must be computed for all pixels in the SAR image. There are $3 \times 10^{12}$ such values. Represented as floating point complex numbers, they amount to be of the order 10 TB of data. Handling this volume of data is a heavy task even for very high performance computers. The computer requirements becomes even more strained by the obvious demand that processing for any practical system should be done in real time, perhaps accepting a processing delay of the order integration time, but certainly with the implication that the processing rate must be on par with the data collection rate. If moreover higher band operation is envisaged, sample densities must be increased, leading to even more severe demands on computational power. In all, while integral (3.39) is a straightforward way of processing small sets of SAR data (typically for test purposes), it is not suitable for massive online processing.

This chapter will specify an approximate method for SAR processing as the SAR path deviates from linear progression due to wind changes and turbulence. Most importantly, in contrast to (3.34), the method is computationally efficient. The method will be referred to as *fast factorized backprojection* (FFBP), in contrast to direct implementation of (3.39), which will be referred to as *global backprojection* (GBP). According to the example above, if the SAR image contains $N$ pixels, the processing burden magnitude for the GBP is $N^3$complex number operations. Adopting FFBP, this number can be reduced to $N^2 \times \log N$, that is, from the order $10^{12}$ to $10^9$. Thereby, the data set in the example can be held within a reasonably large computer RAM, while a compact and affordable computer setup will match the processing burden of a steady-state processing scheme. It is noted the processing burden of an $N \times N$point two-dimensional fast Fourier transform (FFT) is $N^2 \times \log N$. Approximating Hankel transforms by Fourier transforms, the processing burden of spectral inversion (3.32) also becomes $N^2 \times \log N$, that is, the extra computational burden in dealing with nonstraight paths will not be significant.

To be analyzed, the degree of approximation in the method can be kept in control by selecting sampling and pixel densities sufficiently fine. The errors can be further reduced by refined interpolation methods though the analysis will here be restricted to nearest neighbor interpolation.

Only the case of flight geometries deviating from an on average straight and uniform path will be considered. Indeed, a basically nonstraight SAR path, such as the circular, requires knowledge of the ground topography at fractional wavelength accuracy for a fully focused SAR image. The slant range coordinates adopted for the ground, when imaged from a straight path, make the SAR image fully focused independently of topography, given that no instances of fold over or shadowing occurs, as discussed in Chapter 3. For radar platform attempting to fly straight, actual deviations from the straight course will stay relatively small and the defocusing effects of a nonflat ground are normally negligible. The SAR image resulting from the method thus resembles the image that would have been obtained from a straight path, having no deviations from its mean movement. The ground topography still influences the image in causing geometry distortions. Not a subject for this book, ordinary image processing techniques can be applied to interpolate this image into a map-true ground representation by means of a ground elevation data.

The FFBP method has for many years been routinely used in the CARABAS program. It has proved very effective even for quite irregular SAR paths; for instance, the small helicopter used as platform for CARBAS III is severely affected by wind conditions. The SAR images provided in Chapter 3 were registered under conditions of strong side wind gusts. Even so, the FFBP processing technique results in a fully focused SAR images with near ideal RCS sensitivity.

To perform the FFBP processing, the SAR system must be supported by auxiliary aircraft equipment. Methods will be discussed for obtaining the required SAR path geometry parameters from inertial measurement units (IMUs) and/or global navigation satellite systems (GNSS). The basic requirement for fractional wavelength SAR resolution is that the range from each point along the SAR path to each point in the SAR scene must be known within some fraction of a wavelength. While ground topography influence can be neglected, for radar frequencies below 100 MHz, the SAR path shape must be known with an accuracy of a meter or less and with correspondingly higher accuracy at higher frequencies. Knowledge of platform speed is equally critical; relative speed changes must be known to a degree less than the position accuracy divided by SAR path length, that is, must be known to the order $10^{-4}$ or finer for (say) a 10-km path length. Since an IMU basically senses speed and heading changes, the requirements impose stringent requirements on speed and heading drift ratios. A GNSS system senses positions as function of time, and meter and even submeter accuracy falls into what can be met with high-precision GNSS techniques, which, however, are also subjected to drift errors. In the case of GNSS solutions, inertial measurement devices may still be required in addition, since the lever arm from the GNSS antenna to the SAR antenna adds to the SAR path irregularities, as the aircraft maneuvers. CARABAS III adopts this type of navigational technique, combining a phase coherent differential GNSS system (with antennas on the helicopter tail boom) with a simple inertial measurement device, estimating the lever arm movement.

There is also an option of *FFBP autofocus* (called *factorized gradient autofocus*), which relaxes the requirement on navigational accuracy. The main aim in the autofocus development is to obtain an independence of GNSS data, thus making the radar autonomous and entirely relying on the onboard inertial navigation equipment. SAR autofocus is routinely practiced for microwave SAR. In the microwave case, errors in motion measurements impact the SAR image as phase errors in its Fourier transform, on which fact most microwave autofocusing methods are based. For low-frequency SAR, the motion errors cannot be accessed and compensated for in the final image. However, FFBP offers an alternative autofocus option, carrying in itself the possibility for estimation the SAR path geometry parameters, which thus can be determined independently of exterior data. A condition for this autofocusing method to work is that a majority of dominant scatters have been resolved into separate resolution cells. Autofocusing then allows one or two further FFBP iterations, considerably improving resolution. By the relaxed demands of the initially lower resolution and shorter SAR path, the accuracy of standard navigational equipment would suffice.

## 5.2 FFBP Method

Fundamental to the FFBP method is the observation that radar data collected at SAR path points close together determines the low azimuth resolution components of SAR image, whereas the radar data from points wide apart affects the high-resolution SAR image components. From this observation follows the general FFBP idea: the full SAR process can be factorized into an iterative process of merging groups of SAR images given with respect to groups of neighboring linear subapertures. The merged SAR images will have improved resolution corresponding to the merged and thus larger aperture. Factorizing processing allows focusing errors to be coarse initially, and only be refined to the extent that resolution is improving with each iteration stage. As a result, there is a significant computational saving compared to GBP, for which error control is set by resolution of the final SAR image.

Analysis of focusing errors for microwave SAR is generally conducted in terms of phase errors. For ultrawideband imaging, phase has no frequency-independent meaning and analysis ends in requirements for range errors. The analysis itself is conducted for any particular frequency, that is, wavenumber $k = 2\pi/\lambda$ in the radar band. Understand by $\hat{\gamma}_k(r,\phi)$ the ground reflectivity density in slant plane polar coordinates with respect to an origin at the center of a linear SAR path. Let $\gamma_k(y)$ represent the radar data at the position $y$ along the SAR path. Relation (3.22) between reflectivity density and radar data takes in the Fourier domain the simple form

$$\gamma_k(y) = \sum_r \sum_\phi \hat{\gamma}_k(r,\phi) e^{-i2kr_y(r,\phi)} \tag{5.1}$$

Summation is carried out across the SAR image pixels and $r_y(r,\phi)$ is the range from the SAR path point $y$ to the SAR image point $r,\phi$. Given that the subaperture has length $Y$, the SAR image amplitude has according to (3.34) the form (suppressing velocity dependence in (3.34))

$$\chi_k(r,\phi) = \int_{-Y/2}^{Y/2} \gamma_k(y) e^{i2kr_y(r,\phi)} dy \tag{5.2}$$

Represented as (5.2), the SAR image will exhibit angular resolution and manifest in the response of $\chi_k(r,\phi)$ to angular variations $\phi \to \phi'$. In particular, if $Y$ is small, it is expected that $|\phi' - \phi|$ has to be large for $\chi_k(r,\phi')$ to differ significantly from $\chi_k(r,\phi)$. To find the appropriate resolution expression, combine (5.1) and (5.2). For just a single-point reflector, that is, a single term in (5.1)

$$\chi_k(r,\phi') = \hat{\gamma}_k(r,\phi) \int_{-Y/2}^{Y/2} e^{i2k\left[r_y(r,\phi')-r_y(r,\phi)\right]} dy \tag{5.3}$$

From the cosine theorem, the relation $r_y(r,\phi) = \sqrt{r^2 + y^2 - 2ry\sin\phi}$ holds. The difference $\delta r_y(r,\phi) = r_y(r,\phi') - r_y(r,\phi)$ can be estimated first considering $\sqrt{r^2 + y^2 + x}$ for small variations of the variable $x$

$$\delta\sqrt{r^2 + y^2 - x} = \frac{\delta x}{2\sqrt{r^2 + y^2 - x}} \tag{5.4}$$

For $x = 2ry\sin\phi$, variations $\delta x$ due to variations $\delta\phi$ will be small if either $y$ is small compared to $r$ or $\delta\phi$ is small. Both situations will be included here; the case of small $r$ compared to $y$ will not be of interest. Note that even if $\delta\phi$ is as large as 30°, the first-order expansion $\delta\sin\phi = \delta\phi\cos\phi$ remains a fair approximation. In all, if either $y$ or $\delta\phi$ is small, the relation

$$\delta r_y(r,\phi) = -\delta\phi \frac{ry\cos\phi}{r_y(r,\phi)} \approx -y\delta\phi \tag{5.5}$$

approximately holds; the $\cos\phi$ factor has been abolished since angles $\phi$ stay within a ±30° sector.

Insertion of (5.5) into (5.3) yields

$$\chi_k(r,\phi+\delta\phi) \approx 2\hat{\gamma}_k(r,\phi) \int_0^{Y/2} \cos(2ky\delta\phi)\, dy = \hat{\gamma}_k(r,\phi) \frac{\sin(kY\delta\phi)}{k\delta\phi} \tag{5.6}$$

Clearly, (5.5) states the angular spread of the amplitude of a single ground reflector when represented in a SAR image, which has been obtained over an aperture of length $Y$. In particular, considering (5.5) in the limit $\delta\phi \to 0$, the formula states that the amplitude of this ground reflector at the position of the ground reflector is $\chi_k(r,\phi) \approx Y\hat{\gamma}_k(r,\phi)$. Considering angles $\delta\phi \neq 0$, but still not larger than keeping $Y\delta\phi$ small, the below expression follows:

$$\frac{\chi_k(r,\phi+\delta\phi)}{\chi_k(r,\phi)} = \frac{\sin(kY\delta\phi)}{kY\delta\phi} \approx 1 - \frac{1}{6}(kY\delta\phi)^2 \tag{5.7}$$

The formula can be interpreted as stating that the point reflector is represented by a near constant amplitude for angles $\delta\phi$ no larger than to leave $kY\delta\phi$ small. These angles may thus be correspondingly large if the aperture length $Y$ is small. The mean amplitude variation within the angular sector of width $\Phi$ (i.e., $|\delta\phi| \le \Phi/2$) is seen to be $\wp = (kY\Phi/2)^2/18 \approx (Y\Phi/\lambda)^2/2$. The variation becomes the amplitude error introduced if the true values within the sector are approximated by the constant value $\chi_k(r,\phi)$ of the sector center.

The error analysis conducted is fundamental to the FFBP process and implies that the error growth during the FFBP iterations can be kept small by an appropriate choice of angular discretization, that is, bin size (synonymous to pixel size). In particular, it would be possible to perform SAR processing by use of (5.3) in an iterative scheme of merging SAR aperture to obtain increasing lengths $Y$ where in each stage the substitution $\chi_k(r,\phi)/Y \to \hat{\gamma}_k(r,\phi)$ is made. In the early stages, the images can be represented as consisting of very coarse angular sectors, making the computational effort in producing them very small. The subimages have, however, to be obtained for all the short subapertures. In the process of merging apertures, the computational effort for producing each image increases though there will be fewer images to process. As is demonstrated below, the overall computational saving achieved with the method is huge.

For a signal of finite bandwidth, range shifting has a similar defocusing effect as has angular shifting for a signal reconstructed over a finite aperture. Indeed, with $\chi(r,\phi)$ obtained from $\chi_k(r,\phi)$ by a inverse Fourier transform over the bandwidth $B$, from (3.6)

$$\frac{\chi(r+\delta r,\phi)}{\chi(r,\phi)} = e^{i2\omega_c\delta r/c}\frac{\sin 2\pi B\delta r/c}{2\pi B\delta r/c} \approx e^{i2\omega_c\delta r/c}\left[1 - \frac{1}{6}(2\pi B\delta r/c)^2\right] \tag{5.8}$$

Consequently, to keep amplitude errors at the preset acceptable level 0, range discretization should be selected with range bin errors $\wp = (2\pi B\Re/c)^2/72$, that is, bin range widths $\Re = 3c\sqrt{2\wp}/\pi B \approx c\sqrt{2\wp}/B$. In effect, the SAR image range resolution is independent of aperture length and has to be selected the same for small and large apertures. For small apertures, the sectors of nearly constant SAR image amplitude thus have significant angular spread but appear thin in range. For large apertures, the sectors assume an essentially square shape.

To describe the FFBP process in more detail, assume that there are $L + 1$ FFBP iterations. Each stage operates on a data set $X_\ell$; $\ell = 0,\ldots,L$ of SAR image pixel amplitudes with each amplitude $\chi^{\ell}_{m,n,j} \in X_\ell$ labeled by three integer indices. For $j$ fixed, the amplitudes represent a partially resolved SAR image, given in range and angle coordinates and obtained from a subaperture labeled by $j$, as follows from the relations:

$$r = m\Re;\ \phi = 2n\Phi_\ell;\ y = jY_\ell \tag{5.9}$$

For each iteration stage, subaperture length intervals $Y_\ell$ will increase and angular subdivision $\Phi_\ell$ be refined according to the fixed reciprocal relation

$$\Phi_\ell Y_\ell = \lambda\sqrt{2\wp} \tag{5.10}$$

The data input for the first iteration is pulse compressed radar data at the angular resolution of the physical antenna of the SAR aircraft. The data output of the final iteration is the finished image. The initial angular discretization level is typically that of the 60° beam with of the physical SAR antenna, that is, $\Phi_0 = 60° = \pi/3$. The initial along track mean sampling density becomes

$$Y_0 = \lambda\sqrt{2\wp} \tag{5.11}$$

The final angular discretization level will therefore be $\Phi_L = \pi/3N$, with $N$ the number of radar data sample points along the final aperture of length $Y_L$. For a wideband system, $k = 2\pi/\lambda$ in (5.7) is suitably estimated at center frequency. If the bandwidth and center frequencies agrees, it is seen that bin width and along track sample densities agree (i.e., $Y_0 = \Re$). This the case for CARABAS III low band adopting the band 27.5 to 82.5 MHz (i.e., $f_c = B = 55$ MHz). Then, $\Phi_\ell \approx \Re/Y_\ell = 2^{-\ell}$. The CARABAS III high band is set to 137.5 to 357.5 MHz, so in the case $f_c = 247.5$ MHz, while $B = 220$ MHz, so also in this case $Y_0 \approx \Re$, whereas the agreement $\Phi_\ell \approx \Re/Y_\ell = 2^{-\ell}$ remains.

Subaperture merging is assumed to be carried out in groups of $N^{1/L}$ subapertures, where $N^{1/L}$ is rounded to the nearest integer. This integer number will be referred to as the *base* of the merging process. In the merging mean aperture size will increase and angular discretization becomes refined according to

$$Y_\ell = N^{1/L}Y_{\ell-1};\ \Phi_\ell = N^{-1/L}\Phi_{\ell-1} \tag{5.12}$$

In the process, amplitude errors will remain constant according to (5.10). Aperture merging consists of exploiting navigational data to find the appropriate range and angle from each subaperture $j$ to a particular ground point $P$. These will be represented by resolution bin numbers $m(P,j),n(P,j)$ from which $X_\ell$ is computed as the set of amplitudes

$$\chi^{\ell}_{m',n',k'} = \sum_{k=k'}^{k'+N^{1/L}} \chi^{\ell-1}_{m(P,j),n(P,j),j};\ \chi^{\ell-1}_{m(P,j),n(P,j),j} \in X_{\ell-1} \tag{5.13}$$

Let the number of subapertures at iteration level $\ell$ be $A_\ell = N^{(L-\ell)/L}$. For each subaperture, there are $B_\ell = N^{\ell/L}$ beams and thus $NB_\ell = N^{(L+\ell)/L}$ resolution bins. Since each bin amplitude (5.13) is obtained by summing $N^{1/L}$ terms, the processing burden for beam merging at each iteration stage is adding $A_\ell NB_\ell N^{1/L} = N^{(2L+1)/L}$ amplitudes. The total burden for all $L$ stages is consequently $LN^{(2L+1)/L}$ additions. The suitable choice of the number of subapertures to be merged is that which minimizes this number. Note that $N^{1/L} = e^{(\log N)/L}$ and

$$\frac{d}{dL}\left(Le^{(\log N)/L}\right) = 0 \Rightarrow L = \log N \tag{5.14}$$

Since $N$ is large $N^{1/\log N} \approx e$, that is, the minimum processing burden is obtained if subapertures are merged in doublets or triplets, that is, merging base is 2 or 3. There will then be of the order log $N$ iteration stages, whereas the overall SAR processing burden becomes of the order $N^2 \log N$, that is, of the same magnitude as a two-dimensional FFT. The FFT has an exact inverse, whereas the FFBP, even when the SAR path is straight and uniform, remains an approximation of the ground reflectivity distribution.

The processing error for the overall FFBP process is estimated from (5.9). The accumulated amplitude error after $L$ iterations becomes $\wp_{\text{tot}} = 1 - (1 - \wp)^L \approx 1 - e^{L\wp}$. It follows that the FFBP process breaks down entirely if $\wp \approx 1/L$, so $\wp << 1/L$ must be a general requirement. Then, $\wp_{\text{tot}} \approx \wp L$.

Clearly, $\wp^2_{\text{tot}}$ represents the intensity, that is, power processing error. Processing error is one of several multiplicative noise terms contributing to what is known as the *integrated sidelobe level ratio* (ISLR) of the SAR image. The general definition of ISLR is the ratio of multiplicative noise intensity (or power) compared to the mean SAR image intensity (or power). Defocusing, caused by uncompensated motion errors, spreads target responses as a noise floor rather than focusing them to the target footprints. Hence, also motion errors act to increase ISLR. A further strong ISLR source is radio frequency interference notching, since notch sidelobes are very sharp and cause significant sidelobes spreading uniformly across the SAR image. Right/left leakage residuals are further ISLR contributions. Note that ISLR will not necessarily limit SAR sensitivity for a scene, which only contains weak scatterers (i.e., open land not only for the higher frequencies but also forest for frequencies below 100 MHz). The ISLR value will, however, be a limiting factor for detectability of weak targets where there is abundance of much stronger targets. One example is buried structures in cases when the ground surface itself is rough and gives rise to significant back scattering. Another situation is urban environments, since there is an abundance of dihedral scatterers caused by building walls. These constitute very strong scatterers and prevent detectability of targets, which would be detectable under less forbidding conditions.

From practical experience, an ISLR ratio smaller than −15 dB is not to be expected. Requiring processing errors to be less than $\wp^2_{\text{tot}} = -25$ dB, the FFBP method can safely be assumed not to introduce approximation errors of significance. As an example, consider an SAR image consisting of $N \times N = 10^4 \times 10^4$ pixel. Hence, FFBP processing requires $L \approx \log N \approx 10$ iterations of subaperture merging. Aiming for a processing error $\wp^2_{\text{tot}} = -25$ dB, that is, $\wp_{\text{tot}} = 0.06$, a merging error $\wp = 0.006$ is required. Thereby, (5.11) implies $Y_0 = \mathfrak{R} = 0.6$m and $Y_0 = 0.13$m and $\mathfrak{R} = 0.15$m for the CARABAS III low ($\lambda = 5$m at $f_c = 55$ MHz) and high ($\lambda = 1.2$m at $f_c = 247.5$ MHz) bands.

Global backprojection can be considered a special case of FFBP obtained by selecting $L = 1$. The processing burden is (as has been pointed out) $LN^{(2L+1)/L} = N^3$. The selection $L = 2$, often referred to as *local backprojection* (LBP), results in a processing burden $LN^{(2L+1)/L} = N^{5/2}$, that is, it implies a reduction of the processing effort by a factor $\sqrt{N}$ compared to GBP. This reduction is of course quite significant and particularly noteworthy when considering that a fewer number of iteration stages result in smaller processing errors than for the full FFBP. Allowing $\wp_{\text{tot}} \approx 0.06$, as in the example above, for LBP $\wp = 0.03$ is required, compared to $\wp = 0.006$ for FFBP. According to (5.9), discretization of the SAR image into pixels (range and angle combined) can thereby be made five times coarser for LBP than for FFBP (angular and range discretization can be twice as coarse). The FFBP processing effort for the example is $N^2 \log N \approx 10^9$, while, with the allowed coarser discretization, the LBP effort becomes about the same, that is, $(N/\sqrt{5})^{5/2} \approx 10^9$. Therefore, already going from GBP to LBP, the processing effort becomes realizable in a practical computer setup. As will be analyzed in the sequel, the structure of the full FFBP method provides further advantages besides just the processing effort.

The development of the FFBP was initiated by Hellsten et al. [1] by defining the LBP method. Development of the full FFBP possibilities rapidly followed in cooperation between the Swedish Defense Research

Establishment FOI and the University of Linköping (see [2] and [3] for a review). The proprietary rights for FFBP were secured by FOI in several patents. Independently and in parallel, a base 2 FFBP method was developed by McCorkle and Rofheart [4]. In [5], LBP is described as a new method, adopted by MIT Lincoln Laboratory in the US. The paper must not be interpreted as indicating that the method originated at Lincoln Laboratory. Development of the method had commenced at FOI several years before, with the patent US5969662A filed September 21, 1995.

The following section accounts for further developments of FFBP taking place at SAAB and Chalmers University from around 2010 and onward. The section provides a condensed treatment of path irregularities in all three dimensions for base 2 FFBP. The approach opens for autofocusing methods, which are touched upon in Section 5.4. It should also been mentioned that the method has been applied to microwave SAR, as is accounted for in [6–10].

## 5.3 Explicit Treatment of Base 2 FFBP

In an actual SAR processing algorithm (5.25) must be replaced by explicit expressions, which relate radar data to the ground points $P$ adopting SAR path geometry data. If a detailed digital representation of the ground surface is available, this could be used for representing the ground points $P$ in (3.39). As has been pointed out, knowledge of the ground surface shape is not required for a straight path and should not be required for a near straight path either. Generally, a ground slant range representation obtained with respect to the mean heading and speed of the SAR registration suffices in order to obtain the focused SAR image. The present section develops a detailed subaperture merging method based on SAR path geometry data, but not requiring detailed ground topography information. The method restricts merging to subaperture pairs (i.e., base 2 FFBP). Base 3 merging can be set up analogously, while the processing requirements for LBP and GBP are different and the present approach less useful. To keep the book reasonably focused geometry parameter descriptions for LBP and GBP will not be developed. As said, the base 2 FFFB deserves particular attention, since it offers very natural inroads to autofocusing, as will be discussed.

In the iterative scheme of subaperture merging, each merging stage will be evaluated as regards what geometry parameters are required and the accuracy with which they must be known. Since the task of acquiring the navigational data with sufficient accuracy is quite challenging, this analysis, which serves not to overstate the geometry requirements, is clearly important.

Usually, navigational solutions are derived via Kalman filters, which combines inertial and/or GNSS data with their respective error models as well as a platform motion model, to improve navigational accuracy. As will be analyzed in Section 5.5, while crucial for long time navigational accuracy, such Kalman filters hardly affect the path parameters over a minute order SAR registration time. Over short times, a second and less, aircraft dynamics decidedly affects SAR path shape in that aircraft accelerations are low-pass filtered by the aircraft inertia. Hence, straight path deviations stay a negligible fraction of low-frequency SAR wavelength for subsecond subapertures, and motion parameters can be correspondingly sparsely sampled.

Referring to (5.11), for the base 2 FFBP setup, the initial along track number of data samples $N$ should be selected as an even power of 2, that is, $N = 2^L$, with $L + 1$ the number of iteration stages. The SAR path $Q(t)$ can then be represented as a set of affine vectors $Q[(k-1)T_0]Q(kT_0)$; $1 \leq k \leq 2^L$, in which $T_0$ is the sample period at which the pulse compressed radar data has been made available for SAR processing. The path segments will be denoted by $\boldsymbol{Y}_{a1\ldots a_m} = Q[(k-1)T_0]Q(kT_0)$, where $a_1 \ldots a_m$ is the $m$-digit binary representation of the number $k$. Explicitly, such a representation means

$$k = \sum_{j=1}^{m} 2^{(m-j)} a_j \tag{5.15}$$

The binary digits will be denoted "+" for 1 and "−" for 0. Thus, if $L = 10$, there are 1,024 segments, where the 619th one will look like $\boldsymbol{Y}_{+--++-+-++}$. The FFBP approach introduces recursively straightened path segments according to

$$\boldsymbol{Y}_{a_1 \ldots a_m} = \boldsymbol{Y}_{a_1 \ldots a_m -} + \boldsymbol{Y}_{a_1 \ldots a_m +} \tag{5.16}$$

where $1 \leq m \leq L - 1$ and finally

$$\boldsymbol{Y} = \boldsymbol{Y}_{-} + \boldsymbol{Y}_{+} \tag{5.17}$$

It is observed that the end point of $\boldsymbol{Y}_{a_1 \ldots a_m -}$ is the start point of $\boldsymbol{Y}_{a_1 \ldots a_m +}$ and that the endpoint of $\boldsymbol{Y}_{a_1 \ldots -+}$ is the start point of $\boldsymbol{Y}_{a_1 \ldots +-}$. The set of vectors $\boldsymbol{Y}_{a_1 \ldots a_m}$ thereby forms a piecewise linear curve with the linear segments connected in ascending binary order. For each iteration, there is a 50% reduction in the number of segments approximating the SAR path, while at the end of the process, the SAR path has become represented by s single straight path extending between the endpoints of the actual path. Figure 5.1 further illustrates the principle.

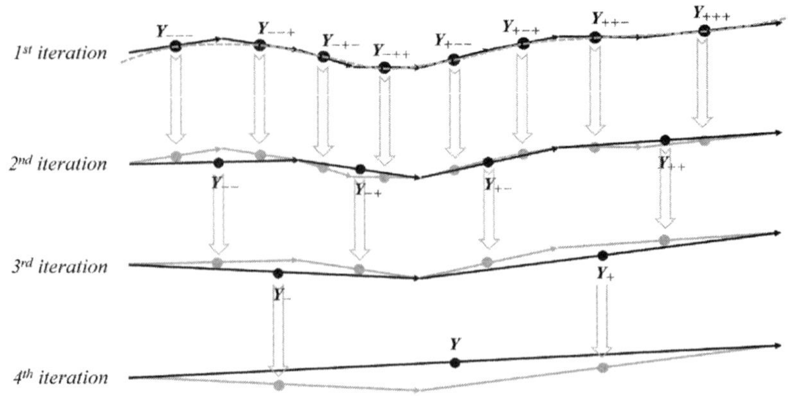


**Figure 5.1** Example of aperture factorization and merging for subaperture mergings in pairs (merging base 2) and for three iterations. The first iteration utilizes along track radar data, sufficiently densely samples so the aircraft track (see the dashed line in the upmost graph) will be smooth in between samples, while the mergings successively straighten the large-scale nonstraight and nonuniformity of the track.

The SAR path is assumed hovering over a ground surface. Let $\boldsymbol{r}$ be the affine vector staring at the midpoint of any segment $\boldsymbol{Y}_{a_1 \ldots a_m}$ and pointing at any ground point. To each $\boldsymbol{Y}_{a_1 \ldots a_m}$, there is attributed the SAR image $\chi_{a_1 \ldots a_m}(\boldsymbol{r})$. The range responses are sampled with a periodicity $T_0$, so initially segments are labeled $\boldsymbol{Y}_{a_1 \ldots a_L}$ and have passage time $T_0$. The final SAR image $\chi(\boldsymbol{r})$ has its origin at the midpoint of $\boldsymbol{Y}$ and corresponds to an overall registration time $2^L T_0$. Angular resolution improves with each recursion step. For the initial slow time range responses, data will not be split into different angular bins and may thus be represented as only depending on range, that is, $\chi_{a_1 \ldots a_L}(\boldsymbol{r}) = \chi_{a_1 \ldots a_L}(|\boldsymbol{r}|)$. It is understood that the right–left discrimination problem has been previously resolved, and that data $\chi_{a_1 \ldots a_L}(|\boldsymbol{r}|)$ stems only from one side of the aircraft (or in fact from the 60° antenna beam pointing to one side of the aircraft).

The subimage merging formulas are

$$\chi_{a_1 \ldots a_m}(\boldsymbol{r}) = \chi_{a_1 \ldots a_m -}(\boldsymbol{r}_-) + \chi_{a_1 \ldots a_m +}(\boldsymbol{r}_+) \tag{5.18}$$

Here, $\boldsymbol{r}_\pm$ and $\boldsymbol{r}$ all point at the same ground point. The explicit transforms $\boldsymbol{r}_\pm \rightarrow \boldsymbol{r}$ required to satisfy this requirement will be developed next.

As mentioned, due to aircraft inertia, the aperture pairs can be assumed to be perfectly straight for the initial aperture mergings. At the stage when the merging process requires compensation for nonuniform flight, the deviations will be assumed to remain fractionally small, but to extend in all directions, that is, to include a possible change of velocity as well horizontal and vertical heading changes. Since ground topography influence is assumed negligible, the ground plane will be assumed flat. The aperture merging parameters will be independent of neighboring apertures and apertures in previous mergings. Hence, the total number of independent SAR path geometry parameters will be that of the total set of geometry parameters for all mergings.

Denote the space positions of the endpoints of the two adjacent aperture endpoints by $Q_1$, $Q_2$, and $Q_3$. Hence, $Y_{a_1\ldots a_m} = Q_1Q_3$, $Y_{a_1\ldots a_m-} = Q_1Q_2$, and $Y_{a_1\ldots a_m+} = Q_2Q_3$. The aperture midpoints, that is, the subimage origins, are denoted by $Q$, $Q_-$, and $Q_+$, respectively. The explicit form of (5.18) should provide formulas enabling the computation of $\boldsymbol{r}_\pm$ for any given $\boldsymbol{r}$ from SAR path geometry data. Thereby, given the merging SAR images, the merged SAR image amplitude can be computed. In vector language, the required transform $\boldsymbol{r}_\pm \rightarrow \boldsymbol{r}$ is simply

$$\boldsymbol{r}_\pm = \boldsymbol{r} - \boldsymbol{D}_\pm \tag{5.19}$$

where the vectors $\boldsymbol{D}_\pm$ go from $Q_\pm$ to $Q$. The plane spanned by the triangle $Q_1Q_2Q_3$, that is, by the subaperture pair to be merged, will be referred to as the *merging plane* and designated $\Pi_{x_1\ldots x_m}$. The parts of $\boldsymbol{r}_\pm$ orthogonal to $\Pi_{a_1\ldots a_m}$ are not affected by (5.19), so the transform (5.19) pertains only to the orthogonal projections $\hat{\boldsymbol{r}}_\pm, \hat{\boldsymbol{r}}$ of $\boldsymbol{r}_\pm$ and $\boldsymbol{r}$ onto this plane, while $\boldsymbol{r}_\pm - \hat{\boldsymbol{r}}_\pm = \boldsymbol{r} - \hat{\boldsymbol{r}}$. It will be convenient to have the SAR images represented with respect to this merging plane, making the required expressions for (3.4) a matter of plane geometry.

The geometry in the merging plane is laid out in Figure 5.2. As seen, the angle at $Q$ in the triangle $PQQ_+$ is $\pi/2 - \hat{\phi} + \delta\phi_{x_1\ldots x_m-}$ and the angle at $Q$ in the triangle $PQ_-Q$ is $\pi/2 + \phi + \delta\phi_{x_1\ldots x_m+}$. Moreover, $|QQ_+| = A_{a_1\ldots a_m-}/2$ and $|QQ_-| = A_{a_1\ldots a_m+}/2$. The cosine theorem determines the modulus of $\hat{\boldsymbol{r}}_\pm$ from $\hat{\boldsymbol{r}}$

$$\hat{r}_\pm = \sqrt{\hat{r}^2 + \frac{1}{4}Y^2_{a_1\ldots a_m\mp} \mp \hat{r}Y_{a_1\ldots a_m\mp}\sin\left(\hat{\phi} \mp \delta\phi_{a_1\ldots a_m\mp}\right)} \tag{5.20}$$

As for angular dependencies of $\hat{\boldsymbol{r}}_\pm$, it is clear from Figure 5.2 that the angle at $Q_+$ in the triangle $PQQ_+$ is $\pi/2 + \hat{\phi}_+ - \Delta\phi_{a_1\ldots a_m}$, while the angle at $Q_-$ the triangle $PQ_-Q_-$ is $\pi/2 - \phi_- - \Delta\phi_{a_1\ldots a_m}$, where $\Delta\phi_{a_1\ldots a_m} = \delta\phi_{x_1\ldots x_m+} + \delta\phi_{x_1\ldots x_m-}$. Hence, by the sine theorem

$$\frac{\cos\left(\hat{\phi} \mp \delta\phi_{a_1\ldots a_m\mp}\right)}{\hat{r}_\pm} = \frac{\cos\left(\hat{\phi}_\pm \mp \Delta\phi_{a_1\ldots a_m}\right)}{\hat{r}} \tag{5.21}$$

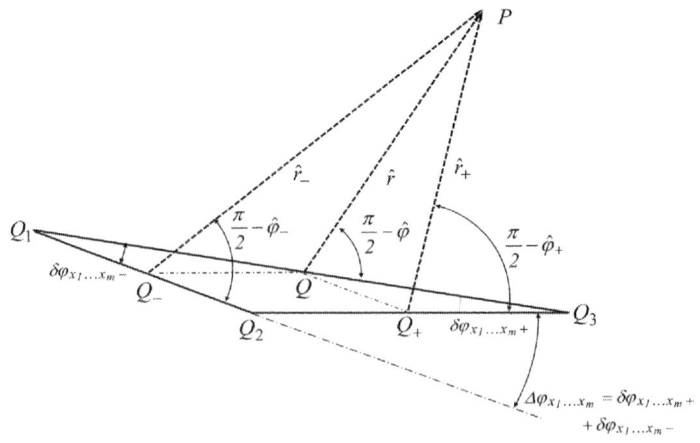

**Figure 5.2** Subaperture merging geometry in the merging plane.

In particular, note the relations

$$\begin{aligned} \hat{r}_{\pm}\cos\left(\hat{\phi}_{\pm} \mp \Delta\phi_{a_1\ldots a_m}\right) &= \hat{r}\cos\left(\hat{\phi} \mp \delta\phi_{a_1\ldots a_m\mp}\right) \\ \hat{r}_{\pm}\sin\left(\hat{\phi}_{\pm} \mp \Delta\phi_{a_1\ldots a_m}\right) &= \hat{r}\sin\left(\hat{\phi} \mp \delta\phi_{a_1\ldots a_m\mp}\right) \mp \frac{1}{2}Y_{a_1\ldots a_m\mp} \end{aligned} \tag{5.22}$$

implied by (5.20) and (5.21). In turn

$$\hat{\phi}_{\pm} = \tan^{-1}\left[\tan\left(\hat{\phi} \mp \delta\phi_{a_1\ldots a_m\mp}\right) \mp \frac{Y_{a_1\ldots a_m\mp}}{2\hat{r}\cos\left(\hat{\phi} \mp \delta\phi_{a_1\ldots a_m\mp}\right)}\right] \pm \Delta\phi_{a_1\ldots a_m} \tag{5.23}$$

Hence, (3.4) is determined upon knowledge of the SAR path parameters $Y_{a_1\ldots a_m\mp}$ and $\delta\phi_{a_1\ldots a_m\mp}$ and $\Delta\phi_{a_1\ldots a_m}$. The angles $\delta\phi_{a_1\ldots a_m\pm}$ are obtained by the sine theorem

$$\frac{\sin\delta\phi_{a_1\ldots a_m\pm}}{Y_{a_1\ldots a_m\mp}} = \frac{\sin\Delta\phi_{a_1\ldots a_m}}{Y_{a_1\ldots a_m}} \Rightarrow \delta\phi_{a_1\ldots a_m\pm} = \sin^{-1}\frac{Y_{a_1\ldots a_m\mp}\sin\Delta\phi_{a_1\ldots a_m}}{Y_{a_1\ldots a_m}} \tag{5.24}$$

Here, $Y_{a_1\ldots a_m}$ are found using the cosine theorem

$$Y_{a_1\ldots a_m} = \sqrt{Y^2_{a_1\ldots a_m-} + Y^2_{a_1\ldots a_m+} + 2Y_{a_1\ldots a_m-}Y_{a_1\ldots a_m+}\cos\Delta\phi_{a_1\ldots a_m}} \tag{5.25}$$

For future reference, the alternative expression to (5.24) is noted as

$$\delta\phi_{a_1\ldots a_m\pm} = \cos^{-1}\frac{Y_{a_1\ldots a_m\pm} + Y_{a_1\ldots a_m\mp}\cos\Delta\phi_{a_1\ldots a_m}}{Y_{a_1\ldots a_m}} \tag{5.26}$$

In all, by means of (5.20) and (5.23) to (5.25), the aperture merging (5.18) can be executed given the lengths $Y_{a_1\ldots a_m\mp}$ of the merging apertures and the angle $\Delta\phi_{a_1\ldots a_m}$ between them. Further simplification of these equations seems not possible as long as obtaining highly accurate trigonometric relations are required. Since $\Delta\phi_{a_1\ldots a_m}$ will be small, $Y_{a_1\ldots a_m\mp}$ nearly equal approximations are of course possible, as will be exemplified in the next section.

The adoption of polar coordinates for merging formulas (5.20) to (5.22) is crucial even though there is an obvious translation of (5.22) into a Cartesian coordinate matrix formulation consisting of $\Delta\phi_{a_1\ldots a_m}$, $\delta\phi_{a_1\ldots a_m\mp}$ rotations, and $Y_{a_1\ldots a_m\mp}/2$ translations. The necessity for polar coordinates rest in the requirement to represent radar data as finely resolved in range and coarsely in angle. The entire computational saving of the FFBP method depends on that this situation is fully exploited and SAR images at each iteration stage are represented at sampling densities corresponding to the angular resolution, which has been attained at this iteration stage.

Since the merging (5.18) pertains only to the orthogonal projections $\hat{\boldsymbol{r}}$ of $\boldsymbol{r}$ onto the merging plane, it would be convenient to express ground points in SAR images by the planar coordinates of this projection. The merging plane may, however, be vertical in which case the coordinate along the normal direction of this is needed for a unique ground point representation. For a general formulation, the merging formulas will be transformed to be valid for planes (to be referred to as *aperture planes*) $\Gamma_{a_1\ldots a_m}$ defined by containing $Y_{a_1\ldots a_m}$ and the horizontal unit base vector $\tilde{\boldsymbol{X}}^B_{a_1\ldots a_m}$ with foot point in $Q$. Let $Y^B_{a_1\ldots a_m}$ be the unit vector along $\boldsymbol{Y}_{a_1\ldots a_m}$ and $\tilde{\boldsymbol{Z}}^B_{a_1\ldots a_m}$ the unit vector orthogonal to the plane $\Gamma_{a_1\ldots a_m}$, both with foot points in $Q$; they are chosen so as to make $\tilde{\boldsymbol{X}}^B_{a_1\ldots a_m}$, $\boldsymbol{Y}^B_{a_1\ldots a_m}$, $\tilde{\boldsymbol{Z}}^B_{a_1\ldots a_m}$ a positively oriented orthonormal base in three-dimensional space with $\tilde{\boldsymbol{Z}}^B_{a_1\ldots a_m}$ pointing upward. Figure 5.3 illustrates the various conventions. The inclination angle $\varepsilon_{a_1\ldots a_m}$ of $Y_{a_1\ldots a_m}$ is also the angle between the aperture plane $\Gamma_{a_1\ldots a_m}$ and the ground. For a near level path, $\varepsilon_{a_1\ldots a_m}$ will be small, just as $\Delta\phi_{a_1\ldots a_m}$ is small for a near straight path. If $\tilde{h}_{a_1\ldots a_m}$ is the distance from $Q$ to the ground along the $\tilde{\boldsymbol{Z}}^B_{a_1\ldots a_m}$ axis, then for the vector pointing to any ground point $P$

$$\mathbf{r} = QP = \tilde{r}\cos\tilde{\phi}\tilde{\mathbf{X}}^B_{a_1\ldots a_m} + \tilde{r}\sin\tilde{\phi}\mathbf{Y}^B_{a_1\ldots a_m} + \left(\tilde{r}\sin\tilde{\phi}\tan\varepsilon_{a_1\ldots a_m} - \tilde{h}_{a_1\ldots a_m}\right)\tilde{\mathbf{Z}}^B_{a_1\ldots a_m} \tag{5.27}$$

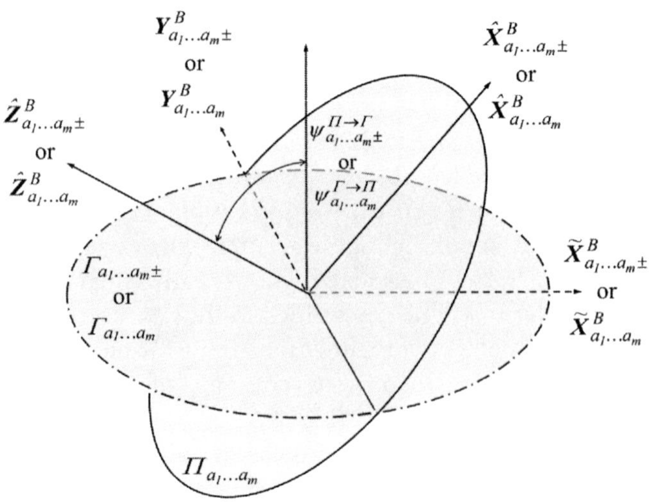


**Figure 5.3** Notational conventions for transforming SAR images between the merging plane $\Gamma_{a_1...a_m}$ and any one of the aperture planes $\Gamma_{a_1...a_m\pm}$, $\Gamma_{a_1...a_m}$. The aperture plane is illustrated as horizontal and the merging plane inclined, although in reality, aperture planes are also inclined. In particular, $\Gamma_{a_1...a_m\pm}$, $\Gamma_{a_1...a_m}$ will not be parallel (with the exception when $\Pi_{a_1...a_m}$ is horizontal).

It follows that with $\tilde{h}_{a_1...a_m}$, $\varepsilon_{a_1...a_m}$ given, the SAR image $\tilde{\chi}_{a_1...a_m}(\boldsymbol{r})$ is determined by the $\tilde{r},\tilde{\phi}$ values alone, thus enabling the simplified notation $\tilde{\chi}_{a_1...a_m}(\tilde{r},\tilde{\phi})$.

For the merging plane $\Pi_{a_1...a_m}$, introduce the unit vector $\hat{X}^B_{a_1...a_m}$, which is orthogonal to $Y_{a_1...a_m}$, by $\hat{X}^B_{a_1...a_m}$, $Y^B_{a_1...a_m}$, $\hat{Z}^B_{a_1...a_m}$, the associated positively oriented orthonormal base with $\hat{Z}^B_{a_1...a_m}$ pointing upward (see Figure 5.3). It is noted that $\Pi_{a_1...a_m}$ share the axis along $Y^B_{a_1...a_m}$ with $\Gamma_{a_1...a_m}$ and the axes along $Y^B_{a_1...a_m\pm}$ with $\Gamma_{a_1...a_m\pm}$. The basis systems for $\Gamma_{a_1...a_m}$, $\Gamma_{a_1...a_m\pm}$, and $\Pi_{a_1...a_m}$ stand in the relations

$$\begin{pmatrix} \tilde{X}^B_{a_1...a_m} \\ Y^B_{a_1...a_m} \\ \tilde{Z}^B_{a_1...a_m} \end{pmatrix} = \begin{pmatrix} \cos\psi^{\Gamma\to\Pi}_{a_1...a_m} & 0 & -\sin\psi^{\Gamma\to\Pi}_{a_1...a_m} \\ 0 & 1 & 0 \\ \sin\psi^{\Gamma\to\Pi}_{a_1...a_m} & 0 & \cos\psi^{\Gamma\to\Pi}_{a_1...a_m} \end{pmatrix} \begin{pmatrix} \hat{X}^B_{a_1...a_m} \\ Y^B_{a_1...a_m} \\ \hat{Z}^B_{a_1...a_m} \end{pmatrix}$$
$$\begin{pmatrix} \tilde{X}^B_{a_1...a_m\pm} \\ Y^B_{a_1...a_m\pm} \\ \tilde{Z}^B_{a_1...a_m\pm} \end{pmatrix} = \begin{pmatrix} \cos\psi^{\Pi\to\Gamma}_{a_1...a_m\pm} & 0 & \sin\psi^{\Pi\to\Gamma}_{a_1...a_m\pm} \\ 0 & 1 & 0 \\ -\sin\psi^{\Pi\to\Gamma}_{a_1...a_m\pm} & 0 & \cos\psi^{\Pi\to\Gamma}_{a_1...a_m\pm} \end{pmatrix} \begin{pmatrix} \hat{X}^B_{a_1...a_m} \\ Y^B_{a_1...a_m} \\ \hat{Z}^B_{a_1...a_m} \end{pmatrix} \quad (5.28)$$

The complete FFB process can be described as a process that for each subaperture during each iteration and for each image coordinate pair $\tilde{r},\tilde{\phi}$ seeks the corresponding image coordinate pairs $\tilde{r}_{\pm},\tilde{\phi}_{\pm}$ by performing the transformations

$$\tilde{r},\tilde{\phi} \rightarrow \tilde{r},\tilde{\phi},\tilde{z} \xrightarrow{\text{rotation}} \hat{r},\hat{\phi},\hat{z} \xrightarrow{\text{merging}} \hat{r}_{\pm},\hat{\phi}_{\pm},\hat{z} \xrightarrow{\text{rotation}} \tilde{r}_{\pm},\tilde{\phi}_{\pm},\tilde{z}_{\pm} \rightarrow \tilde{r}_{\pm},\tilde{\phi}_{\pm} \tag{5.29}$$

The rotations are according to (5.28) and the merging according to (3.6) to (3.8). For the rotation, it is necessary to complement the SAR image with the $\tilde{z}$ coordinate of the ground pointing vector $\boldsymbol{r}$ according to (5.27). The value $\hat{z}$ obtained after the initial rotation is retained during the merging, while affecting $\tilde{r}_{\pm},\tilde{\phi}_{\pm}$ in the final rotation stage.

In polar coordinates, the rotation $\Pi \rightarrow \Gamma$ takes the form

$$\begin{pmatrix} \hat{r}\cos\hat{\phi} \\ \hat{r}\sin\hat{\phi} \\ \hat{z} \end{pmatrix} = \begin{pmatrix} \cos\psi^{\Gamma\rightarrow\Pi}_{a_1\ldots a_m} & 0 & -\sin\psi^{\Gamma\rightarrow\Pi}_{a_1\ldots a_m} \\ 0 & 1 & 0 \\ \sin\psi^{\Gamma\rightarrow\Pi}_{a_1\ldots a_m} & 0 & \cos\psi^{\Gamma\rightarrow\Pi}_{a_1\ldots a_m} \end{pmatrix} \begin{pmatrix} \tilde{r}\cos\tilde{\phi} \\ \tilde{r}\sin\tilde{\phi} \\ \tilde{z} \end{pmatrix}$$
$$\Rightarrow \begin{cases} \hat{r} = \sqrt{\left(\tilde{r}\cos\tilde{\phi}\cos\psi^{\Gamma\rightarrow\Pi}_{a_1\ldots a_m} - \tilde{z}\sin\psi^{\Gamma\rightarrow\Pi}_{a_1\ldots a_m}\right)^2 + \tilde{r}^2\sin^2\tilde{\phi}} \\ \hat{\phi} = \sin^{-1}\dfrac{\tilde{r}\sin\tilde{\phi}}{\hat{r}} \\ \hat{z} = \tilde{r}\cos\tilde{\phi}\sin\psi^{\Gamma\rightarrow\Pi}_{a_1\ldots a_m} + \tilde{z}\cos\psi^{\Gamma\rightarrow\Pi}_{a_1\ldots a_m} \end{cases} \tag{5.30}$$

Similarly (the merging can be completed having found $\tilde{r}_{\pm}$ and $\tilde{\phi}_{\pm}$ and expressions for $\tilde{z}_{\pm}$ are not required)

$$\begin{aligned} \tilde{r}_{\pm} &= \sqrt{\left(\hat{r}\cos\hat{\phi}\cos\psi^{\Pi\rightarrow\Gamma}_{a_1\ldots a_m\pm} - \hat{z}\sin\psi^{\Pi\rightarrow\Gamma}_{a_1\ldots a_m\pm}\right)^2 + \hat{r}^2\sin^2\hat{\phi}} \\ \tilde{\phi}_{\pm} &= \sin^{-1}\frac{\hat{r}\sin\hat{\phi}}{\tilde{r}_{\pm}} \end{aligned} \tag{5.31}$$

Consider the suitable bin (or pixel) sizes in which the SAR image amplitudes are assumed constant in the numerical representation of the SAR images. Since the aperture planes are nearly parallel to the ground plane, the bin shapes can be set to agree in the two planes. Imaging ground at the tilted incidence exploited in SAR, it is observed that ground range resolution will be coarser than range resolution. Aperture bin width can therefore safely be set by the slant range values agreed on in the preceding section, that is, $\mathfrak{R} = c\sqrt{2\varepsilon}/B$. To simplify matters and based on the discussion of the preceding section, it will be assumed that bin width and initial sampling density agrees (i.e., $Y_0 = \mathfrak{R} = \lambda_c\sqrt{2\varepsilon} = c\sqrt{2\varepsilon}/B$). Angular bin widths are selected according to (5.9), that is, $\Phi_\ell \approx \mathfrak{R}/Y_{a_1\ldots a_m} = 2^{-\ell}$.

For the merging plane, discretization will depend on its inclination. In the extreme case of responses infinitely far away or operating at zero altitude, vertical platform position will not influence radar range at all. Correspondingly, discretization in the $\hat{X}^B_{a_1\ldots a_m}$ direction of a vertical merging plane could be arbitrarily coarse. In the opposite case that the merging plane tends is horizontal, resolution is the same as in the aperture plane. Again, the simplest approach is to let bin size be set by range resolution and thus agree with aperture plane bin size, thus selecting $\hat{\mathfrak{R}} = \tilde{\mathfrak{R}}$. The relation $\hat{\Phi}_\ell = 2^{-\ell}$ follows.

For the aperture planes, discretization in the $\tilde{z}$ direction may be set to agree with $\mathfrak{R}$. Since the aperture planes are nearly horizontal, the number of $\tilde{z}$ bins across which the images spreads will be low. The number of $\hat{z}$ bins, being the result of the rotation $\tilde{r},\tilde{\phi},\tilde{z} \rightarrow \hat{r},\hat{\phi},\hat{z}$, can be large if the merging plane tends to become vertical. Computationally this is compensated by a small number of bins in the $\hat{x}$ direction for near vertical merging planes.

While the merging algorithm depends only on the triples of apertures $Y_{a_1\ldots a_m\pm}$ and $\boldsymbol{Y}_{a_1\ldots a_m}$, these must represent an underlying continuous SAR path for the resulting SAR subimages to constructively contribute in the processing. A principle that guarantees this continuity is that next iteration level aperture geometry parameters are derived from the parameters determining the apertures at the present level, with initial apertures obtained from the actual platforms positions during data collection. Basic to the aperture geometry are the parameters $\tilde{h}_{a_1\ldots a_m}$, $Y_{a_1\ldots a_m}$, and $\varepsilon_{a_1\ldots a_m}$ defining each $\boldsymbol{Y}_{a_1\ldots a_m}$. Consequently, these should be determined from the corresponding parameters $\tilde{h}_{a_1\ldots a_m\pm}$, $Y_{a_1\ldots a_m\pm}$, and $\varepsilon_{a_1\ldots a_m\pm}$ of the apertures $Y_{a_1\ldots a_m\pm}$ merging into $Y_{a_1\ldots a_m}$. In addition, the angle $\Delta\phi_{a_1\ldots a_m}$ between the apertures $\boldsymbol{Y}_{a_1\ldots a_m\pm}$ is required in order to settle the merging geometry. Given this angle $\delta\phi_{a_1\ldots a_m\pm}$ and $Y_{a_1\ldots a_m}$ follow from (5.24) and (5.25).

From the requirement that $Y_{a_1\ldots a_m}$, $Y_{a_1\ldots a_m\pm}$ should form a triangle, that is, $Y_{a_1\ldots a_m} = Q_1Q_3$, $\boldsymbol{Y}_{a_1\ldots a_m} = Q_1Q_2$, and $\boldsymbol{Y}_{a_1\ldots a_m+} = Q_2Q_3$, the end points of $\boldsymbol{Y}_{a_1\ldots a_m\pm}$ should join, and particularly so in the vertical direction. It follows that:

$$
\begin{aligned}
&2h_{a_1\ldots a_m+} - Y_{a_1\ldots a_m+}\sin\varepsilon_{a_1\ldots a_m+} = 2h_{a_1\ldots a_m-} + Y_{a_1\ldots a_m-}\sin\varepsilon_{a_1\ldots a_m-} \\
&\qquad \Rightarrow Y_{a_1\ldots a_m+}\sin\varepsilon_{a_1\ldots a_m+} + Y_{a_1\ldots a_m-}\sin\varepsilon_{a_1\ldots a_m-} = 2\left(h_{a_1\ldots a_m+} - h_{a_1\ldots a_m-}\right)
\end{aligned}
\tag{5.32}
$$

The height of the merged aperture $\boldsymbol{Y}_{a_1\ldots a_m\pm}$ follows that:

$$h_{a_1\ldots a_m} = \frac{h_{a_1\ldots a_m+} + h_{a_1\ldots a_m-}}{2} + \frac{Y_{a_1\ldots a_m+}}{4}\sin\varepsilon_{a_1\ldots a_m+} - \frac{Y_{a_1\ldots a_m-}}{4}\sin\varepsilon_{a_1\ldots a_m-} \tag{5.33}$$

The vertical extension of $Y_{a_1\ldots a_m} = Q_1Q_3$ may alternatively be represented as the combined vertical extension of $Y_{a_1\ldots a_m-} = Q_1Q_2$ and $Y_{a_1\ldots a_m+} = Q_2Q_3$, so

$$\varepsilon_{a_1\ldots a_m} = \sin^{-1}\frac{Y_{a_1\ldots a_m+}\sin\varepsilon_{a_1\ldots a_m+} + Y_{a_1\ldots a_m-}\sin\varepsilon_{a_1\ldots a_m-}}{Y_{a_1\ldots a_m}} \tag{5.34}$$

An arbitrary selection of $\tilde{h}_{a_1\ldots a_m\pm}$, $Y_{a_1\ldots a_m\pm}$, and $\varepsilon_{a_1\ldots a_m\pm}$ may violate (5.32) and will thus not correspond to a continuous SAR path. Adhering to the principle that the parameters are the result of previous mergings (5.32) will be satisfied, however.

To express the angles $\psi^{\Pi\to\Gamma}_{a_1\ldots a_m}$ and $\psi^{\Gamma\to\Pi}_{a_1\ldots a_m\pm}$ in the variables $\Delta\phi_{a_1\ldots a_m}$, $Y_{a_1\ldots a_m\pm}$, and $\varepsilon_{a_1\ldots a_m\pm}$, consider first the vertical projection (i.e., the ground plane projection) $Q_1^\perp Q_2^\perp Q_3^\perp$ of the triangle $Q_1Q_2Q_3$. Since $\tilde{\boldsymbol{X}}^B_{a_1\ldots a_m}$ is horizontal as well as orthogonal to $\boldsymbol{Y}^B_{a_1\ldots a_m}$, it is at right angles to the vertical plane through $\boldsymbol{Y}^B_{a_1\ldots a_m}$ and consequently with the ground plane projection of $\boldsymbol{Y}^B_{a_1\ldots a_m}$. The geometry of Figure 5.4 follows, in which $\tilde{\boldsymbol{X}}^B_{a_1\ldots a_m}$ is at right angles to both $Q_1Q_3$ and $Q_1^\perp Q_3^\perp$ just as $\tilde{\boldsymbol{X}}^B_{a_1\ldots a_m-}$ is at right angles to both $Q_1Q_2$ and $Q_1^\perp Q_2^\perp$ and $\tilde{\boldsymbol{X}}^B_{a_1\ldots a_m-}$ to $Q_2Q_3$ and $Q_2^\perp Q_3^\perp$.

The sides of $Q_1^\perp Q_2^\perp Q_3^\perp$ are given by

$$Y^\perp_{a_1\ldots a_m} = Y_{a_1\ldots a_m}\cos\varepsilon_{a_1\ldots a_m} \tag{5.35}$$

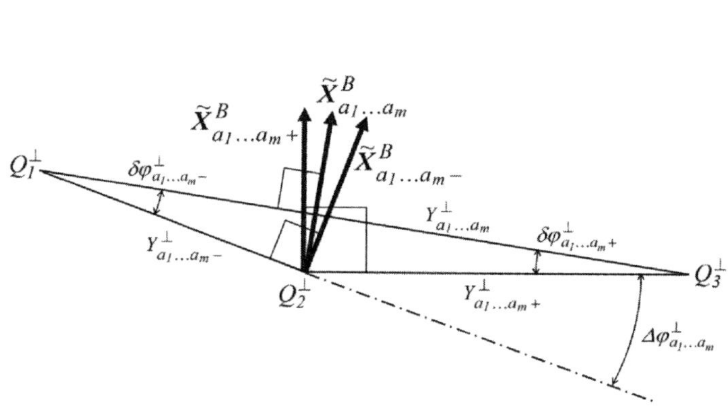


**Figure 5.4** Angular relations pertaining to the ground projected triangle $Q_1^\perp Q_2^\perp Q_3^\perp$.

The angles of $Q_1^{\perp}Q_2^{\perp}Q_3^{\perp}$ follow from the sine and cosine theorems. From the cosine theorem

$$\cos\Delta\phi^{\perp}_{a_1\ldots a_m} = \frac{{Y^{\perp}_{a_1\ldots a_m}}^2 - {Y^{\perp}_{a_1\ldots a_m+}}^2 - {Y^{\perp}_{a_1\ldots a_m-}}^2}{2Y^{\perp}_{a_1\ldots a_m+}Y^{\perp}_{a_1\ldots a_m-}} \tag{5.36}$$

According to (5.32) and (5.33)

$$\Delta\phi^{\perp}_{a_1\ldots a_m} = \cos^{-1}\frac{\cos\Delta\phi_{a_1\ldots a_m} - \sin\varepsilon_{a_1\ldots a_m+}\sin\varepsilon_{a_1\ldots a_m-}}{\cos\varepsilon_{a_1\ldots a_m+}\cos\varepsilon_{a_1\ldots a_m-}} \tag{5.37}$$

which determines $\Delta\phi^{\perp}_{a_1\ldots a_m}$ as a function of $\Delta\phi_{a_1\ldots a_m}, \varepsilon_{a_1\ldots a_m\pm}$. For (5.36) to remain meaningful, the argument of the inverse cosine function must be less than unity; that is, variables of the argument must be subjected to the constraint

$$\Delta\phi_{a_1\ldots a_m} \geq \left|\varepsilon_{a_1\ldots a_m+} - \varepsilon_{a_1\ldots a_m-}\right| \tag{5.38}$$

The inequality remains strict as long as the merging plane is at an oblique angle, while it becomes an equality when the merging plane is vertical.

Use the sine theorem to relate the angles $\delta\phi^{\perp}_{a_1\ldots a_m\pm}$ to $\Delta\phi^{\perp}_{a_1\ldots a_m}$

$$\frac{\sin\delta\phi^{\perp}_{a_1\ldots a_m\pm}}{Y^{\perp}_{a_1\ldots a_m\mp}} = \frac{\sin\Delta\phi^{\perp}_{a_1\ldots a_m}}{Y^{\perp}_{a_1\ldots a_m}} \Rightarrow \delta\phi^{\perp}_{a_1\ldots a_m\pm} = \sin^{-1}\frac{Y^{\perp}_{a_1\ldots a_m\mp}\sin\Delta\phi^{\perp}_{a_1\ldots a_m}}{Y^{\perp}_{a_1\ldots a_m}} \tag{5.39}$$

From (5.25), (5.34), and (5.37), $\delta\phi^{\perp}_{a_1\ldots a_m\pm}$ are obtained in the independent variables $\Delta\phi^{\perp}_{a_1\ldots a_m}$, $Y_{a_1\ldots a_m\pm}$, and $\varepsilon_{a_1\ldots a_m\pm}$. Explicit expressions for $\delta\phi^{\perp}_{a_1\ldots a_m\pm}$ are lengthy so $\delta\phi^{\perp}_{a_1\ldots a_m\pm}$ are better computed using (5.38), with $Y^{\perp}_{a_1\ldots a_m\pm}$, $Y^{\perp}_{a_1\ldots a_m}$, and $\cos\Delta\phi^{\perp}_{a_1\ldots a_m}$ computed separately. The case of a vertical merging plane is noted to correspond to $Y^{\perp}_{a_1\ldots a_m} = Y^{\perp}_{a_1\ldots a_m+} + Y^{\perp}_{a_1\ldots a_m-}$, while $\Delta\phi^{\perp}_{a_1\ldots a_m} = 0$, so $\delta\phi^{\perp}_{a_1\ldots a_m\pm} = 0$ as expected.

Consider $\psi^{\Gamma\to\Pi}_{a_1\ldots a_m+}$, that is, the angle of rotation from $\tilde{\boldsymbol{X}}^{B}_{a_1\ldots a_m\pm}$ to $\hat{\boldsymbol{X}}^{B}_{a_1\ldots a_m\pm}$, around the direction $\boldsymbol{Y}^{B}_{a_1\ldots a_m+}$. Now, $\boldsymbol{Y}^{B}_{a_1\ldots a_m-} \times \boldsymbol{Y}^{B}_{a_1\ldots a_m+}$ is directed along $\hat{\boldsymbol{Z}}^{B}_{a_1\ldots a_m}$, so $\boldsymbol{Y}^{B}_{a_1\ldots a_m+} \times (\boldsymbol{Y}^{B}_{a_1\ldots a_m-} \times \boldsymbol{Y}^{B}_{a_1\ldots a_m+})$ is directed along $\hat{\boldsymbol{X}}^{B}_{a_1\ldots a_m+}$. Hence, there is a relation

$$\begin{aligned}\cos\psi^{\Gamma\to\Pi}_{a_1\ldots a_m+} &= \frac{\tilde{\mathbf{X}}^{B}_{a_1\ldots a_m+} \bullet \left[\mathbf{Y}^{B}_{a_1\ldots a_m+} \times \left(\mathbf{Y}^{B}_{a_1\ldots a_m-} \times \mathbf{Y}^{B}_{a_1\ldots a_m+}\right)\right]}{\left|\mathbf{Y}^{B}_{a_1\ldots a_m+} \times \left(\mathbf{Y}^{B}_{a_1\ldots a_m+} \times \mathbf{Y}^{B}_{a_1\ldots a_m-}\right)\right|} \\ &= \frac{\tilde{\mathbf{X}}^{B}_{a_1\ldots a_m+} \bullet \mathbf{Y}^{B}_{a_1\ldots a_m-}}{\sin\Delta\phi_{a_1\ldots a_m}}\end{aligned} \tag{5.40}$$

The scalar product $-\tilde{\boldsymbol{X}}^B_{a_1\ldots a_m+} \bullet \boldsymbol{Y}^B_{a_1\ldots a_m-}$ is assessed as the ground projection $\boldsymbol{Y}^{B\perp}_{a_1\ldots a_m-}$ projected onto $-\tilde{\boldsymbol{X}}^B_{a_1\ldots a_m+}$. According to the orthogonality relations of Figure 5.4, the angle between $\boldsymbol{Y}^{B\perp}_{a_1\ldots a_m-}$ and $-\tilde{\boldsymbol{X}}^B_{a_1\ldots a_m+}$ is $\pi/2 - \Delta\phi^{\perp}_{a_1\ldots a_m}$, that is, $-\tilde{\boldsymbol{X}}^B_{a_1\ldots a_m+} \bullet \boldsymbol{Y}^B_{a_1\ldots a_m-} = \sin\Delta\phi^{\perp}_{a_1\ldots a_m}\cos\varepsilon_{a_1\ldots a_m-}$. Similarly, arguing regarding $\psi^{\Gamma\to\Pi}_{a_1\ldots a_m-}$

$$\psi^{\Gamma\to\Pi}_{a_1\ldots a_m\pm} = \cos^{-1}\frac{\sin\Delta\phi^{\perp}_{a_1\ldots a_m}\cos\varepsilon_{a_1\ldots a_m\mp}}{\sin\Delta\phi_{a_1\ldots a_m}} \tag{5.41}$$

Coming to $\psi^{\Pi\to\Gamma}_{a_1\ldots a_m}$, this is the angle of rotation from $\hat{\boldsymbol{X}}^B_{a_1\ldots a_m}$ to $\tilde{\boldsymbol{X}}^B_{a_1\ldots a_m}$ around the direction $\boldsymbol{Y}^B_{a_1\ldots a}$. It is clear that $\boldsymbol{Y}_{a_1\ldots a_m} \times (\boldsymbol{Y}^B_{a_1\ldots a_m-} \times \boldsymbol{Y}^B_{a_1\ldots a_m+})$ is directed along $\hat{\boldsymbol{X}}^B_{a_1\ldots a_m}$. Invoking (5.16)

$$\cos\psi^{\Pi\to\Gamma}_{a_1\ldots a_m} = \frac{\tilde{\boldsymbol{X}}^B_{a_1\ldots a_m} \bullet \left[\left(\boldsymbol{Y}^B_{a_1\ldots a_m-} + \boldsymbol{Y}^B_{a_1\ldots a_m+}\right)\times\left(\boldsymbol{Y}^B_{a_1\ldots a_m-} \times \boldsymbol{Y}^B_{a_1\ldots a_m+}\right)\right]}{\left|\left(\boldsymbol{Y}^B_{a_1\ldots a_m-} + \boldsymbol{Y}^B_{a_1\ldots a_m+}\right)\times\left(\boldsymbol{Y}^B_{a_1\ldots a_m-} \times \boldsymbol{Y}^B_{a_1\ldots a_m+}\right)\right|}$$
$$= \frac{\begin{matrix}\tilde{\boldsymbol{X}}^B_{a_1\ldots a_m} \bullet \boldsymbol{Y}^B_{a_1\ldots a_m-}\left(Y_{a_1\ldots a_m+} + Y_{a_1\ldots a_m-}\cos\Delta\phi_{a_1\ldots a_m}\right) \\ -\tilde{\boldsymbol{X}}^B_{a_1\ldots a_m} \bullet \boldsymbol{Y}^B_{a_1\ldots a_m+}\left(Y_{a_1\ldots a_m-} + Y_{a_1\ldots a_m+}\cos\Delta\phi_{a_1\ldots a_m}\right)\end{matrix}}{Y_{a_1\ldots a_m}\sin\Delta\phi_{a_1\ldots a_m}} \tag{5.42}$$

Again referring to Figure 5.4, the angles between $\pm\boldsymbol{Y}^{B\perp}_{a_1\ldots a_m\pm}$ and $-\tilde{\boldsymbol{X}}^B_{a_1\ldots a_m}$ are seen to be $\pi/2 - \delta\phi^{\perp}_{a_1\ldots a_m\pm}$, that is, $\tilde{\boldsymbol{X}}^B_{a_1\ldots a_m} \bullet \boldsymbol{Y}^B_{a_1\ldots a_m\pm} = \sin\delta\phi^{\perp}_{a_1\ldots a_m\pm}\cos\varepsilon_{a_1\ldots a_m\pm}$. Also using (5.26)

$$\psi^{\Pi\to\Gamma}_{a_1\ldots a_m} =$$
$$\cos^{-1}\frac{\sin\delta\phi^{\perp}_{a_1\ldots a_m-}\cos\delta\phi_{a_1\ldots a_m+}\cos\varepsilon_{a_1\ldots a_m-} + \sin\delta\phi^{\perp}_{a_1\ldots a_m+}\cos\delta\phi_{a_1\ldots a_m-}\cos\varepsilon_{a_1\ldots a_m+}}{\sin\Delta\phi_{a_1\ldots a_m}} \tag{5.43}$$

In (5.41), $\Delta\phi^{\perp}_{a_1\ldots a_m}$ is computed from (5.37), and in (5.43), $\delta\phi_{a_1\ldots a_m+}$ is computed from (5.39). Hence, $\psi^{\Gamma\to\Pi}_{a_1\ldots a_m\pm}$, just as $\Delta\phi^{\perp}_{a_1\ldots a_m}$, depends on the three independent variables $\Delta\phi^{\perp}_{a_1\ldots a_m}$, $\varepsilon_{a_1\ldots a_m\pm}$, while $\psi^{\Pi\to\Gamma}_{a_1\ldots a_m}$, just as $\delta\phi^{\perp}_{a_1\ldots a_m\pm}$, depends on all five variables $\Delta\phi_{a_1\ldots a_m}$, $Y_{a_1\ldots a_m\pm}$, $\varepsilon_{a_1\ldots a_m\pm}$. Just to illustrate the formulas, assume $\Delta\phi^{\perp}_{a_1\ldots a_m} = 5°$, $Y_{a_1\ldots a_m-} = 1.2$ km, $Y_{a_1\ldots a_m+} = 1$ km, $\varepsilon_{a_1\ldots a_m-} = -1°$, and $\varepsilon_{a_1\ldots a_m+} = 3°$. Then $Y_{a_1\ldots a_m} = 2.1979$ km and $\varepsilon_{a_1\ldots a_m} = 0.8184°$, while $\psi^{\Gamma\to\Pi}_{a_1\ldots a_m-} = 53.1388°$, $\psi^{\Gamma\to\Pi}_{a_1\ldots a_m+} = 53.0864°$, and $\psi^{\Pi\to\Gamma}_{a_1\ldots a_m} = 53.1410°$. The near equality of the angles between aperture and merging planes is expected since the subapertures are nearly colinear, as indeed is the situation for actual SAR paths.

The formulas derived define the parameters required for aperture merging. They provide the full three-dimensional representation of an arbitrary nonlinear aircraft trajectory broken down into subapertures, determined by geometry variables, which can be acquired by suitable aircraft positioning equipment. Positioning techniques will be treated in Section 5.5.

## 5.4 Sensitivity to Motion Errors

Before discussing the positioning techniques, the sensitivity to geometry parameter errors needs to be investigated. It is noted that splitting SAR processing in iterative subaperture mergings, as in the FFBP method, provides an efficient inroad in studying path errors arising at different scales, that is, both the short-term irregularity errors and longtime large aperture positioning errors. As will be a topic in Section 5.5, time and distance are crucial factors for both inertial and GNSS measurement accuracy.

As is assumed, the aim is to fly a linear trajectory making a linear path, parallel to the ground, a fair approximation of the actual path. To first order the geometry, measurement errors have the same defocusing effect when processing for a nonstraight path, as measurement errors of the same magnitude when processing for a perfectly straight path. Thereby, the error analysis can be carried out by applying the FFBP merging formulas derived on first-order deviations from a linear path.

The transformation $\tilde{r},\tilde{\phi} \to \tilde{r}_{\pm},\tilde{\phi}_{\pm}$ given by (5.20) and (5.23) assumes that there are no measurement errors. If the same transformation is applied when measurement errors prevail, the result will be that in the merging process, the coordinates $\tilde{r}_{\pm},\tilde{\phi}_{\pm}$ obtained correspond to values $\tilde{r} + \Delta_{\pm}\tilde{r}$, and $\tilde{\phi} + \Delta_{\pm}\tilde{\phi}$, which are offset by displacements $\Delta_{\pm}\tilde{r},\Delta_{\pm}\tilde{\phi}$ from the original values $\tilde{r},\tilde{\phi}$. If it happens that $(\Delta_{+}\tilde{r},\Delta_{+}\tilde{\phi}) = (\Delta_{-}\tilde{r},\Delta_{-}\tilde{\phi})$, such displacements will cause the merged image to distort, but not to defocus. As long image distortions remain small they are acceptable; the influence of topography will anyway introduce such a distortion (but no defocusing) as was concluded in Chapter 3. The requirement for path measurement error control will rather be that these errors must not give rise to defocusing. Accordingly, the SAR image amplitudes $\tilde{r}_\pm, \tilde{\phi}_\pm$ must be separated no further than what is acceptable as regards image amplitude errors in the merging process:

$$\begin{aligned} &\tilde{\phi}_+ \to \phi + \Delta_+\tilde{\phi};\ \phi_- \to \tilde{\phi} + \Delta_-\tilde{\phi};\ D\tilde{\phi} = \Delta_+\tilde{\phi} - \Delta_-\tilde{\phi}, \left| D\tilde{\phi} \right| \le \Phi_\ell \\ &\tilde{r}_+ \to \tilde{r} + \Delta_+\tilde{r};\ \tilde{r}_- \to \tilde{r} + \Delta_-\tilde{r};\ D\tilde{r} = \Delta_+\tilde{r} - \Delta_-\tilde{r}, \left| D\tilde{r} \right| \le \Re \end{aligned} \tag{5.44}$$

The parameters $D\tilde{r}$ and $D\tilde{\phi}$ are the focusing range and angular errors arising in the merging process, respectively. Consider first (5.44) restricted to the merging plane. From Figure 5.2, the following inverse representations of (5.20) and (5.23) are evident:

$$\begin{aligned} \hat{r} &= \sqrt{\hat{r}_\pm^{\,2} + \frac{1}{4} Y_{a_1 \ldots a_m \mp}^{\;\;2} \pm \hat{r}_\pm Y_{a_1 \ldots a_m \mp} \sin\left( \phi_\pm \mp \Delta\phi_{a_1 \ldots a_m} \right)} \\ \hat{\phi} &= \tan^{-1}\left[ \tan\left( \hat{\phi}_\pm \mp \Delta\phi_{a_1 \ldots a_m} \right) \pm \frac{Y_{a_1 \ldots a_m \mp}}{2\hat{r}_\pm \cos\left( \hat{\phi}_\pm \mp \Delta\phi_{a_1 \ldots a_m} \right)} \right] \pm \delta\phi_{a_1 \ldots a_m \mp} \end{aligned} \tag{5.45}$$

Expressions for $\hat{r} + \Delta_\pm \hat{r}, \hat{\phi} + \Delta_\pm \hat{\phi}$ are sought as a first-order series expansion of (5.45) with speed and heading changes along the merging subapertures given by

$$\begin{aligned} Y_{a_1 \ldots a_m \pm} &= \frac{Y_{a_1 \ldots a_m} \pm \Delta Y_{a_1 \ldots a_m \pm}}{2} \\ \delta\phi_{a_1 \ldots a_m +} = \delta\phi_{a_1 \ldots a_m -} &= \frac{\Delta\phi_{a_1 \ldots a_m}}{2} \end{aligned} \tag{5.46}$$

where $|\Delta Y_{a_1 \ldots a_m \pm}| << Y_{a_1 \ldots a_m \pm}$ and $\Delta\phi_{a_1 \ldots a_m} << 1$ now will be considered as speed and heading measurement errors for a fundamentally straight and uniform SAR path. Hence, $\tilde{r}_\pm, \tilde{\phi}_\pm$ are given by (5.20) and (5.23) restricted to a straight path. It follows that $\Delta_\pm \tilde{r}, \Delta_\pm \tilde{\phi}$ can be approximated as:

$$\begin{aligned} \Delta_\pm \hat{r} &= \frac{\Delta Y_{a_1 \ldots a_m \mp}}{2} \left( \frac{\partial \hat{r}}{\partial Y_{a_1 \ldots a_m \mp}} \right)_0 + \Delta\phi_{a_1 \ldots a_m} \left( \frac{\partial \hat{r}}{\partial \hat{\phi}_\pm} \right)_0 \\ \Delta_\pm \hat{\phi} &= \frac{\Delta Y_{a_1 \ldots a_m \mp}}{2} \left( \frac{\partial \hat{\phi}}{\partial Y_{a_1 \ldots a_m \mp}} \right)_0 \mp \Delta\phi_{a_1 \ldots a_m} \left[ \left( \frac{\partial \hat{\phi}}{\partial \hat{\phi}_\pm} \right)_0 - \frac{1}{2} \right] \end{aligned} \tag{5.47}$$

where the "0" index understands the straight path conditions, that is, $Y_{a_1\ldots a_m\pm} = Y_{a_1\ldots a_m}/2$ and $\Delta\phi_{a_1\ldots a_m} = 0$. To first order in the range and angular focusing errors, (5.49) becomes

$$
\begin{aligned}
D\hat{r} &= \frac{\Delta Y_{a_1\ldots a_m+}}{2}\left(\frac{\partial \hat{r}}{\partial Y_{a_1\ldots a_m-}}\right)_0 - \frac{\Delta Y_{a_1\ldots a_m-}}{2}\left(\frac{\partial \hat{r}}{\partial Y_{a_1\ldots a_m-}}\right)_0 + \Delta\phi_{a_1\ldots a_m}\left[\left(\frac{\partial \hat{r}}{\partial \hat{\phi}_+}\right)_0 - \left(\frac{\partial \hat{r}}{\partial \hat{\phi}_-}\right)_0\right] \\
D\hat{\phi} &= \frac{\Delta Y_{a_1\ldots a_m+}}{2}\left(\frac{\partial \hat{\phi}}{\partial Y_{a_1\ldots a_m-}}\right)_0 - \frac{\Delta Y_{a_1\ldots a_m-}}{2}\left(\frac{\partial \hat{\phi}}{\partial Y_{a_1\ldots a_m-}}\right)_0 - \Delta\phi_{a_1\ldots a_m}\left[\left(\frac{\partial \hat{\phi}}{\partial \hat{\phi}_+}\right)_0 + \left(\frac{\partial \hat{\phi}}{\partial \hat{\phi}_-}\right)_0 - 1\right]
\end{aligned}
\tag{5.48}
$$

It will be useful to refine $\Delta Y_{a_1\ldots a_m\pm}$ into mean and gradient velocity errors $\Delta_M Y_{a_1\ldots a_m}$ and $\Delta_G Y_{a_1\ldots a_m}$, thus requiring

$$
\begin{aligned}
\Delta_M Y_{a_1\ldots a_m} &= \frac{\Delta Y_{a_1\ldots a_m+} + \Delta Y_{a_1\ldots a_m-}}{2} \\
\Delta_G Y_{a_1\ldots a_m} &= \frac{\Delta Y_{a_1\ldots a_m+} - \Delta Y_{a_1\ldots a_m-}}{2}
\end{aligned}
\Rightarrow
\begin{aligned}
\Delta Y_{a_1\ldots a_m+} &= \Delta_M Y_{a_1\ldots a_m} + \Delta_G Y_{a_1\ldots a_m} \\
\Delta Y_{a_1\ldots a_m-} &= \Delta_M Y_{a_1\ldots a_m} - \Delta_G Y_{a_1\ldots a_m}
\end{aligned}
\tag{5.49}
$$

The derivatives computed from (5.45) become

$$
\begin{aligned}
\left(\frac{\partial \hat{r}}{\partial Y_{a_1\ldots a_m\mp}}\right)_0 &= \frac{Y_{a_1\ldots a_m} \pm 4\hat{r}_\pm \sin\hat{\phi}_\pm}{8\hat{r}} \\
\left(\frac{\partial \hat{r}}{\partial \hat{\phi}_\pm}\right)_0 &= \pm\frac{\hat{r}_\pm Y_{a_1\ldots a_m}\cos\hat{\phi}_\pm}{4\hat{r}} \\
\left(\frac{\partial \hat{\phi}}{\partial Y_{a_1\ldots a_m\mp}}\right)_0 &= \pm\frac{8\hat{r}_\pm \cos\hat{\phi}_\pm}{16\hat{r}_\pm^2\cos^2\hat{\phi}_\pm + \left(4\hat{r}_\pm \sin\hat{\phi}_\pm \pm Y_{a_1\ldots a_m}\right)^2} \\
\left(\frac{\partial \hat{\phi}}{\partial \hat{\phi}_\pm}\right)_0 &= \frac{16\hat{r}_\pm^2 \pm 4\hat{r}_\pm Y_{a_1\ldots a_m}\sin\hat{\phi}_\pm}{16\hat{r}_\pm^2\cos^2\hat{\phi}_\pm + \left(4\hat{r}_\pm \sin\hat{\phi}_\pm \pm Y_{a_1\ldots a_m}\right)^2}
\end{aligned}
\tag{5.50}
$$

The parameters $\tilde{r}_\pm, \tilde{\phi}_\pm$ can be reexpressed in $\tilde{r}, \tilde{\phi}$ by means of (5.20) to (5.23) restricted to $Y_{a_1\ldots a_m\pm} = Y_{a_1\ldots a_m}/2$ and $\Delta\phi_{a_1\ldots a_m} = 0$. Thereby, (5.50) simplifies considerably to

$$\left(\frac{\partial \hat{r}}{\partial Y_{a_1\ldots a_m\mp}}\right)_0 = \frac{\sin\hat{\phi}}{2}$$
$$\left(\frac{\partial \hat{r}}{\partial \phi_\pm}\right)_0 = \pm\frac{Y_{a_1\ldots a_m}\cos\hat{\phi}}{4}$$
$$\left(\frac{\partial \hat{\phi}}{\partial Y_{a_1\ldots a_m\mp}}\right)_0 = \pm\frac{\cos\hat{\phi}}{2\hat{r}} \tag{5.51}$$
$$\left(\frac{\partial \hat{\phi}}{\partial \hat{\phi}_\pm}\right)_0 - \frac{1}{2} = \frac{1}{2} \mp \frac{Y_{a_1\ldots a_m}\sin\hat{\phi}}{4\hat{r}}$$

From insertion of (5.49) and (5.51) into (5.48), the following first-order expressions are obtained for range and angular focusing errors:

$$D\hat{r} = \Delta_G Y_{a_1\ldots a_m}\frac{\sin\hat{\phi}}{2} + \Delta\phi_{a_1\ldots a_m}\frac{Y_{a_1\ldots a_m}\cos\hat{\phi}}{2}$$
$$D\hat{\phi} = \Delta_M Y_{a_1\ldots a_m}\frac{\cos\hat{\phi}}{2\hat{r}} - \Delta\phi_{a_1\ldots a_m} \tag{5.52}$$

To capture the basic requirements, for motion measurement accuracy, (5.52) is applied to the cases of horizontal and vertical merging planes. First consider the horizontal case in which $\psi^{\Gamma\to\Pi}_{a_1\ldots a_m\pm} = -\psi^{\Pi\to\Gamma}_{a_1\ldots a_m} = 0$, that is, $\hat{r} = \tilde{r}, \hat{\phi} = \tilde{\phi}$, $\hat{r}_\pm = \tilde{r}_\pm, \hat{\phi}_\pm = \tilde{\phi}_\pm$ and $D\hat{r} = D\tilde{r}$ , $D\hat{\phi} = D\tilde{\phi}$ . Consider for simplicity that bandwidth and center frequency agree, as is the situation at least approximately for CARABAS III high and low bands. As was established in Section 5.2, $\Phi_\ell = \Re/Y_{a_1\ldots a_m}$. Also assuming the standard situation $\tilde{\phi} \le 30°$, (5.52) implies the following velocity and heading error tolerances for both range and angle focusing:

$$\text{range focusing:} \quad \left|\Delta_G Y_{a_1\ldots a_m}\right| \le 4\Re, \qquad \left|\Delta\phi_{a_1\ldots a_m}\right| \le 2\frac{\Re}{Y_{a_1\ldots a_m}}$$
$$\text{angular focusing:} \quad \left|\Delta_M Y_{a_1\ldots a_m}\right| \le 2\frac{\Re\hat{r}}{Y_{a_1\ldots a_m}}, \quad \left|\Delta\phi_{a_1\ldots a_m}\right| \le \frac{\Re}{Y_{a_1\ldots a_m}} \tag{5.53}$$

Note that for subapertures short compared to stand of range, that is, during initial stages of the FFBP process, the mean speed across the subaperture is not critical, but any speed change is. Only when subaperture length approaches standoff range will the mean speed become important, and is the governing speed requirement in the final merging step. Not surprisingly, heading errors are of greater importance for angular than for range focusing.

With the assumption of a horizontal merging plane, the defocusing heading errors are horizontal. A corresponding analysis is made for vertical heading errors, spanning a vertical merging plane. A vertical merging plane corresponds to $\Delta\phi^{\perp}_{a_1\ldots a_m} = \delta\phi^{\perp}_{a_1\ldots a_m} = 0$ according to (5.37) and (5.39). From (5.40) and (5.43), it follows that $\psi^{\Gamma\to\Pi}_{a_1\ldots a_m\pm} = -\psi^{\Pi\to\Gamma}_{a_1\ldots a_m} = \pi/2$. Speed errors do not upset the straightness of the SAR path and will thus be independent of the merging plane orientation. It will therefore not be necessary to consider them anew, that is, $Y_{a_1\ldots a_m+} = Y_{a_1\ldots a_m-} = Y_{a_1\ldots a_m}/2$ is assumed.

Again, the analysis starts with transformation $\tilde{r},\tilde{\phi} \to \tilde{r}_{\pm},\tilde{\phi}_{\pm}$ given by (5.20) and (5.23) for the case of no measurement errors, that is, a straight and uniform horizontal aperture. The case of a vertical merging plane encompasses that due to measurement errors, the merging apertures are assigned nonzero inclination angles $\varepsilon_{a_1\ldots a_m\pm}$. For this analysis, it is noted that focusing does not depend on the merged aperture inclination $\varepsilon_{a_1\ldots a_m}$. This is actually a special case of the analysis of Chapter 4 that focusing is independent of ground topography. Since the merging subapertures are of equal length $\varepsilon_{a_1\ldots a_m} = (\varepsilon_{a_1\ldots a_m+} + \varepsilon_{a_1\ldots a_m-})/2$, so focusing only depends on the angle $\Delta\phi_{a_1\ldots a_m}$ between the subapertures, presently satisfying the relation $\Delta\phi_{a_1\ldots a_m} = \varepsilon_{a_1\ldots a_m+} - \varepsilon_{a_1\ldots a_m-}$. Hence, range and angular defocusing parameters become

$$D\tilde{r} = \varepsilon_{a_1\ldots a_m+}\left(\frac{\partial\tilde{r}}{\partial\varepsilon_{a_1\ldots a_m+}}\right)_0 - \varepsilon_{a_1\ldots a_m-}\left(\frac{\partial\tilde{r}}{\partial\varepsilon_{a_1\ldots a_m-}}\right)_0 = \Delta\phi_{a_1\ldots a_m}\left(\frac{\partial\tilde{r}}{\partial\varepsilon_{a_1\ldots a_m\pm}}\right)_0$$

$$D\tilde{\phi} = \varepsilon_{a_1\ldots a_m+}\left(\frac{\partial\tilde{\phi}}{\partial\varepsilon_{a_1\ldots a_m+}}\right)_0 - \varepsilon_{a_1\ldots a_m-}\left(\frac{\partial\tilde{\phi}}{\partial\varepsilon_{a_1\ldots a_m-}}\right)_0 = \Delta\phi_{a_1\ldots a_m}\left(\frac{\partial\tilde{\phi}}{\partial\varepsilon_{a_1\ldots a_m\pm}}\right)_0 \tag{5.54}$$

Note that the expressions refer to the aperture planes rather than the merging plane as in (5.44). Indeed, for the dependence on $\varepsilon_{a_1\ldots a_m\pm}$ to be fully included in the merging, also the rotations (5.28) must be incorporated. To keep the notational burden low when invoking these, expressions will be reduced to first-order approximations in $\varepsilon_{a_1\ldots a_m\pm}$. Then inserting (5.27) into (5.28) for $\psi^{\Gamma\to\Pi}_{a_1\ldots a_m\pm} = \pi/2$

$$\begin{pmatrix} \hat{x}_{\pm} \\ \hat{y}_{\pm} \\ \hat{z}_{\pm} \end{pmatrix} = \begin{pmatrix} 0 & 0 & 1 \\ 0 & 1 & 0 \\ -1 & 0 & 0 \end{pmatrix} \begin{pmatrix} \tilde{x}_{\pm} \\ \tilde{y}_{\pm} \\ \tilde{y}_{\pm} \varepsilon_{a_1 \ldots a_m \pm} - \tilde{h}_{a_1 \ldots a_m} \end{pmatrix} \tag{5.55}$$

Combining with (3.8), it is noted that that the latter can be restated in Cartesian coordinates

$$\begin{pmatrix} \hat{x} \\ \hat{y} \\ \hat{z} \end{pmatrix} = \begin{pmatrix} 1 & \mp \frac{\Delta\phi_{a_1 \ldots a_m}}{2} & 0 \\ \pm \frac{\Delta\phi_{a_1 \ldots a_m}}{2} & 1 & 0 \\ 0 & 0 & 1 \end{pmatrix} \times \left[ \begin{pmatrix} 1 & \pm\Delta\phi_{a_1 \ldots a_m} & 0 \\ \mp\Delta\phi_{a_1 \ldots a_m} & 1 & 0 \\ 0 & 0 & 1 \end{pmatrix} \begin{pmatrix} \hat{x}_{\pm} \\ \hat{y}_{\pm} \\ \hat{z}_{\pm} \end{pmatrix} \pm \frac{1}{4} \begin{pmatrix} 0 \\ Y_{a_1 \ldots a_m} \\ 0 \end{pmatrix} \right] \tag{5.56}$$

Finally, there is the rotation back to the merged aperture plane through the angle $-\psi^{\Pi \to \Gamma}_{a_1 \ldots a_m} = \pi/2$

$$\begin{pmatrix} \tilde{x} \\ \tilde{y} \\ \tilde{z} \end{pmatrix} = \begin{pmatrix} 0 & 0 & -1 \\ 0 & 1 & 0 \\ 1 & 0 & 0 \end{pmatrix} \begin{pmatrix} \hat{x} \\ \hat{y} \\ \hat{z} \end{pmatrix} \tag{5.57}$$

Combining the three transformations and then reexpressing the $\tilde{x}$ and $\tilde{y}$ coordinates in $\tilde{r}$ and $\tilde{\phi}$ coordinates

$$\left. \begin{array}{l} \tilde{x} = \tilde{x}_{\pm} \\ \tilde{y} = \tilde{y}_{\pm} \pm \dfrac{Y_{a_1 \ldots a_m} + \Delta\phi_{a_1 \ldots a_m} \tilde{h}_{a_1 \ldots a_m}}{4} \end{array} \right\} \Rightarrow$$

$$\tilde{r} = \sqrt{\tilde{r}_{\pm} + \frac{Y_{a_1 \ldots a_m}^{2}}{16} + \frac{\Delta\phi_{a_1 \ldots a_m} \tilde{h}_{a_1 \ldots a_m} Y_{a_1 \ldots a_m}}{8} \pm \tilde{r}_{\pm} \sin\tilde{\phi}_{\pm} \frac{Y_{a_1 \ldots a_m} + \Delta\phi_{a_1 \ldots a_m} \tilde{h}_{a_1 \ldots a_m}}{2}}$$

$$\tilde{\phi} = \tan^{-1}\left[ \tan\tilde{\phi}_{\pm} \pm \frac{Y_{a_1 \ldots a_m} + \Delta\phi_{a_1 \ldots a_m} \tilde{h}_{a_1 \ldots a_m}}{4\tilde{r}_{\pm} \cos\tilde{\phi}_{\pm}} \right] \tag{5.58}$$

It is observed that the expressions depend only on $\varepsilon_{a_1 \ldots a_m \pm}$ in the combination $\Delta\phi_{a_1 \ldots a_m} = \varepsilon_{a_1 \ldots a_m +} - \varepsilon_{a_1 \ldots a_m -}$, that is, a vertical heading change between the merging subapertures. The derivatives become

$$\left(\frac{\partial \tilde{r}}{\partial \varepsilon_{a_1 \ldots a_m \pm}}\right)_0 = \pm \frac{\tilde{h}_{a_1 \ldots a_m}}{4\tilde{r}}\left(\tilde{r}_\pm \sin\tilde{\phi}_\pm \pm \frac{Y_{a_1 \ldots a_m}}{4}\right)\left(\frac{\partial \Delta\tilde{\phi}}{\partial \varepsilon_{a_1 \ldots a_m \pm}}\right) = \frac{\tilde{h}_{a_1 \ldots a_m} \sin\tilde{\phi}}{4}$$

$$\left(\frac{\partial \tilde{\phi}}{\partial \varepsilon_{a_1 \ldots a_m \pm}}\right)_0 = \frac{\pm \dfrac{\tilde{h}_{a_1 \ldots a_m}}{4\tilde{r}_+ \cos\tilde{\phi}_\pm}\left(\dfrac{\partial \Delta\tilde{\phi}}{\partial \varepsilon_{a_1 \ldots a_m \pm}}\right)}{1 + \left(\tan\tilde{\phi}_\pm \pm \dfrac{Y_{a_1 \ldots a_m} + \Delta\phi_{a_1 \ldots a_m} \tilde{h}_{a_1 \ldots a_m}}{4\tilde{r}_\pm \cos\tilde{\phi}_\pm}\right)^2} = \frac{\tilde{h}_{a_1 \ldots a_m} \cos\tilde{\phi}}{4\tilde{r}} \tag{5.59}$$

The accuracy requirements for measuring vertical heading changes are consequently

$$\begin{aligned} &\text{range focusing:} && \left|\Delta\phi_{a_1 \ldots a_m}\right| \leq 8\frac{\Re}{h_{a_1 \ldots a_m}} \\ &\text{angular focusing:} && \left|\Delta\phi_{a_1 \ldots a_m}\right| \leq 4\frac{\tilde{r}\Re}{Y_{a_1 \ldots a_m} \tilde{h}_{a_1 \ldots a_m}} \end{aligned} \tag{5.60}$$

For short apertures, the range requirement is the more stringent, whereas finally when aperture length is close to standoff range, the angular focusing requirements become more stringent. As for vertical heading errors, it is implied by (5.60) that these remain critical (since range focusing errors are not reduced) also for short apertures. In contrast, horizontal heading errors are relaxed for short apertures. Overall, vertical heading error influence is smaller than horizontal, in particular since altitude generally is smaller than SAR path length and standoff range.

An obvious reflection is that the first-order expansions of the merging formulas also may be used for SAR processing, given that the aircraft path is sufficiently straight. The objective would be simplified merging formulas making computations speedier. From the viewpoint of computational burden, simplifications (even if possible) are not motivated in determining the angles $\psi^{\Gamma \to \Pi}_{a_1 \ldots a_m \pm}$ and $\psi^{\Pi \to \Gamma}_{a_1 \ldots a_m}$, which only have to be evaluated once for every subaperture image (rather than for each pixel in the image). As for merging formulas (5.24) and (5.25), a first-order expansion does not relax the relatively time-consuming task of performing square root and trigonometric coordinate calculations for all image pixels. However, for a perfectly linear path, the merging transformations are all the same for all subapertures, within each FFBP iteration. They can thus be realized by precalculated lookup tables. First-order path nonlinearities will be represented as linear deviations from these lookup tables. These deviations are more readily calculated than recalculating the lookup values from the exact coordinate expressions involving new trigonometric evaluations, so indeed some computational savings are possible.

Evidently, depending on the degree of nonlinearity of the trajectory, the linear approximations will introduce errors. The coordinate errors $D\tilde{r}$ and $D\tilde{\phi}$ are obtainable by forming the difference between the exact merging solution derived in the preceding section and the first-order approximation of this section. The error tolerance is determined by the range and angular resolution just as for the analyzed accuracy requirement for the path parameters. These investigations, on the validity of first-order path approximations, are suitably carried out numerically on a case-to-case basis. A gross result is that first-order approximations may be sufficient for fixed wing platforms, which only to small extent deviates from linear motion. The path of a helicopter platform under conditions of wind gusts may be far from linear (e.g., speed changes of up to 50%), making the linear approximation of the merging formulas far from sufficient. In contrast, the exact merging formulas has proved to provide well-focused SAR images also in these difficult cases, as was exemplified in Chapter 3.

In favor of exact solutions, it is noted that highly efficient means exist for fast trigonometric evaluation. The CORDIC method is an old and well-known method, which appears very well suited for FFBP (its standard realm of application today is in computer graphics). The CORDIC method finds trigonometric function values through shift register calculations performing an iterative refinement of the function argument. Very likely, the CORDIC method allows a reuse of lookup tables by adding the angular adjustments required for nonlinear SAR paths, in ways similar to the situation for a first-order approximation. However, with CORDIC, the solution to the exact formulation would be computed instead, allowing SAR paths far from linear.

## 5.5 Motion Estimation Methods

Concluding the account of FFBP processing, methods for measuring the SAR path geometry parameters are outlined. Basically, three fundamental technologies offer themselves:

- Inertial measurements based on onboard IMUs;
- Satellite positioning, GNSS;
- Autofocusing, based on the collected radar data itself.

Each of these technologies can in turn be used in different ways; in particular, they either aim at a complete solution or provide input to a combination of methods. Examples will be provided. What particular method is selected depends on many factors. One is obviously the required measurement accuracy, which will increase with system size. Contrary to this trend, a small system installed on a small aircraft moves in a less stable way, and needs a more detailed registration of the short-term aircraft motion. A different aspect is that there sometimes is an advantage of not having to rely on external sources for establishing the SAR path. For the GNSS, reliance has to be put on the availability of GNSS satellite signals. Moreover, advanced use of the GNSS (such as *precision point positioning* (PPP); see the text below) also requires availability of GNSS related data on the Internet. In a military conflict, GNSS services might be jammed or spoofed. But also operating in remote places, gathering data from the Internet might prove difficult. For these reasons, depending on the application at hand, inertial methods might be preferred over the GNSS.

IMUs are well suited for establishing the SAR path geometry apart from falling short fully reaching the required speed measurement accuracy. Autofocus is not to be regarded as a stand-alone method of determining the SAR path geometry. However, it may well suit as a complement to IMU measurements. Also, GNSS-based methods for path determination may require the support of IMUs, since the SAR path is that of the SAR antenna. Thus, the motion of the lever arm between the GNSS antennas (typically on top of the aircraft) and the SAR antennas (at the bottom) has to be established. Performance requirements for the IMUs required for this particular task are, however, modest, so the IMUs are low cost.

The basic working principle of an IMU is to measure the rate of change of inertial motion, which can be broken down into six parameters comprising angular and velocity changes along three orthogonal axes. Continuously integrating these with respect to time, the aircraft speed, heading and attitude can be established, and integrating a second time, the overall motion of the aircraft is obtained. Crucial for the method is the resulting speed, attitude and heading errors (see [11] for concise treatment). The overall time behavior of these errors is as a random walk error growth, consisting partly of a zero-mean diffusive error growth, with expectation values for parameter magnitudes increasing as the square root of time. Added to the diffusive errors there are also bias (or offset) errors, characterized by having expectation values growing linearly in time. The bias errors are minimized by accurate IMU calibration, though calibration has limits in precision and repeatability. For instance, since the IMU can never be levelled perfectly, there will be a residue of Earth gravitation in what the IMU senses as horizontal direction, leading to a horizontal velocity bias growing linearly in time. Bias residues are characteristics of each type of IMU, and are stated along with the diffusive errors in the IMU performance sheet. According to the analysis of the preceding section, the FFBP requirement for heading accuracy increases as aperture lengths and thus aperture times are increased in the process of aperture merging, Due to the time dependencies of the diffusive and bias errors, there will be a time or aperture length limit set by the quality of the IMU measurements, beyond which the accuracy of these measurements does not meet the focusing requirements.

In an approximate order of increasing cost and performance, IMUs can be categorized by their different ways of implementing the required gyro function. Microelectromechanical systems (MEMS) IMUs are small micromechanical devices. Rather than realizing the gyro by means of a rotating reference, they exploit a vibrating miniature body. When the body is subjected to rotation, the resulting Coriolis force yields the rotational rate. In the category of gyros adopting a rotating reference, besides the now outdated mechanical rotating gyros, there are ring laser gyros (RLGs) and fiber optic gyros (FOGs). Both gyro types are based on light rays traveling in opposite directions around a closed loop (arranged by mirrors or optical fiber). By means of the so-called Sagnac effect (corresponding to Doppler effect for linear motion), the rate of rotation of the loop can be measured from the level of interference between the two rays.

Table 5.1 provides performance figures for three "tactical grade" IMUs, using the three different gyro principles. The roughly similar performance

**Table 5.1**
Some IMU Performance Examples

| IMU | | Rotation | | Velocity | |
|---|---|---|---|---|---|
| **Method** | **Type** | **Bias residual** | **Diffusive** | **Bias residual** | **Diffusive** |
| MEMS | SD1500 | 1.5°/h | $0.025°/\sqrt{h}$ | 3.6 m/s/h | $0.06\ m/s/\sqrt{h}$ |
| FOG | KVH1750 | 0.1°/h | $0.012°/\sqrt{h}$ | 1.8 m/s/h | $0.07\ m/s/\sqrt{h}$ |
| RLG | K14801 | 0.5°/h | $0.05°/\sqrt{h}$ | 14 m/s/h | $0.05\ m/s/\sqrt{h}$ |

figures are noted. The Systron Donner inertial MEMS gyro, which is one of the examples, is quite high end compared to MEMS gyros in general. The Kearfott RLG IMU figures provided exemplifies the lower cost segment of RLG IMUs. Also, the fiber optic KVH gyro has high performance for its kind. The three IMUs seem to represent a realistic compromise between affordability (and indeed availability in the open market) and the relatively high performance requirements for meter-wave SAR. The different physical units adopted in their data sheets have been converted into common units in Table 5.1. Doing so, it is noted that the velocity bias has the physical dimension of speed drift per unit time, that is, the physical dimension of acceleration. Mostly, it is expressed in [$\mu$g], but in Table 5.1, this unit has been converted into the presently convenient unit [m/s/h]. The velocity diffusive error is expressed in [m/s/$\sqrt{\text{h}}$], which is noted to have the same physical dimension as [$\mu$g/$\sqrt{\text{Hz}}$ ]. The latter is the unit usually found in IMU data sheets.

The main alternative to IMU motion measurement is the use of navigation satellite systems such as GNSS. *Ephemeris*, that is, the predictable motion of heavenly objects, has been exploited for navigation throughout the history of humanity. The key to GNSS positioning is the ephemeris of GNSS satellites, which is obtainable in several ways. Standard ephemeris is coded into the signals transmitted from the satellites, whereas enhanced quality GNSS ephemeris is available on the Internet. Some services offer this data in near real time. The GNSS signals also carries time stamps, imposed by precise onboard clocks. Receiving at least four satellite signals at some point $P$ on or above the Earth surface, the Earth coordinates at $P$ can be determined from the satellite position knowledge. Obviously, the suitably of GNSS to the present application depends on that submeter measurement accuracy can be achieved. The process of GNSS position determination is quite complex, with the position error budget having a wide range of contributions. The important error sources are listed in Table 5.2 [12]. The different errors will influence the position measurement quite differently depending on how the GNSS measurement is carried out. As will be discussed, it is possible to improve GNSS accuracy far beyond the meter order errors in the table, which pertains to the standard GNSS positioning using a single stand-alone receiver. With these errors juxtaposed, the resulting accuracy is around 6m, the value which only marginally depends on the resolution code. The coarser C/A code is openly available, whereas the finer P(Y) code is available to a large albeit restricted community.

Fundamental to the advanced uses of GNSS is that the only error, which has the character of white noise, is the receiver time resolution. The remaining error appears as biases to true values changing only slowly with time and location. In the course of time, and with satellites moving (thus with signals

**Table5.2**
GNSS Error Budget

| | | |
|---|---|---|
| 1. | Standard ephemeris error | 2.5m |
| 2. | Satellite clock error | 2m |
| 3. | Ionosphere delay error | 5m |
| 4. | Tropospheric delay error | 0.5m |
| 5. | Receiver time resolution C/A code | 3m |
| 6. | Receiver time resolution P(Y) code | 0.3m |

propagating through different parts of the atmosphere), propagation speed will change, resulting in a position error drift set by the atmospheric errors. Similarly, ephemeris and clock errors create range errors increasing with the time between measurements. In the course of a few minutes to several hours, the drift grows to agree with the errors of Table 5.2. In the course of even longer periods (days), the satellite clocks are adjusted and variations atmospheric conditions average out, making the mean drift over such periods tending to zero. To obtain accurate Earth positions, averaging over a very long time is practiced in geodetic applications.

Positioning by a stand-alone GNSS receiver is commonly referred to as *single point* positioning. An important approach to reduce errors is *differential* GNSS in which GNSS measurements of a receiver at unknown position (called the rover) are compared to measurements conducted in parallel by a nearby reference station located at a fixed known position. The two receivers are experiencing essentially the same atmospheric ephemeris and clock errors, so the method allows accuracy to be improved to that of receiver time resolution.

Neither a single point nor a differential GNSS offers in their standard form performance compliant with the demands of the present SAR applications. The accuracy attainable with a nondifferential GNSS is too coarse. Differential GNSS accuracy based on P(Y) code suffices on some occasions, but has to rely on a nearby reference receiver. There is an alternative set of methods, which provides order of magnitudes higher position accuracy, and can be used without having access to a P(Y) code or a reference station. The methods are based on retrieving signal carrier phase by correlation to the carrier predicted for the nearest integer number of wavelengths. Phase determination can be carried out with a granularity of about 1% of the wavelength. Since a GNSS adopts a carrier frequency of 1.5 GHz, that is, 0.2m wavelength, a range accuracy of ±0.002m is attainable in principle.

Using phase for refined range measurements requires addressing the ambiguity problem; the range up to the nearest integer number of wavelengths must be known a priori. Certainly, range is known with the standard GNSS accuracy of 6m. With differential measurements, accuracy is improved and the ambiguity problem is simplified. Form the difference between range ambiguously determined (up to an unknown number of wavelengths) from phase and the unambiguous albeit inaccurate range determined from code. The standard method of resolving the ambiguity is by least-square minimization of this difference, collected over the time a particular satellite is present in received data. There are several variants of this procedure, using either the P(Y) and C/A codes.

Two main precision GNSS methods based on phase measurements stand out as well developed and openly available. They are known as *real-time kinematics* (*RTK*) [13] and *PPP* [14]. The first is a differential method relying on a reference receiver fixed at a known position, while the position (or motion) of the rover receiver is to be determined. The RTK performance is affected by the GNSS errors tabulated above, which act to reduce the degree of correlation between the GNSS signals at the two receivers, as they become increasingly separated. Typically, the separation between the two receivers has to stay within 20 km without ionospheric modeling, while it can be extended to more than 50 km if such is included. The second method PPP is more recent. It assumes a single receiver. Ambiguity resolution becomes more critical than for a differential method, but relies instead on that the satellite data extracted from the GNSS signals is further corrected by ionospheric models. These corrections are available from services, found on the Internet. Further enhancement of these methods is attained by two-frequency receivers (receiving both L1 and L2 signals, with carrier frequency separated). Use of two frequencies improves compensation for ionospheric dispersion and increases robustness of ambiguity resolution.

It is also observed that the stringent requirements on range accuracy are for determining Earth ground coordinates in the absolute sense. To determine position changes equally accurate, resolution of the ambiguity problem is not necessary. Forming the difference between adjacent phase measurements, the Doppler shift of the satellite signal is measured. The Doppler shift exhibits a certain bandwidth reflecting that any nonlinearity in rover movements must be encompassed within a phase difference less than 180°. In fact, given rover speed uncertainty $\Delta v$, the repetition rate must be higher than $f_{\rm rep} > B_{\rm Doppler} = \Delta v/\lambda$. Since $\Delta v = 2a/f_{\rm rep}$ for accelerations limited to $a$, for wavelength $\lambda$, it follows that $f_{\rm rep} > \sqrt{2a/\lambda}$. With rover acceleration limited to $a$ = 10 m/s$^2$, a repetition rate of 10 Hz is thereby sufficient. While GNSS signals have a C/A code repetition rate of 1 kHz [15], commercially available GNSS receivers operate at significantly lower rates. A 10-Hz reception rate is possible with high-end receivers. Having secured sufficient reception bandwidth, Doppler shift and thus range shift rate is acquired unambiguously. Integration of the range shift rate in the course of time provides the net range shift. Including several satellites, the rover net displacement and its three dimensional detailed shape are established. Unless exterior corrections are included, net displacements must be limited to 20 km (the reference /rover separation for uncorrected RTK) and occur over at most a few minutes or less, in order that GNSS drift errors should not degrade the measurement.

Clearly, horizontal rigid rotations or translations of the SAR path are just changes of the image content and will not affect image focusing. Also recall that rigid rotations of the path in a vertical plane or changes of path altitude do not upset SAR image focusing, though they may cause distortions to an image given in aperture plane or slant range coordinates. Hence, for the present application of determining the SAR path, it is obvious that the just-mentioned method of measuring displacements is sufficient. In fact, this particular method seems very well suited for low-frequency SAR motion determination, since the allotted period and size of displacement fits well with typical path lengths in this case. The practical (and quite decisive) concern is that this method of displacement measurement seems not developed and available as of today. It may be that there has been a lack of demand, or possibly that such a method exists but not in any open version. Within the CARABAS development, available RTK and PPP methods have therefore been resorted to.

The following is a summary of the practical experiences obtained with RTK and PPP for the CARABAS III system (for details, see the master thesis by Helsten et al. [16]). In CARABAS III, the update rate for GNSS data is $f_{\text{rep}}$ = 10 Hz. Position measurements are found to vary randomly within 0.01m. For gentle nonturbulent flight, CARABAS III helicopter accelerations would be limited to $a$ = 1 m/s$^2$. Position uncertainty due to unknown movement is thereby limited to $\Delta r = a/2f_{\text{rep}}^2$ = 0.01 m, that is, of the same magnitude as the measurement fluctuations. For turbulent flight, the position uncertainty becomes larger than the measurement error. Hence, for CARABAS III, the GNSS data cannot be further improved by filtering with respect to platform inertia, but represents the most likely helicopter position at the intervals of measurement. For obtaining position data in between GNSS registrations linear interpolation between adjacent registrations becomes the reasonable method.

Added to the uncorrelated error, there is a three-dimensionall position drift, arising due to the GNSS errors developing in time. Modeled as a diffusive process, a drift rate of the order 0.003 m/$\sqrt{s}$ is expected. It will thus yield a 0.03m drift for a 100s SAR path. For all reasonable lengths of low-frequency SAR paths, position uncertainty would stay within 0.1m. In the preceding section, it was established that the required position error in any direction must be smaller than some multiple of bin size $\Re$. According to Section 5.2, suitable $\Re$ values for CARABAS III low and high bands were found to be $\Re = 0.6$m and $\Re = 0.15$m. Note that systems larger than CARABAS III requires one or two more iteration steps in the FFBP process, implying that marginally smaller bins are required to keep the same level of processing noise. As is evident from (5.53) and (5.60), the accuracy requirements are most pressing in the final FFBP stages, in which the aperture length is approaching standoff range. The governing heading accuracy requirement is $\Delta\phi = \Re/Y$, whereas the governing relative speed accuracy requirement becomes $\Delta Y = Y\Delta V/V = 2\Re$, with $V$ being the aircraft ground speed. It can thereby be concluded that relying on suitable GNSS techniques in both the high and low band cases, motion parameter errors have been demonstrated to fulfill the requirements.

Consider the alternative of using IMUs instead of GNSS for SAR path measurements. IMUs are different from GNSS in primarily determining speed and heading rather than position. As for speed accuracy, $\Delta Y = Y\Delta V/V = 2\Re$ can be restated as $\Delta V = \Re/T$, keeping the convention of $T$ representing half the SAR integration time (i.e., $Y = 2VT$). For CARABAS III, a path length 2.5 km is covered in 100s. Consequently, for low-band operation ($\Re = 0.6$m), heading and speed linear and diffusive drifts are demanded to stay within 1.7°/h, 0.7 m/s/h and 0.2°/$\sqrt{\text{h}}$, 0.09 m/s/$\sqrt{\text{h}}$, respectively. The IMU performance figures tabulated in Table 5.1 are seen to satisfy these requirements with the exception of speed linear drift; the velocity bias residuals for the cited systems are larger than (or at best of the same order as) the required 0.7 m/s/h. For larger systems operating at larger standoff ranges, SAR paths are longer (say, 10 km), though the installation will be on faster aircrafts. Typical integration times are still of the order 100s, so IMU speed requirements remain unchanged. Due to the longer aperture, the heading requirements sharpen and become 0.4°/h, 0.05°/$\sqrt{\text{h}}$, which is still within the bounds of IMU achievable performance. High-band ($\Re = 0.15$m) heading requirements for the short CARABAS III aperture become the same as for the large system at low band. The high-band speed requirements become 0.2 m/s/h and 0.02 m/s/$\sqrt{\text{h}}$. The diffusive error is outside the performance bounds of the tabulated IMUs. Evidently, at high band, large aperture systems would be outside the cited IMU performance bounds in essentially every aspect.

Note the following difference between GNSS and IMUs. The geometry accuracy requirements (5.53), (5.60) translate into that position accuracy $\approx \mathfrak{R}$ is required for all subaperture lengths in the FFBP merging process. Applicability of a specific GNNS system is thereby decided by whether it produces this position accuracy or not. For speed and heading, which are sensed by IMUs, (5.53) and (5.60) imply that error levels increase with aperture length. Therefore, for the IMU total error (the net error of all iteration combined, as discussed in Section 5.2) to remain the same as the total GNSS error, it may be allowable to let $\mathfrak{R}$ increase to even twice the value adopted for a GNSS system for the final mergings. Balancing this additional error growth, the IMU measurement errors would be considerably smaller than those for the GNSS system, during the initial iteration stages.

Also note what was mentioned in Section 5.2 that with the alternative LBP method, $\mathfrak{R}$ can be made twice as large as for base 2 FFBP, correspondingly relaxing drift requirements. IMU compliance regarding speed accuracy improves and may be acceptable below 100 MHz. For the high-band IMU motion determination, accuracy remains far from sufficient.

Navigation systems often combine an IMU with a single point GNSS receiver, using a Kalman filter to derive an optimal joint navigational solution. The GNSS receiver basically senses position, and thus serves to make velocity and heading errors approach zero over large time spans. The Kalman filter, operating on joint GNSS and IMU data, improves the navigational solution also for shorter times, than the time it takes for IMU velocity and heading drifts to become detectable by the GNSS measurements. In effect, GNSS data serves to suppress the IMU bias errors. The possible improvement is limited by the nonsystematic, that is, diffuse IMU errors. Ideally, the filtering removes the IMU bias errors entirely. However, depending on the degree of irregularity of the SAR path and the GNSS data quality, the navigational solution may be far from that mark.

Since bias errors grow linearly with time, whereas the diffuse errors grow with square root of time, diffuse errors will dominate for short time intervals. For longer times, the Kalman filter controls and limits the errors. The time lapse it takes for the diffuse and bias errors to balance is thus the minimum time lapse for IMU speed and heading estimates to be improved by GNSS data. For speed errors in Table 5.1, these time lapses are 1s, 5s, and 0.5s, respectively, for the MEMS, FOG, and RLG examples. For heading errors, the corresponding values are 1s, 51s, and 36s. The two last figures approaches the typical 100s SAR integration time, implying that the Kalman filter improves SAR path geometry estimation relatively little, or not at all depending on circumstances. For the remaining figures, a significant improvement is possible. In particular, a filtered solution would be a beneficial improvement of standalone IMU data for estimating SAR path speed.

As already have been mentioned, also a phase measuring GNSS system may need augmentation with an IMU, since the relative motion (lever arm motion) between GNSS and radar antenna phase center must be accurately determined. As an example on how this can be done in a relatively straightforward manner, CARABAS III uses a combination of high-performance phase retrieving GNSS and a low-grade IMU for measuring the lever arm motion. The GNSS system includes twin GNSS antennas mounted on the helicopter tail boom (see Figure 4.8). These interferometrically retrieve the tail boom orientation. A purposely designed Kalman filter weights the IMU data with this data, thereby controlling the IMU drift.

Apart from IMU and GNSS methods for motion determination, a third method of SAR autofocus offers itself. Autofocus is routinely practiced in microwave SAR, and is in fact quite necessary in view of the exacting motion determination requirements, which prevail for these high-resolution systems. However, in this case, the antenna beam is narrow, which allows radar data to be considered as linear rather than circular mean values across the radar image. Thereby, the impact of irregular motion can be represented by phase shifts in the image Fourier domain, while such a representation is not possible with the wide angle antenna adopted for low-frequency SAR. Consequently, the two established microwave autofocus methods called map drift and phase gradient [3.1], do not work for low-frequency SAR.

For low-frequency SAR, there is a quite different circumstance that enables autofocus. Since wavelength order resolution is attained and objects only scatter effectively if of wavelength order size, a typical feature of low-frequency SAR is that image resolution cells contain just a single scatterer. The scattering from resolution cells will thus be isotropic. Consequently, if a SAR scene is imaged from two different flight paths, the resulting SAR images $\chi(P)$ and $\chi'(P)$, represented with respect to ground points $P$, will have proportional envelopes $|\chi(P)| = C|\chi'(P)|$, with a constant scale factor $C$ determined by imaging conditions such as standoff range. For microwave SAR, the situation is different; at the level of resolution cells, the amplitude fluctuates strongly if the SAR path is altered, a phenomenon known as *speckle*. Speckle is routinely suppressed by *multilook*, that is, forming SAR images $\chi(P)$, $\chi'(P)$, ... from a sequence of neighboring SAR paths and from these constructing an image given by the averaged amplitude envelope $\langle|\chi(P)|\rangle$.

Consider a pair of low-frequency SAR images $\tilde{\chi}_{\mp}(P)$ to be merged by the base 2 FFBP process. Form the quotient

$$\aleph = \frac{\left[ \int\limits_{\text{image}} \left| \hat{\chi}_-\left(\hat{r}_-, \hat{\phi}_-\right) \right| \left| \hat{\chi}_+\left(\hat{r}_+, \hat{\phi}_+\right) \right| r dr d\phi \right]^2}{\int\limits_{\text{image}} \left| \hat{\chi}_+\left(\hat{r}_+, \hat{\phi}_+\right) \right|^2 r dr d\phi \int\limits_{\text{image}} \left| \hat{\chi}_-\left(\hat{r}-, \hat{\phi}-\right) \right|^2 r dr d\phi} \leq 1 \qquad (5.61)$$

with $\hat{r}_\pm, \hat{\phi}_\pm$ determined by means of (5.20) and (5.23) and knowledge of the merging geometry. Equality occurs only if $| \hat{\chi}_\pm(\hat{r}_\pm, \hat{\phi}_\pm) |$ are proportional. For images isotropic in the sense described, $\aleph$ may consequently approach unity. This is, however, the situation if and only if the merging geometry parameters are correctly set. In consequence, autofocusing could be carried out as subimage merging under different assumptions on the geometry parameters, choosing the parameter set that yields the highest $\aleph$ value.

Images will not be isotropic and the criterion will not work in either of two cases: 1) there are several wavelength sized scatters within the resolution cells and 2) there is an abundance of objects extending over several resolution cells (building walls and fences are examples). Due to the first requirement, autofocusing can only work in the later FFBP stages, where resolution has become sufficiently fine to bring at least the majority of scatterers into single resolution cells. Moreover, extend objects must not be dominant; autofocusing will not work for imaging a city, for instance. The ideal scene content is forest with trees as sparse as the resolution of the merging images. Also note that (5.61) refers to the amplitude envelope; in the underlying images, the phase as well as the spectral support (as defined in Chapter 3) varies rapidly with any change in the SAR path, that is, complex SAR images will never be isotropic.

Crucial for the usefulness of the indicated autofocusing method is that it applies locally to the merging of each subaperture pair. The optimization of $\aleph$ is thus reduced to the variation of the geometry parameters $Y_{a_1 \ldots a_m \pm}$, $\varepsilon_{a_1 \ldots a_m \pm}$ (aperture length and inclination), characterizing either subaperture or the angle $\Delta\phi_{a_1 \ldots a_m}$ between them in all five parameters at a time. Inspecting the merging formulas of Section 5.3, the subaperture heights $\tilde{h}_{a_1 \ldots a_m \pm}$ have no final influence on merging and thus focusing. Among $Y_{a_1 \ldots a_m \pm}$, $\varepsilon_{a_1 \ldots a_m \pm}$, $\Delta\phi_{a_1 \ldots a_m}$, it is only $\Delta\phi_{a_1 \ldots a_m}$ that comes into the merging as an independent parameter. Both $Y_{a_1 \ldots a_m}$ and $\varepsilon_{a_1 \ldots a_m}$ are derived in the merging process though at lower accuracy than required for the next iteration. Autofocusing thus amounts to fine-tuning of $Y_{a_1 \ldots a_m \pm}$, $\varepsilon_{a_1 \ldots a_m \pm}$. Consider, for instance, the final aperture length $Y$. According to (5.53), this must be known with the precision $\Delta Y = 2\Re$, that is, $\Delta Y_\pm \approx \sqrt{2}\Re$ is required for the final merging. The apertures $Y_\pm$ are given with path length uncertainty $\Delta Y_{\pm} = \Re$, if previous mergings have been carried out correctly. The uncertainty is consequently insufficient for the final merging. Improved length estimates can be obtained by autofocusing. The merging process would assess the hypotheses $Y_{\pm} + n_{\pm}\Re \rightarrow Y_{\pm}$, where $n_{\pm} = -1,0,1$. In all, there are six such hypotheses, among which $n_{+}$ and $n_{-}$ are selected so that $\aleph$ is optimized. The uncertainty of $Y_{a_1 \ldots a_m \pm}$ would improve to $\sqrt{2}\Re$, so $Y$ accuracy becomes $2\Re$ as required. This method of speed accuracy improvement is of course of particular interest in combination with IMU measurements, where accurate determination of the speed parameter was found to be a challenge.

Optimization of (5.61) may similarly be carried out, varying vertical and horizontal subaperture headings. In this process, it is crucial that focusing and thus (5.61) only depend on changes in the heading between the merging subapertures. Hence, focusing can be conducted locally for each merging subaperture pair by variation of vertical and horizontal headings as well as variation of the mean velocity and the velocity gradient in all variations of four parameters. The further overall tilting or rotation required to fit the resulting subaperture into the next level merging will not upset the focusing.

Development of the autofocusing method just described started at Saab in Sweden around 2010 [17], while further development and also successful applications to real CARABAS data are accounted for a doctoral thesis of Torgrimsson et al. [18]. This work can also be accessed through [19–21].

## References


[1] Hellsten, H., et al. "Development of CARABAS II SAR," *Proceedings of Radar Sensor Technology*, SPIE AeroSense Conference, Orlando, FL, Vol. 2747, April 8–9, 1996, pp. 48–60.

[2] Ulander, L.M.H., H. Hellsten, and G. Stenström, "Synthetic Aperture Radar Processing Using Fast Factorized Backprojection," *IEEE Transactions on Aerospace and Electronic Systems*, Vol. 39, No. 3, 2003, pp. 760–766.

[3] Nilsson S. "Application of Fast Backprojection Techniques for Some Inverse problems in Integral Geometry," Ph. D. Dissertation 499, Linköpings University, Linköping, Sweden, 1997.

[4] McCorkle, J., and M. Rofheart, "An Order N2log(N) Backprojector Algorithm for Focusing Wide Angle, Wide Bandwidth Arbitrary Motion Synthetic Aperture Radar," SPIE AeroSense Conference, Orlando, FL, Vol. 2747, April 8–9, 1996, pp. 25–36.

[5] Yegulalp, A.F., "Fast Backprojection Algorithm for Synthetic Apertures Radar," *Proceedings of the 1999 IEEE Radar Conference*, Waltham, MA, April 20–22, 1999, pp. 60–65.

[6] Brandfass, M., and L.F. Lobianco, "Modified Fast Factorized Backprojection as Applied to X-Band Data for Curved Flight Paths," Seventh European Conference on Synthetic Aperture Radar (EUSAR), Graf-Zeppelin-Haus, Friedrichshafen, Germany, 2008, pp. 1–4.

[7] Frölind, P.O., and L.M.H. Ulander, "Evaluation of Angular Interpolation Kernels in Fast Back-projection SAR Processing," *IEEE Proceedings of Radar, Sonar and Navigation*, Vol. 153, No. 3, 2006, pp. 243–249.

[8] Ulander, L.M.H., P.O. Frölind, and D. Murdin, "Fast Factorized Back-projection for Processing of Microwave SAR Data," Sixth European Conference on Synthetic Aperture Radar (EUSAR), Dresden, Germany, 2006.

[9] Rodriguez-Cassola, M., et al., "Efficient Time-Domain Image Formation with Precise Topography Accommodation for General Bistatic SAR Configurations," *IEEE Transactions on Aerospace and Electronic Systems*, Vol. 47, No. 4, 2011, pp. 2949–2966.

[10] Woodman, O.J., "An introduction to Inertial Navigation," Technical Report Number 696, University of Cambridge Computer Laboratory, UK, 2007.

[11] Wikipedia:https://en.wikipedia.org/wiki/Error_analysis_for_the_Global_Positioning_System.

[12] Navipedia: http://www.navipedia.net/index.php/Real_Time_Kinematics.

[13] Navipedia: http://www.navipedia.net/index.php/Precise_Point_Positioning.

[14] Kaplan, E., and C. Hegarty, *Understanding GPS: Principles and Applications*, Norwood, MA: Artech House, 2005.

[15] Arvidsson J., "GPS Based Flight Motion Capture—Development and Evaluation of a High Performance Standalone GPS Navigation Solution for an Airborne Radar System," Master's Thesis, Chalmers, Göteborg, Sweden, 2016.

[16] Hellsten, H., P. Dammert, A. Åhlander, "Autofocus in Fast Factorized Backprojection for Processing of SAR Images when Geometry Parameters are Unknown," *Proceedings of the 2010 IEEE Radar Conference*, Washington D.C., USA, 2010.

[17] Torgrimsson, J., "Factorized Geometrical Autofocus for Synthetic Aperture Radar Processing," Ph. D. Dissertation, Chalmers, Göteborg, Sweden, 2015.

[18] Torgrimsson, J., et al., "Factorized Geometrical Autofocus for Syntetic Apertuer Radar Processing," *IEEE Trans. on Geoscience and Remote Sensing*, Vol. 52, No. 10, 2014, pp. 6674–6687.

[19] Torgrimsson J., et al., "Autofocus and Analysis of Geometrical Errors within the Framework of Fast Factorized Backprojection," *Proceedings of SPIE*, Baltimore, USA, Vol. 9093, 2014.

[20] Torgrimsson J., et al., "An Efficient Solution to the Factorized Geometrical Autofocus Problem," *IEEE Trans. on Geoscience and Remote Sensing*, Vol. 54, No. 8, 2016, pp. 4732–4748.

# 6

# Multidata Target Detection

## 6.1 Introduction

The concept of a target is understood as any class of objects for which the electromagnetic scattering characteristics are broadly known and which may be potentially present in a radar: in particular, an SAR image. Target detection comprises analyzing the radar/SAR image for features corresponding to the target scattering characteristics, thereby establishing the presence and location of the target.

Target detection is straightforward when the target RCS is larger than the RCS for any clutter response (stemming from objects outside the target class). Targets are then detected and positioned by thresholding the radar image with respect to the target class minimum RCS. Intensity threshold detection is of course the standard detection method for air surveillance radar. To an extent, threshold detection also works for man-made structures in microwave SAR images, since they distinguish themselves from natural ones by an abundance of flat surfaces, straight edges, and right angels. Specular double bounce scattering, that is, corner reflector scattering, thereby act to make man-made RCS values dominate over those of natural objects. For meter-wave SAR, as was analyzed in Chapter 2, backscattering from natural objects also involves double bounces, albeit the scattering occurs in the Rayleigh or resonance regions for both man-made and natural objects. Hence, in contrast to microwaves, there is no systematic difference in the resulting scattering magnitudes, and intensity thresholding will no longer be viable for distinguishing man-made objects.

Apart from intensity thresholding, detection may also be based on target shape characteristics. For instance, targets can be detected by scrutinizing the SAR image for any pattern agreeing with the target footprint dimensions. Characteristics of the target shadow can sometimes be an important source for finding the target. Shape detection requires an image resolution substantially better than target dimensions. It is the obvious target detection method for optical images. For low-frequency SAR, the method is less useful, since targets may be of wavelength size and thus not larger than the possible wavelength limited resolution. It is noted that shape extracting target detection involves significant element of image interpretation, making automatic detection routines at best time consuming, but for many situations also nonrobust.

Further SAR detection methods are possible by exploiting several channels of independent data, enabling targets to be discriminated from clutter in the multidimensional data space obtained. Examples are data gathered using different frequency bands, polarizations, or flight paths. In these specific cases, target extraction is based on differences in the frequency, polarization, or angular response between targets and natural clutter. Among the many choices of combining such multiple data sets, there are a few ones that stand out as singularly effective for low-frequency SAR target detection, which will be analyzed here in detail. One of the foremost is change detection, which combines data gathered over a particular scene from two distinct times. Stationary targets are extracted and positioned on the hypothesis that they have entered, changed position, or vanished from the scene between the two registration instances. As will be discussed in detail, a unique feature of low-frequency SAR, not shared with microwave SAR (or optical sensors for that matter), is the stationarity of clutter between such registrations. Following essentially parallel although not identical flight paths during repeated flights with a time span ranging from minutes to several months, high-performance change detection is possible. Also, low-frequency SAR resolution is relatively coarse, so surveillance rate can be high and vast areas surveyed on a regular basis. With such registrations serving as reference, there is a standby capability for finding targets by a renewed registration, anywhere in the surveyed area, whenever a need to do so arises.

Two further cases of multidata detection will also be analyzed, namely, DPCA detection of moving targets and polarimetric detection of buried targets. As described in Chapter 3, the former case entails conjunct registration of two SAR images, time shifted a fraction of a second, making a moving target appear as phase shifted, whereas the stationary ground clutter exhibits no phase shift. Subtracting the images the stationary ground clutter will cancel, leaving phase-shifted residues representing the moving targets. A mathematically similar detection method can be adopted for subsurface objects, since, according to the account of the bare ground SPM theory (in Chapter 2), conjunctly registered vertically and horizontally polarized SAR images can be weighted to make the ground surface clutter cancel. Subsurface targets will thereby appear as residues. In contrast to the discussion of the physics of buried and moving targets in Chapters 2 and 3, respectively, this common clutter removal method will here be developed into a statistical detection scheme. As in any form of detection, the statistical uncertainties arise for the reasons of incomplete knowledge of the scattering properties of the target and the surrounding terrain. The uncertainty is further augmented by thermal noise causing additive errors, as well as unknown and thus uncompensated transfer characteristics, causing multiplicative errors. The theoretical developments of subsurface and moving target imaging in Chapters 2 and 3 must be turned into methods, taking these uncertainties into account.

There is still another important meter-wave SAR multidata application, namely, the construction of elevation maps for ground below trees. An alternative and fitting name to this application is *bald Earth mapping*. The case is that trees tend to grow to reach a common height. Thus, trees will be short on ridges or hilltops and high in valleys in between. In particular, in the tropics, where growth rate is large, this can be a very pronounced phenomenon, preventing topographic maps of such terrain to be obtained by conventional stereo photography. Scanning laser ranging (LiDAR) is an alternative since the laser ray can hit the ground in gaps in the foliage. However, the method requires vegetation not to be too dense, and is therefore less suitable for tropical forest. Meter-wave SAR offers a possibility, where in fact the tree stem returns provide the required topographic information. As was analyzed in Chapter 2, the tree returns have as their phase center where the tree stem meets the ground surface. Thus, obtaining SAR images along two separate paths, the differences in the slant range projections of the undulating ground enable a height function of the latter to be reconstructed. The method has been exploited by the NASA GeoSAR system, having UHF (P-band) SAR antennas at the wingtips of a business jet [2] of Chapter 1. Also, the VHF band has been successfully exploited, using data from two separate flights along parallel trajectories [1]. Indeed, in view of that the main application is for very dense tropical rain forest, the use of frequencies below 100 MHz is presumably preferable. As mentioned in Section 6.1, the experimental and theoretical development for this application has not yet been brought up to the level of the other applications here addressed. Definite statements about achievable performance and best procedures and for this application are therefore postponed to a later occasion.

The assessment and compensation for different system transfer characteristics is vital for all forms of multidata target detection. System calibration is therefore an important concern. For the three cases to be discussed, calibration must achieve that for the pair of SAR images collected, the ground surface clutter appears equal so that targets can be discriminated by a process of linear combination of the multichannel data. Absolute calibration of the SAR images is not necessary; what is crucial is that equalization can be performed to achieve equality in representation for the ground clutter, without destroying the target signatures. The importance of channel equality is well known from microwave radar, in which array antenna performance critically depends on equalization between antenna elements. For microwaves, the basic method to achieve transfer function calibration as well as equalization is by adopting a corner reflector reference. For meter-wave radar, anechoic chambers and similar methods to avoid diffuse clutter returns do not work. The one way to isolate corner reflector returns from surrounding clutter is by SAR processing exploiting significant bandwidth and aperture angle. Hence, corner reflector measurements will not provide transfer characteristics for single frequencies or single target bearings, making them less applicable for calibration and equalization in the meter-wave case.

For the achieving equalization, there is an alternative to calibrated point targets, namely, using clutter returns. The STAP method used for microwave array antenna GMTI (as mentioned in chapter 3) is an example. Loosely corresponding to STAP, clutter-based equalization is very useful for low-frequency SAR, and the three detection techniques to be developed all utilize clutter-based equalization. The change detection method does so with the purpose of obtaining the probability of whether any particular image point contains a target; targets are then found by thresholding the image for a sufficiently high probability for this to be the case (say 99%). The probabilistic foundation of this method is the Bayes theorem. Crucial for the method is the formation of a likelihood quotient between the complementary hypotheses of there being a target versus there being no target in any given image position. It shares this concept of likelihood quotient with standard detection methods, which, however, relies on an a priori assumption of the clutter distribution (e.g., Gaussian) rather than an actual estimate from the SAR image itself.

The provided moving and subsurface detection methods enable a preliminary target screening of the SAR data. Full detection schemes may involve further actions. For moving targets, an obvious step is target tracking, improving detection certainty by considering the character of the track (e.g., discriminating between human and animal movers by the pattern of movement). For subsurface targets, change detection is a possibility, for example, decimating newly laid mines from permanent subsurface clutter, such as stones and tree roots. Generally, however, for mine detection, complementary detection steps of an entirely nonradar character (use of metal detectors and so on) are required. Once a full detection scheme has been set up, the Bayesian method of providing target probabilities will be applicable. Such probability analyses are outside the scope of the book and will not be further commented upon.

## 6.2 Bayesian Change Detection

Change detection performs detection of targets having relocated (including vanished from or entered) between two radar registrations. Methods for change detection have been around for many years [2]. For CARABAS, the first version was developed in the mid-nineties to late nineties [3]. Generally, SAR change detection entails three basic steps, namely 1) the equalization of reference and update (also known as mission) SAR images, 2) the formation of a measure stating the likelihood of a target, given the reference and update amplitudes, and 3) setting up a filter based on this measure, which distills out the targets.

Change detection can be coherent, in which the target likelihood measure depends on image phases, or incoherent, in which the measure only depends on the modulus of the amplitudes. Coherent methods exhibit a greater sensitivity. However, such enhanced sensitivity is not always an advantage; it may also make the comparison fluctuate for reasons not controllable or relevant. From the investigations done by FOI, little evident advantage has been found in using coherent methods. The conclusion is for change detection pertaining to detection of stationary surface targets with meter-wave radar, for the ground penetration and GMTI methods to be discussed coherency is crucial, and may well remain so extending these cases to also include change detection.

For noncoherent change detection, to obtain equality between reference and update images, the radar transceiver must be calibrated so that transceiver transfer parameters become as equal as possible. Low-frequency antenna patterns are highly robust, so if the antenna installation on the radar platform remains unaltered, significant antenna pattern variations are unlikely. Coherent detection schemes to be described are different in this sense, for equalization to match the higher sensitivity, the schemes that will be developed are adaptive, and involve fine-tuning of transfer characteristics on a case-to-case basis.

Equalization also includes the geometrical alignment of the reference and update images. Georeferencing of SAR images to subresolution accuracy is suitably implemented as a standard routine, since motion path accuracy anyhow must be very precise, as explained in the preceding chapter. This alignment precision has proved sufficient for noncoherent change detection (coherent change detection may require a still higher precision, depending on the particular way in which it is attempted). An image alignment precondition is that the two flights follow similar headings. Incidence angle variation to different standoff range can be compensated, relying on ground projected SAR images, as exploited in the FFBP process of the last chapter.

With the conditions stated, viz., transceiver calibration, image georectification, and alignment of SAR paths, the high degree of equality of low-frequency SAR clutter images follows, at least for their modulus or intensity representations. The route to powerful change detection schemes is thereby open. In the detailed design of such schemes, there are, however, three further observations, which ought to be weighted in the following.

1. Meter-wave terrain clutter statistics is significantly more dependent on scene content than what is the case for microwaves. The fundamental reason for this to be case is that meter-wave backscattering from a resolution cell generally stems from a single object, whereas for microwave, it is better regarded as the sum of backscattering from multiple scatterers. Microwave clutter backscatter is thus effectively a random process and becomes well represented by a standard probability distribution (normal, log normal, *K*, ...). As will be evident in the sequel, for meter-wave radar, the clutter distribution is a matter of the actual scene content, down to the presence of individual objects (e.g., actual targets, buildings, and fences). Estimation of clutter statistics from the actual SAR images thus turns out as an important prerequisite for an efficient change detection scheme.
2. Also the clutter variability between the reference and update images has a character heavily dependent on image content at hand. An open field will, even though it is covered by grass, yield essentially zero reflectivity at VHF. A significant portion of the SAR image pixels thus may have zero reflectivity and show no variability between images. VHF tree clutter apart from being weak is also very deterministic; it consists mainly of returns from tree trunks and since there are isotropic, they prove very similar response between the reference and update images. In contrast, cultural clutter from houses, power cables,

and metal fencing is very strong and nonisotropic. As a result, such structures cause relatively strong fluctuations between reference and update images.

3. The statistics of reflectivity variations differ if caused by targets appearing/disappearing or just being the unavoidable variations between repeated registrations of an unchanged scene. Obviously, it is this very difference, which enables the discrimination of targets in the midst of clutter. However, the estimation of clutter change statistics from the SAR image pairs becomes significantly erroneous unless this difference is taken into account. This requirement is accomplished by letting target presence be assigned a probability. High probability targets are considered preliminary detections and are excluded in a refined estimation of clutter changes. If the target probabilities thereby improve, the target detections are considered as valid, while further targets can be identified, removed, and clutter estimation further refined, in an iterative process.

An extensive work on change detection has been carried out by FOI and others (see [4–8]; note that [6] identifies change detection with a challenge problem for which data at least up to 2016 has been available from the U.S. AFRL WPAFB website). The approaches evaluated in the references have in common the use of a *likelihood quotient*, measuring the change significance by adopting a generalized likelihood quotient tests (GLRTs). Though in part similar to the present method, the approach here is new and unique in explicitly respecting observations 1–3. This approach, developed at Saab around 2013, has been demonstrated highly effective also in difficult situations of relatively small vehicle targets under heavy foliage or vehicles located in the immediate vicinity of strong cultural clutter.

Starting the formal development of this change detection scheme, let $a_U(\boldsymbol{x})$ and $a_R(x)$ be update and reference magnitudes (modulus of complex image amplitude) at the same pixel (the two images are aligned), whereas $\boldsymbol{x} = \boldsymbol{x}_T$ denotes the event that the update image contains a target at $\boldsymbol{x}$, which was not there in the reference image The quantity stating the target probability in this setting is the conditional probability density $p\langle \boldsymbol{x} = \boldsymbol{x}_T | a_U(\boldsymbol{x}), a_R(x)\rangle$, written in short form $p\langle \boldsymbol{x}_T | a_U, a_R\rangle$. While it would be difficult to assess $p\langle \boldsymbol{x}_T | a_U, a_R\rangle$ directly, according to the Bayes theorem

$$p\left\langle \boldsymbol{x}_T \middle| a_U, a_R \right\rangle = \frac{p\left\langle a_U \middle| \boldsymbol{x}_T, a_R \right\rangle p\left\langle \boldsymbol{x}_T \middle| a_R \right\rangle}{p\left\langle a_U \middle| a_R \right\rangle} \tag{6.1}$$

Below the probability densities, $p\langle a_U|x_T,a_R\rangle$ and $p\langle a_U|a_R\rangle$ will be derived. The probability density $p\langle \boldsymbol{x} = \boldsymbol{x}_T|a_R\rangle$ for target location is conditional and allowed to depend on reference image magnitude. For instance, a high $a_R$ magnitude can be taken to indicate a location already occupied and thus a less likely position for an emerging target. No such a priori assumption of target positions is, however, necessary. As targets are nonoverlapping, a simple assumption is that if there are $N$ pixels in the SAR image, $k$ targets, and each target is $M$ pixel large, then $p\langle \boldsymbol{x}_T|a_R\rangle = p\langle \boldsymbol{x}_T\rangle = Mk/N$.

The denominator of (6.1) is the probability density for an update magnitude given the reference magnitude. With ideal imaging conditions and with no targets, emerging $p\langle a_U|a_R\rangle$ would be zero unless $a_U = a_R$. For actual images, there are two mutually exclusive cases affecting $p\langle a_U|a_R\rangle$ corresponding to the two statements "the magnitudes $a_U,a_R$ are got, but there is no target at the pixel" and "the magnitudes are got and there is a target at the pixel." For the former statement, there pertains the pure clutter distribution $p\langle a_U|\boldsymbol{x}_C,a_R\rangle$ and for the latter the pure target distribution $p\langle a_U|\boldsymbol{x}_T,a_R\rangle$. Hence

$$p\langle a_U|a_R\rangle = p\langle a_U|\boldsymbol{x}_T,a_R\rangle p\langle \boldsymbol{x}_T|a_R\rangle + p\langle a_U|\boldsymbol{x}_C,a_R\rangle\left(1 - p\langle \boldsymbol{x}_T|a_R\rangle\right) \tag{6.2}$$

By insertion of the above parameter into (6.1), with $(1 - p\langle \boldsymbol{x}_T|a_R\rangle) \approx 1$, the following formula is obtained for the hypothesis of $K$ targets:

$$p_k\langle \boldsymbol{x}_T|a_U,a_R\rangle = \frac{1}{1 + \dfrac{N}{Mk}\dfrac{p\langle a_U|\boldsymbol{x}_C,a_R\rangle}{p\langle a_U|\boldsymbol{x}_T,a_R\rangle}} \tag{6.3}$$

From (6.3), it follows that $p\langle \boldsymbol{x}_T|a_U,a_R\rangle \leq 1$. Apart from the number of targets assumed, $p_k\langle \boldsymbol{x}_T|a_U,a_R\rangle$ is determined by the ratio $\eta(\boldsymbol{x}) = p\langle a_U|\boldsymbol{x}_T,a_R\rangle / p\langle a_U|\boldsymbol{x}_C,a_R\rangle$, which is the present correspondent to the likelihood ratio, mentioned above. For target probability to be high, $\eta$ must be high and vice versa. Target nominees are thus obtained by finding the strongest local maxima for $\eta$. Rather than attempting the present form of detection, which also assesses the number of targets in the scene, GLRT-related methods usually perform detection by applying a common threshold value across the image. The threshold is set to meet a requirement for false alarm rate in pure clutter. Detection performed in this way is known as *constant false alarm rate* (CFAR). From target collection trials, it is possible to erect so-called *receiver operating characteristic* (ROC) *curves* plotting the fraction of the true targets detected versus the number of false alarms for different threshold settings.

The target probability distribution $p\langle a_U|\boldsymbol{x}_T, a_R\rangle$ will be based on a target imaging model, developed below. For $p\langle a_U|\boldsymbol{x}_C, a_R\rangle$, a histogram or frequency count in the data will be adopted. Targets must be avoided in this count, adopting an iterative method of target excision, as postulated in observation 3. The specific numerical settings assumed below have been found effective for change detection of stationary vehicles under dense tree cover, using the CARABAS III low band 27.5 to 82.5 MHz. Evidently, the settings may need to be adjusted for other situations, like operating at other frequencies, across other types of terrain, or for other target classes.

Consider as given a SAR image pair $S_0$. One image constitutes the reference and the other the update. An association between pixel pairs and ground positions $\boldsymbol{x}$ is assumed. Let $S$ denote some subset of $S_0$, which is supposed free from targets, that is, just containing ground clutter stemming from objects, which have remained unchanged between the reference and the update images. Hence, $S$ can be used to estimate $p_k\langle a_U|\boldsymbol{x}_C, a_R\rangle$, where it is understood that the reduction of $S_0$ to $S$ excludes $k$ targets. The targets are supposed to be signified by the requirement that $p_K\langle \boldsymbol{x} = \boldsymbol{x}_T|a_U(\boldsymbol{x}), a_R(x)\rangle$ remains high not just at isolated points $\boldsymbol{x}$ but consistently within some area, representing minimum target footprint. Set this area to $2.5 \times 2.5$m, suitably agreeing with the CARABAS III low-band resolution. With the pixel size of $0.5 \times 0.5$m, there will be $M = 25$ pixels within the target footprint, so $p\langle \boldsymbol{x}_T\rangle = 24k/N$, where $N$ is the number of pixel pairs within $S_0$. By inserting $p_k\langle a_U|\boldsymbol{x}_C, a_R\rangle$ and $p\langle \boldsymbol{x}_T\rangle$ into (6.3), $p_k\langle \boldsymbol{x} = \boldsymbol{x}_T|a_U, a_R\rangle$ is computed. The result depends nonlinearly on $a_U, a_R$ and can therefore be expected to fluctuate strongly even for pixel-to-pixel changes of $\boldsymbol{x}$, in spite of the fact that pixel size is significantly finer than SAR resolution. In fact, fluctuations are inevitable, since outlayer clutter amplitudes weigh heavily in calculating $p_k\langle \boldsymbol{x} = \boldsymbol{x}_T|a_U, a_R\rangle$, while these are represented by very few image samples, the probability estimates will thereby be noisy. The remedy is to average $p_k\langle \boldsymbol{x}_T|a_U, a_R\rangle$ with respect to the target footprints. Denote such a footprint, centered on $\boldsymbol{x}$, by $\Lambda_T(\boldsymbol{x})$ and the corresponding average of $p_k\langle \boldsymbol{x}_T|a_U, a_R\rangle$ around $\boldsymbol{x}$ by $p_k\langle \boldsymbol{x} = \boldsymbol{x}_T|\Lambda_T(\boldsymbol{x})\rangle$. Reflecting the nonlinearity of the probability fluctuations, median rather than an arithmetic mean performs better and will be adopted.

The calculation of $p_k\langle \boldsymbol{x} = \boldsymbol{x}_T|\Lambda_T(\boldsymbol{x})\rangle$ is turned into an iterative detection scheme by repeated reselections of $S$. Thus, the first iteration considers that there is a single target in the scene, setting $k = 1$. The full SAR image $S_0$ serves as a temporary substitute for the target free subset $S$. The position of the first target is found by computing the clutter probability $p_0\langle a_U|x_C, a_R\rangle$ and thus $\eta_0(\boldsymbol{x})$, and based on that, $S_0$ represents target free clutter. Given $\eta_0(\boldsymbol{x})$ and $k = 1$, it is possible to compute a temporary estimate $\hat{p}_1\langle \boldsymbol{x}_T|\Lambda_T\rangle$. The position of the target is obtained as the position $\boldsymbol{x}_1$, which maximizes $\hat{P}_1\langle \boldsymbol{x} = \boldsymbol{x}_T | \Lambda_T(\boldsymbol{x})\rangle$. Once the potential target has been located to $\boldsymbol{x}_1$, a region around $\boldsymbol{x}_1$ is removed in order to make an improved estimate of the target presence. Denote by $\Lambda(\boldsymbol{x}_1) \subset S_0$ this region, consisting of 30 × 30 pixels (i.e., 15 × 15m), centered on $\boldsymbol{x}_1$. The size of $\Lambda(\boldsymbol{x}_1)$ has been selected to totally encompass single targets, without touching targets in the vicinity. The improved estimate $p_1\langle x_T | \Lambda_T(\boldsymbol{x}_0)\rangle$ is then obtained, basing $p_1\langle a_U | \boldsymbol{x}_C, a_R\rangle$ on the statistics of $S_1 = S_0 - \Lambda(\boldsymbol{x}_0)$. It is noted that in a number of instances of $a_U, a_R$ amplitudes characteristic of the region $\Lambda(\boldsymbol{x}_0)$ are removed in the new histogram count $p_1\langle a_U | \boldsymbol{x}_C, a_R\rangle$. Thereby, $p_1\langle a_U(\boldsymbol{x}) | \boldsymbol{x}_C, a_R a_U(\boldsymbol{x})\rangle < \hat{P}_0\langle a_U(\boldsymbol{x}) | \boldsymbol{x}_C, a_R(\boldsymbol{x})\rangle$ within $\Lambda(\boldsymbol{x}_0)$, whereupon it follows that $p_1\langle \boldsymbol{x}_0 = \boldsymbol{x}_T | \Lambda_T(\boldsymbol{x}_0)\rangle > \hat{P}_1\langle \boldsymbol{x}_0 = \boldsymbol{x}_T | \Lambda_T(\boldsymbol{x}_0)\rangle$. For a target to be detectable, $p_1\langle \boldsymbol{x}_0 = \boldsymbol{x}_T | \Lambda_T(\boldsymbol{x}_0)\rangle$ should assume a value approaching unity. If that is not the case, there may be more targets in the scene, or alternatively, there are no detectable targets at all.

To test for further targets, find a second target nominee by obtaining the preliminary probability estimate $\hat{P}_2\langle \boldsymbol{x} = \boldsymbol{x}_T | \Lambda_T(\boldsymbol{x})\rangle$ and find the position $\boldsymbol{x}_2$ within $S_1$, which maximizes this. Then form $S_2 = S_1\ \Lambda(\boldsymbol{x}_2)$ from which $p_2\langle \boldsymbol{x}_2 = \boldsymbol{x}_T | \Lambda_T(\boldsymbol{x}_2)\rangle$ is obtained. Again, $p_2\langle \boldsymbol{x}_2 = \boldsymbol{x}_T | \Lambda_T(\boldsymbol{x}_2)\rangle > \hat{P}_2\langle \boldsymbol{x}_2 = \boldsymbol{x}_T | \Lambda_T(\boldsymbol{x}_2)\rangle$. The iteration progresses in this way, proceeding until $\hat{P}_K\langle \boldsymbol{x}_K = \boldsymbol{x}_T | \Lambda_T(\boldsymbol{x}_K)\rangle \approx p_K\langle \boldsymbol{x}_K = \boldsymbol{x}_T | \Lambda_T(\boldsymbol{x}_K)\rangle$for some integer $K$. Thereby, $K - 1$ targets have been found. Each is given with the position $\boldsymbol{x}_k$; $k = 1,\ldots,K - 1$, and the probability $p_{K-1}\langle \boldsymbol{x}_K = \boldsymbol{x}_T | \Lambda_T(\boldsymbol{x}_K)\rangle$; $k = 1,\ldots,K - 1$ of being a target.

The task of determining the distribution $p\langle a_U | \boldsymbol{x}_T, a_R\rangle$ will now be tackled. The simple scattering model on which it is built entails neglecting the reflectivity coupling between the target and the background. Hence, the complex amplitude scattered from the update image at the position of the target is that of the reference image at the target location with the amplitude of the target superimposed. Effects like multiple scattering and shadowing are not incorporated.

An obvious choice for target statistics is that targets within a backscattering magnitude interval $a_{\min} \le a \le a_{\max}$, characteristic for the vehicle class considered, all appear with equal probability. Correspondingly, the magnitude probability density would be constant for all magnitudes in this interval. As it turns out, with this choice, elliptic integrals are encountered when computing $p\langle a_U | \boldsymbol{x}_T, a_R\rangle$. To avoid this complication, the close alternative of choosing the complex probability density constant will be adopted. The difference in detection performance between the two choices is inconsequential.

Denote complex probability densities by symbols $\wp$ to discern them from magnitude probability densities, denoted by a plain $p$. For probabilities not depending on phase, $p(a) = 2\pi a \wp(a)$. Hence, the assumption of constant complex probability density implies a magnitude probability density increasing linearly with magnitude. For the constant complex probability density, its actual form is determined by normalization as

$$\wp_T\left(ae^{i\phi}\right) = A \Rightarrow 1 = \oint \int_{a_{1-\min}}^{a} \wp_T\left(ae^{i\phi}\right)a'\,da'\,d\phi = A\pi\left(a_{\max}^2 - a_{\min}^2\right)$$
$$\Rightarrow \wp_T\left(ae^{i\phi}\right) = \begin{cases} 0 \Leftarrow a \le a_{\min} \\ \dfrac{1}{\pi\left(a_{\max}^2 - a_{\min}^2\right)} \Leftarrow a_{\max} \ge a > a_{\min} \\ 0 \Leftarrow a > a_{\max} \end{cases} \tag{6.4}$$

According to the assumption of no coupling between the target and the background, the update complex image amplitude is $a_U e^{i\phi_U} = a_T e^{i\phi_T} + a_C e^{i\phi_C}$, where $a_C e^{i\phi_C}$ is the complex update amplitude for only clutter and $a_T e^{i\phi_T}$ is the complex target amplitude. Since the phases are fully random

$$\wp\left\langle a_U \middle| x_T, a_R\right\rangle = \oint\int_0^\infty \wp_T\left(a_U - a_C e^{i\phi}\right)\wp\left\langle a_C \middle| x_C, a_R\right\rangle a_C\,da_C\,d\phi \tag{6.5}$$

The dispersion of $p\langle a_U|x_C,a_R\rangle$, that is, the expected spread of $|a_U - a_R|$ for clutter only, must be significantly smaller than the corresponding spread for $p\langle a_U|x_T,a_R\rangle$, that is, for targets; otherwise, change detection cannot work. Even when small, this clutter dispersion still influences $p\langle a_U|a_R\rangle$ and (6.1), since $p\langle a_T|a_R\rangle$ is small and $1 - p\langle a_T|a_R\rangle$ is large in (6.2). However, dispersion effects in (6.5) have a second-order influence on $p\langle a_U|a_R\rangle$ and can be neglected. With no dispersion, $\wp\langle a_C|x_C,a_R\rangle$ is just a delta function $B\delta(a_C - a_R)$. The constant $B$ is determined by normalization as

$$\oint\int_0^\infty \wp\left\langle a \middle| x_C, a_R\right\rangle a\,da\,d\phi = \oint\int_0^\infty B\delta\left(a - a_R\right)a\,da\,d\phi = 2\pi a_R B = 1 \Rightarrow B = \frac{1}{2\pi a_R} \tag{6.6}$$

Thereby

$$\int_0^\infty \wp\left\langle a_C \middle| x_C, a_R\right\rangle \wp_T\left(\left|a_U - a_C e^{i\phi}\right|\right) a_C\,da_C$$
$$= \frac{1}{2\pi^2\left({a_{\max}}^2 - {a_{\min}}^2\right)} \Leftarrow a_{1-\min} < \left|a_U - a_R e^{i\phi}\right| \le a_{1-\max} \tag{6.7}$$

For other cases of $|a_U - a_R e^{i\phi}|$, the integral is zero. It follows that (6.7) can be restated as:

$$\wp\left\langle a_U \middle| x_T, a_R \right\rangle = \frac{\Delta\phi\left(a_U, a_R\right)}{2\pi^2\left(a_{\max}^2 - a_{\min}^2\right)} \tag{6.8}$$

The meaning of the angular span $\Delta\delta(a_U, a_R)$ is as the angle of the intersection of the circle given by $|a_U - a_R e^{i\phi}|$, for varying $\phi$, with the nonzero support of $p_T$, which is the situation illustrated in Figure 6.1. The angular span can be computed analytically. Seemingly, the computation would have to consider several separate cases, depending on whether the circle meets either the inner or outer or both boundaries of the spectral support, or even that it is disjoint with these. As it turns out, the computation leads to second-order algebraic equations for which the real part determines where the circle meet the inner and outer spectral boundaries (as illustrated in Figure 6.1). The following common expression is obtained for the angles of these positions with respect to the imaginary axis:

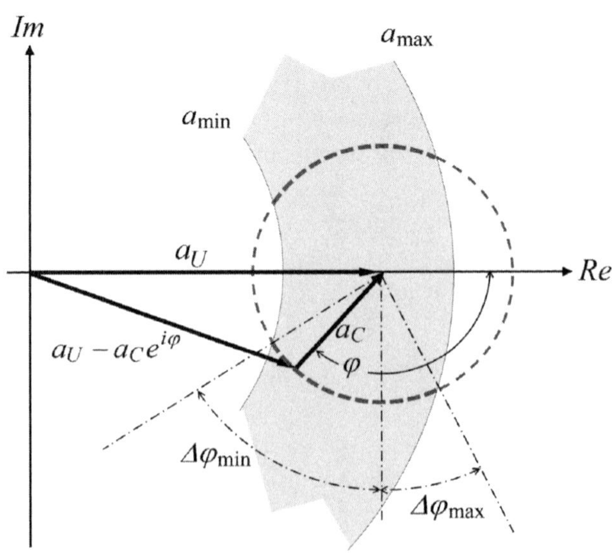


**Figure 6.1** Nonzero contributions to in the complex amplitude plane are within the arc being the intersection between possible target amplitudes and the clutter amplitudes leading to the measured update amplitude.

$$\Delta\phi_{\max/\min}\left(a,a_0\right)=\tan^{-1}\frac{a_{\max/\min}^2-a^2-a_0^2}{\mathrm{Re}\sqrt{a_{\max/\min}^2-\left(a-a_0\right)^2}\,\mathrm{Re}\sqrt{\left(a+a_0\right)^2-a_{\max/\min}^2}} \tag{6.9}$$

Zero real parts in the denominator of the $\tan^{-1}$ argument mean that $\Delta\phi_{\max/\min}=\pm\pi/2$. It follows that the different cases all become wrapped up in the expression for arc length:

$$\Delta\phi\left(a,a_0\right)=2\left|\Delta\phi_{\max}\left(a,a_0\right)-\Delta\phi_{\min}\left(a,a_0\right)\right| \tag{6.10}$$

Whenever the circle does not touch the inner or outer radii of the $p_T$ spectral support, the formula gives $\Delta\phi_{\max/\min}=\pm\pi/2$, that is, the upper and lower arc halves in Figure 6.1 close. Equation (6.10) thereby provides the correct arc length also in these situations, according to which there is an exact analytic expression for (6.8). Being an exact expression, (6.9) is normalized as can be verified numerically. A graphical representation of the magnitude probability density $p\langle a_U|x_T,a_R\rangle$, following from (6.11), is provided in the plate section (Color Plate 4).

Given (6.11), the remaining outstanding issue for the application of (6.3) is the details of the histogram estimation of $p_K\langle a_U|x_C,a_R\rangle$ for a given clutter distribution, from the SAR image pair sequence $S_K$; $K=0,1,\ldots$. The methodology for this task will now be treated. As always, this histogram analysis has to overcome the dilemma of finding a balance between the requirement for small bin sizes to obtain a finely graded statistics and the statistical errors caused by too few samples when bins become small. To attain minimum impact from these shortcomings, two measures will be taken:

1. Reflecting the $a_U,a_R$ correlation, $p_K\langle a_U|x_C,a_R\rangle$ is concentrated along some line $a_R\cos\alpha-a_U\sin\alpha=0$ in the $a_U,a_R$-plane. With radar imaging conditions being equal for the reference and update images, the correlation line would bisect the $a_U,a_R$-axes. In practice, standoff range and transceiver attenuation (reflecting variations in the radio-frequency interference environment) are generally not the same between registrations, whereby $\alpha$ may deviate from the 45° tilt. Given that a representative angle $\alpha$ has been found, it is possible to replace the update magnitude by magnitude change (in the direction orthogonal to this correlation line) by the substitution $a_U\rightarrow Da=a_U\cos\alpha-a_R\sin\alpha$. The advantage is that all $a_U,a_R$ events become concentrated

into a narrow $Da,a_R$ region. Hence, the $Da,a_R$ representation is more suitable than the $a_U,a_R$ representation, for reducing the histogram to as few bins as possible.

2. The distribution $p\langle Da|x_C,a_R\rangle$ thins when $Da$ and $a_R$ approach their extreme values. Hence, unless bin size is coarse, these situations will become represented by few, coincidentally even zero, events. Estimation errors will grow correspondingly large. To attain more evenly distributed estimation errors and recognizing that $p\langle Da|x_C,a_R\rangle$ is expected to approximately exhibit an exponential behavior, bin size is to be selected variable, allowing an exponential increase for large $Da,a_R$ values. It is recognized that the only cases where targets appearing in the update image become large are when $Da > 0$, even though Color Plate 4 of the plate section makes it evident that a nonzero probability $p\langle Da|x_T,a_R\rangle$ stretches out into negative $Da$, that is, destructive target clutter interference. Restricting the probability for target just to the case $Da > 0$, the $Da,a_R$ magnitudes maps onto one-to one fashion magnitudes $Db,b_R$, where the $b$-variables are suitable logarithms of the $a$-variables. Choosing the bin size constant in this $b$-scale, it will increase exponentially in the $a$-scale, as was required.

The angle $\alpha$ can be assumed the same for all iterations $k = 0,1,\ldots$. It can be determined from covariance matrix eigenvalues based on the initial image pair $S_0$. The covariance matrix technique was adopted in Chapter 4 for eliminating the nadir return. The present application is similar, though the context is different. Presently, denote $\boldsymbol{a} = (a_U,a_R)$ and seek the mean direction in the $a_U,a_R$-plane for all $a_U,a_R$-plane events. By the same argument as providing the nadir direction in Chapter 4, this is given by the maximum eigenvector of the covariance matrix $\boldsymbol{M} = \langle \boldsymbol{a}^\dagger \boldsymbol{a}\rangle$. The angle $\alpha$ is the angle of this eigenvector with respect to the $a_U$ axis, while it is the angle with respect to the $a_R$ axis for obtaining the $Da$ amplitudes.

Use of a logarithmic magnitude scale for clutter statistics is similar to considering amplitude ratios rather than amplitude differences in the original linear scale. Amplitude ratios seem the normal way for the various change detection schemes that have been suggested in the microwave regime (and stated to be a way of suppressing influence of speckle in that case) [7, 9, 10]. They have also been applied to low-frequency SAR in the FOI GLRT, when assuming a log-normal distribution.

A convenient rule for establishing the logarithmic $b$-scale is that unity bin size $\Delta b = 1$ corresponds to a bin size $\Delta a$ of half the magnitude $a$ (this canon applied to both $a_R$ and $Da$). The rule can be restated as $da/db = 0.5a$

(i.e., $a = Ce^{0.5b}$ with $C$ an integration constant). Referring to Chapter 3 (and the experience of CARABAS III high and low bands), low-frequency SAR images span reflectivity intervals never exceeding 60 dB. Thus, $b = 0$ can be set to correspond to $a_{\min} = 0.001a_{\max}$. Then

$$a = 0.001a_{\max} \times e^{0.5b} \Rightarrow b = 14, a = 1.0967a_{\max} \tag{6.11}$$

The $a$-scale will correspond to 15 $b$-magnitude bins, each of unity size. It is noted that

$$b = 2\frac{3 + {}^{10}\log\left(a/a_{\max}\right)}{{}^{10}\log e} \tag{6.12}$$

Cumulative probabilities are a simple and reliable way of assessing probabilities from histograms, whereas attempting this estimation by probability densities is difficult. To compute the cumulative probability $p_k\langle \le Db | x_C, b_R \rangle$ relational conditions are applied to $S_k$, selecting event subsets with reference amplitudes in any particular $b_R$-bin. Assume there are $N(b_R)$ such elements for each $b_R$-bin. Apply a second set of conditions on these, subselecting the events with change amplitudes $\le Db$. Assuming there are $N(<Db, b_R)$ elements in each subselection, the cumulative probability is $p_k\langle \le Db | x_C, b_R \rangle = N(<Db, b_R)/N(b_R)$.

While it is suitable to establish the clutter change probability in logarithmic scale, the target change probability is known a priori in the linear scale by (6.11). Transitioning this into logarithmic scale is of course feasible, but the coarse bins in the logarithmic scale will coarsen its dynamic precision for large amplitudes. This is an unwanted consequence, since in contrast to high change amplitudes being rare (i.e., of low probability), large target change amplitudes is what is to be expected (i.e., will have high probabilities). Thus, the error imposed by an unnecessarily coarse amplitude scale will be serious. The error can be avoided simply by interpolating clutter probabilities back to linear scale, after the histogram probabilities have been determined, where it is noted that the cumulative probability $P\langle \le Db | x_C, b_R \rangle$ equals the corresponding cumulative probabilities $P_k\langle \le Da(Db) | x_C, a_R(b_R) \rangle$. Higher order interpolation scheme (e.g., cubic) is noted to possibly break the monotonic decrease of probabilities with increasing amplitudes and is therefore not appropriate, so linear interpolation is used in the transition. Color Plate 3 and Color Plate 6 show probability densities obtained in this manner.

With these final remarks on estimating the clutter distribution, the complete procedure for attaining $p_k\langle x_T | a_U, a_R \rangle$ by means of (6.3) has been accounted for. The method will now be illustrated by two fairly diverse examples of target change detection in foliage.

The first example is tests done with CARABAS III in June 2014. The test site GåraG is a sparsely inhabited area, offering a variety of concealment conditions, up to and including what can be expected in cases like tropical rain forest, even though the site is located in the south of Sweden. Variations come from the terrain being hilly and forest a mixture of conifers and deciduous trees, mainly oak, birch, and spruce. Spruce, which is a dominant tree species in northern temperate forests, has relatively thin branches, causing moderate attenuation of the low-frequency radar signal. By contrast, oak has thick and long branches stretching out in nearly horizontal directions. Typical tropical tree species exhibit similar geometry, making the issue of low-frequency radar signal attenuation comparable for oak and tropical trees.

The test involved registrations at both the CARABAS III high and low bands 137.5 to 357.5 MHz and 27.5 to 82.5 MHz. The test purpose was to investigate radar performance with respect to the type of forest concealment as well as target size. Standoff range was moderate, around 1000m to the scene center of 1000m-wide SAR images. Helicopter altitudes for the high and low bands were 250m and 350m, respectively. The results would scale to larger standoff ranges by increasing platform altitude and possibly transmit power, according to the radar design principles laid down in Chapter 4.

High- and low-band SAR imagery of the Gåra test is provided in Color Plates 1 and 2 of the plate section. They contain two target locations, offering one difficult and one simpler detection case. The difficult one consisted of two fairly small vehicle targets (a Land Rover and a midsize personal car), parked right up to a stand of adult oak trees, obscuring the signal path from the CARABAS helicopter. The vehicles were oriented along the SAR path, aligned with each other but separated some 10m. Detection and separation of the two the targets were successful at the low band, but failed at the high band. That the higher band still offers some foliage penetration performance was demonstrated by the less demanding test, using a large dumper truck located in a spruce tree stand. This target was not available for the low band, nor was it necessary in view of the success for the significantly more difficult first site.

This test outcome is in accordance with the analysis of vegetation attenuation and tree clutter scattering in Chapter 2. In particular, it provides the experimental evidence that while oak trees cause significant attenuation at frequencies above 100 MHz, the attenuation effects diminish and become acceptably small below 100 MHz.

Color Plates 1, 2, and 7 (as discussed below) are a way to compactly illustrate the potential of the SAR imagery to perform change detection. Each figure is obtained as an equal blend of a reference (before target deployment) and update (after target deployment) image. The reference image is painted green and the update image red. Any image point not changing magnitude becomes yellow in the composite image, with a magnitude increase causing a tendency toward red and a magnitude reduction a tendency toward green. In particular, for change detection of emerging targets to be successful, these must stand out as red against a generally yellow clutter background. Evidently, the two vehicle targets at the first site appear in the low-band image in exactly this way, while the dumper is traceable in the high-band image in a similar fashion. It is noted that the low-band image exhibits a remarkably uniform color, that is, there is little in the way of clutter changes competing with the appearance of actual targets (as subsequent statistical analysis fully confirms). For the high band, coloring is more ambiguous. For instance, for the two vehicles at the first site, traces of their broadsides can be anticipated as vague reddish stripes (somewhat stronger for the Land Rover). However, there is an abundance of reddish (and indeed greenish) influences in the image implying that for the high band, the two vehicles will not be reliably detected. It should be noted, however, that the original dynamics of the SAR images has become significantly flattened in the process of graphical illustration; thus, relatively small intensity differences in the illustrations indicate significant magnitude differences in actual data.

In applying the Bayesian method for the clutter estimation considered, the test area is narrowed to 500m width, an image of that size contains a sufficient pixel amount for reliable clutter estimates. Moreover, for the purpose of testing the complex amplitudes of the footprint of the personal car has been cut out and added to three other randomly selected locations within the image, obtaining three further (now artificial) targets. The added copies is attenuated with factors 1, 0.75, and 0.5, respectively (yielding an RCS reduction with 0, 2.5, and 6 dB). The full span of different target sizes provides some indication on when targets become too weak for reliable detection.

Applying the Bayesian iterations to the image pair, all targets but the weakest become reliably detected and correctly located. The fall out of target probabilities during the six required iterations is provided in Table 6.1. As $k = 6$, the iteration stop condition $p_k\langle \boldsymbol{x}_k = \boldsymbol{x}_T | \Lambda_T(\boldsymbol{x})\rangle \approx p_{K+1}\langle \boldsymbol{x}_K = \boldsymbol{x}_T | \Lambda_T(\boldsymbol{x}_K)\rangle$ occurs. The effect of gradual removal of target responses out of the clutter probability density $p_k\langle Da | \boldsymbol{x}_C, a_R\rangle$ is depicted in Color Plate 3. The initially nonzero probability for large clutter variations $Da$ at low reference magnitude $a_R$ goes to zero as targets are excised out of the data. At the same time, there is a nonzero target probability density $p\langle Da | \boldsymbol{x}_T, a_R\rangle$ for these same events, in accordance with Color Plate 4. Color Plate 6 provides the final probability image $p_5\langle \boldsymbol{x} = \boldsymbol{x}_T | \Lambda_T(\boldsymbol{x})\rangle$. Note in particular the very marked contrast between the four targets appearing with near 100% probability of being targets and points in general with probabilities of being a target of the order −50 dB or less. The weakest target, being rejected out of the class of likely targets, is only assigned a probability of 8%, reflecting that the target mix with natural clutter changes, as is actually seen to be the case in Color Plate 3.

Table 6.1 can be summarized in a relation between what might be termed *detection rate* (the number of targets acknowledged compared to the overall number of targets) (DR) and probability of false alarms. The relation resembles the aforementioned ROC curve. Given $p\langle \boldsymbol{x} = \boldsymbol{x}_T\rangle$, it is noted that the proposition "there is a target at $\boldsymbol{x}$" is false with the probability $1 - p\langle \boldsymbol{x} = \boldsymbol{x}_T\rangle$. In other words, $1 - p\langle \boldsymbol{x} = \boldsymbol{x}_T\rangle$ is false alarm probability for a target at $\boldsymbol{x}$. The probability for the $k$ strongest target nominees to all be targets is the product $\Pi p\langle \boldsymbol{x} = \boldsymbol{x}_T\rangle$ for these. The probability that any of the $k$ strongest nominees is a false alarm is then $p\langle FA\rangle = 1 - \Pi p\langle \boldsymbol{x} = \boldsymbol{x}_T\rangle$. Thus, if detection statistics is obtained by there being $N$ true targets, the false alarm probability $p\langle FA\rangle$ holds for the detection rate $n/N$. From Table 6.1 and considering that there are five targets in all, the following relations are obtained: $p\langle FA\rangle = 0.36\%$ for $DR = 20\%$, $p\langle FA\rangle = 0.75\%$ for $DR = 40\%$, $p\langle FA\rangle = 1.18\%$ for $DR = 60\%$, $p\langle FA\rangle = 1.65\%$ for $DR = 80\%$, and $p\langle FA\rangle = 91.86\%$ for $DR = 100\%$. Of course, the choice in acknowledging the four strongest but not the weakest target nominee as true targets is obvious.

The Bayesian approach has the advantage of enabling false alarm probabilities to be computed rather than established by actually encountering false alarms and counting the frequency of their occurrence. To determine false

**Table 6.1**
Target Probabilities Found With the Bayesian Iteration Scheme in the Five-Target Detection Test

| Iteration | First target | Second target | Third target | Fourth target | Fifth target | Sixth target |
|---|---|---|---|---|---|---|
| 0 | 6.90% | | | | | |
| 1 | 10.80% | 7.49% | | | | |
| 2 | 99.63% | 99.59% | 2.2% | | | |
| 3 | 99.63% | 99.59% | 3.88% | 3.20% | | |
| 4 | 99.63% | 99.59% | 18.38% | 16.13% | 3.37% | |
| 5 | 99.64% | 99.60% | 99.57% | 99.52% | 8.28% | 5.68% |
| 6 | 99.64% | 99.60% | 99.57% | 99.52% | 8.27% | 5.68% |

alarm probabilities by frequency count requires very extensive experiments. It tends to become practically impossible as these probabilities (as is the case) become very small.

The method of Bayesian iteration was developed at Saab about three years ago, and test opportunities are still too few for it to have been thoroughly tested. However, apart from the test accounted for, another critical test were made in an area "Roxtuna," with a Land Rover parked in the concealment of spruce in the vicinity of strong cultural clutter from metal fencing and buildings (an abandoned jail) (see Color Plates 6 and 7). The jail fence is clearly traceable in Color Plate 7. The Land Rover becomes correctly detected with a probability 99.38%, whereas second and third target nominees are assigned probabilities 0.09% and 0.07%, respectively, allowing them to be relegated to random clutter fluctuations. The purely red indications in the middle three-quarters down in Color Plate 7 are weak responses (recall the dynamical flatness of the image) and turn out with target probabilities identically zero. The reason is that being purely red, the reference magnitude is effectively zero, and with small difference magnitude, these events rest in the $p\langle a_U|x_T,a_R\rangle = 0$ triangle in the bottom-left corner in Color Plate 4.

To investigate that the Roxtuna outcome was not a matter of coincidence, numerical experiments have been performed in which the scene artificially rearranged in various ways (e.g., the reference and update images were split into two and the halves were interchanged). On no occasion did the manipulations result in obviously erroneous probability assignments, giving a high target probability to nontarget occurrences.

The success of these Roxtuna detection experiments point to an important feature of the scheme, namely, the capability to separate reference to update magnitude changes due to emerging targets from large changes due to the presence of strong reflectors (like those of cultural clutter). In these situations, traditional CFAR schemes increase detection threshold values, making the detection schemes insensitive to small targets like vehicles. Some scrutinization of Color Plate 6 shows that CFAR is certain to fail in the Roxtuna case.

A first account of the Roxtuna and Gåra tests was given in [11]. It should be noted that the probability assignments in Table 6.1 slightly differ from [11], due to a slight variation in parameter settings. Moreover, the artificial targets have been put in new (arbitrarily selected) positions to illustrate the robustness of the scheme. Certainly, repositioning will have some effect on probability assignments, but the effect should be small, which proves to be the case. The Bayesian method has also been tested, with data from FOI, obtained with the older CARABAS II system operating against targets hidden by spruce forest. The method performed flawlessly also in these investigations [12].

## 6.3 Covariance Moving Target Extraction

Above, covariance methods were used for the equalization of the reference and update image. Using the covariance of the collected data to obtain channel equalization will now be applied to the DPCA technique for moving target detection and the polarimetric technique for subsurface target detection. In these applications, equalization will be performed on complex rather than magnitude data, as was the case of change detection. The objective prevails of extracting targets as residuals by identifying and removing a common dominant data term from the multiple data sets. In moving target detection, just as for change detection, the dominant component is the stationary ground response. Detecting subsurface targets, the ground surface response is the dominating and (up to a magnitude factor) equal component between vertical and horizontal polarizations. By its suppression, the response from the subsurface objects can be discriminated.

It is recalled from Chapter 3 that the DPCA technique exploited (at least) two longitudinally displaced but otherwise equally responding antennas, each creating a SAR image. The SAR images become shifted in time due to the separation between antennas. The time shift has no impact on the stationary ground image components, which are thus equal in the two images. For a moving target, the time shift imposes a phase shift between the target response in either image. Subtracting one image from the other the stationary ground components cancel, while the target responses do not. The possibilities for detecting the moving target rest in the magnitude of this residue, compared to errors in the equality between the antennas as well as additive noise.

For meter-wave DPCA, the antennas will have very weak directivity, as was discussed at length in Chapter 4. Thereby, the transfer characteristics (expressed in any of the alternative terms antenna diagram, transmissivity, or gain), become influenced by the antenna location on the airframe. Since fractional bandwidth is large, the influence will also vary across the frequency band. Without compensating for this difference, GMTI performance will be poor, as will be demonstrated. The required equalization can be attempted in two ways. The obvious method is by calibration, assuming antenna transfer characteristics to remain unchanging. Ideally, calibration determines calibration adjustment factor for each spectral point in the two-dimensional Fourier transform of the SAR images. However, as described above, it is difficult by any means to establish the transfer characteristics in the meter wave case. One way, not overly accurate, is computational, relying on an electromagnetic solver applied to a CAD mesh of the airframe structure. There is, however, a further difficulty in that the aircraft generally overfly the SAR scene with some yaw offset due to side winds, which not only affects the directional dependencies in the antenna radiation patterns but also changes the displacement between the antenna phase centers, causing additional forward/aft phase differences. The offset may vary during an ongoing registration. It becomes intertwined in the SAR processing motion compensation, which turns the Fourier domain multiplication of calibration data into a hard-to-define convolution filter. For these good reasons, equalization from calibration cannot be expected to perform well. The alternative and remaining way of equalization is data driven (i.e., adaptive), with eigenvalue decomposition of the covariance matrix being the tool, as will be discussed in detail now.

For the covariance method itself, the forward and rear antenna transfer properties must be expressible as two constants. Presently, allowance must be made for transfer characteristics to vary within the radar frequency band and antenna beam, so the following particular way of applying the method suggests itself. According to Chapter 3, SAR image frequency wavenumbers $k_x$ and $k_y$ have an approximate correspondence to radar frequency and antenna angular characteristics, namely, denoting frequency $f$ and azimuth radiation direction (off the broadside) $\theta$, and the relations $k = \sqrt{k_x^2 + k_y^2}$ and $\theta = \tan^{-1}(k_y/k_y)$ hold, where $k \approx \pi f \sin\psi_0/c$, with $\psi_0$ being the mean incidence angle of the SAR images. Applied to the DPCA case, the forward and rear gains translate into complex valued transfer functions $a^{\pm}(k_y,k_y)$, modifying the two-dimensional Fourier transforms $\hat{\gamma}^{\pm}_{k_x,k_y}$ of the ideal radiometrically calibrated forward and aft SAR images $\hat{\gamma}^{\pm}(x,y)$ according to $\hat{\gamma}^{\pm}_{k_x,k_y} \rightarrow \gamma^{\pm}_{k_y,k_y} = a^{\pm}(k_y,k_y)\,\hat{\gamma}^{\pm}_{k_x,k_y}$. The antenna transmissivity variations with respect to the incidence combine with range dependence and also enter into $a^{\pm}(k_y,k_y)$, though the influence is less direct. In all, equalization based on the covariance can be achieved by splitting up the image spectrum into spectral segments $\Omega$, sufficiently small that $a^{\pm}(k_y,k_y)$ can be considered constant within these. Choosing the segments $\Omega$ very small will result in very deep cancellation between the forward and rear SAR images, but at the price of having noise fluctuations affecting the transfer characteristics estimation. As this occurs, the mover response becomes degraded and eventually destroyed. On the other hand, too coarse segments results in poor equalization with associated errors (creating false alarms as will be seen).

In order to determine the appropriate segment size, represent the forward and aft SAR images by the model

$$\gamma^{\pm}_{k,\theta} = a^{\pm}_{\Omega}\left(\hat{f}_{k,\theta} + e^{i\alpha_{\pm}}\hat{b}_{k,\theta} + n^{\pm}_{k,\theta}\right) \tag{6.13}$$

Here, the SAR image spectrum is represented by the polar coordinates $k$ and $\theta$, equivalently to $k_x$ and $k_y$. Moreover, $\hat{h}_{k,\theta}$ is the calibrated noise-free response from the mover footprint in the SAR image and $\hat{f}_{k,\theta}$ is the calibrated noise-free stationary response of the terrain surrounding the mover, while $n^{\pm}(k,\theta)$ represent exterior noise appearing with the same variance but uncorrelated in the forward and aft antennas. It is important that the radar image data is calibrated as well as possible (e.g., from simulated antenna behavior) by which $a^{\pm}(k_y,k_y)$ will be close to unity. With zero clutter at the mover position $\alpha_+ = -\alpha_-$ (according to the theory developed in Chapter 3), whereas if the clutter overwhelms the mover, $\alpha_+ = \alpha_-$. The energy stemming from $\hat{f}_{k,\theta}$ is assumed much larger than that from $\hat{h}_{k,\theta}$. According to Parseval's theorem, the signal-to-noise ratio is the same in the image spectrum as in the SAR image. The possible suppression can obviously not improve on this, but may still be sufficient to detect the mover. This is so because even though the mover energy may be much smaller than the noise energy, it is focused to the very small mover footprint in the SAR image.

For each spectral segment, the covariance matrix is obtained by computing

$$M = \left\langle ZZ^{\dagger} \right\rangle_{\Omega} = \begin{pmatrix} \left\langle \left| \gamma_{k,\theta}^{+} \right|^{2} \right\rangle_{\Omega} & \left\langle \gamma_{k,\theta}^{+} \gamma_{k,\theta}^{-*} \right\rangle_{\Omega} \\ \left\langle \gamma_{k,\theta}^{-} \gamma_{k,\theta}^{+*} \right\rangle_{\Omega} & \left\langle \left| \gamma_{k,\theta}^{+} \right|^{2} \right\rangle_{\Omega} \end{pmatrix} \tag{6.14}$$

As indicated, averaging is performed over $(k_y,k_y) \in \Omega$. It is observed that $\boldsymbol{M}$ is Hermitian, implying that its eigenvalues, if nonzero, are orthogonal and real (the scalar product being defined as $\boldsymbol{x} \bullet \boldsymbol{y} = x_1^* y_1 + x_2^* y_2$ since vectors are complex). That any of the eigenvalues are zero corresponds to det $\boldsymbol{M} = 0$, that is, $|F_+|^2|F_-|^2 = |\langle F_-^* F_+ \rangle|^2$. However, the Cauchy inequality states

$$\left\langle \left| \gamma_{k,\theta}^{+} \right|^{2} \right\rangle_{\Omega} \left\langle \left| \gamma_{k,\theta}^{+} \right|^{2} \right\rangle_{\Omega} \geq \left| \left\langle \gamma_{k,\theta}^{+} \gamma_{k,\theta}^{-*} \right\rangle_{\Omega} \right|^{2} \tag{6.15}$$

with equality if and only if $\gamma_{k,\theta}^{+} = X\gamma_{k,\theta}^{-}$ for some constant $X$. Thus, equality prevails if $\hat{h}_{k,\theta} = n^{\pm}(k,\theta) = 0$ and transfer characteristics are constant within $\Omega$ in which case $X = a_-/a_+$.

Consider the eigenvalue problem for any arbitrary Hermitian matrix

$$\begin{pmatrix} a & z \\ z^* & b \end{pmatrix} \begin{pmatrix} \chi_1 \\ \chi_2 \end{pmatrix} = \lambda \begin{pmatrix} \chi_1 \\ \chi_2 \end{pmatrix} \Leftrightarrow \hat{H} \begin{pmatrix} \chi_1 \\ \chi_2 \end{pmatrix} = \begin{pmatrix} a-\lambda & z \\ z^* & b-\lambda \end{pmatrix} \begin{pmatrix} \chi_1 \\ \chi_2 \end{pmatrix} = 0 \tag{6.16}$$

The equation system is solvable if and only if det$\hat{\boldsymbol{H}}$ = 0 with solutions determined up to proportionality. Selecting normalization to be $|\chi_2|^2 + |\chi_1|^2 = 1$, where $\chi_2$ is real, the solution to (6.16) is unique and becomes

$$\chi_1 = \frac{-z}{\sqrt{(a-\lambda)^2 + |z|^2}};\ \chi_2 = \frac{a-\lambda}{\sqrt{(a-\lambda)^2+|z|^2}} \tag{6.17}$$

The requirement det$\hat{\boldsymbol{H}}$ = 0 is satisfied by considering det$\hat{\boldsymbol{H}}$ = 0 as the equation for the eigenvalues. Hence

$$(a-\lambda)(b-\lambda)-|z|^2 = 0 \Rightarrow \lambda_{\text{max/min}} = \frac{a+b}{2} \pm \frac{1}{2}\sqrt{4|z|^2 + (a-b)^2} \tag{6.18}$$

The eigendirections pertaining to the maximum and minimum eigenvalues are referred to as the maximum and minimum eigendirections. With the matrix entries of (6.14), the eigenvalues takes the form

$$\lambda_{\text{max/min}} = \frac{\left\langle\left|\gamma_{k,\theta}^{+}\right|^2\right\rangle_\Omega + \left\langle\left|\gamma_{k,\theta}^{-}\right|^2\right\rangle_\Omega \pm \sqrt{4\left|\left\langle\gamma_{k,\theta}^{+}\gamma_{k,\theta}^{-\,*}\right\rangle_\Omega\right|^2 + \left(\left\langle\left|\gamma_{k,\theta}^{+}\right|^2\right\rangle_\Omega - \left\langle\left|\gamma_{k,\theta}^{-}\right|^2\right\rangle_\Omega\right)^2}}{2} \tag{6.19}$$

For the moment, neglect the noise term in (6.13). Since $\hat{f}_{k,\theta}$ represent the terrain surrounding the mover, the terrain and mover frequency responses are independent whereby

$$\begin{aligned} \left\langle\left|\gamma_{k,\theta}^{\pm}\right|^2\right\rangle_\Omega &= \left|a_\Omega^{\pm}\right|^2\left(\left\langle\left|\hat{f}_{k,\theta}\right|^2\right\rangle_\Omega + \left\langle\left|\hat{h}_{k,\theta}\right|^2\right\rangle_\Omega\right) \\ \left\langle\gamma_{k,\theta}^{+}\gamma_{k,\theta}^{-\,*}\right\rangle_\Omega &= a_\Omega^{+}a_\Omega^{-\,*}\left[\left\langle\left|\hat{f}_{k,\theta}\right|^2\right\rangle_\Omega + e^{i(\alpha_+-\alpha_-)}\left\langle\left|\hat{h}_{k,\theta}\right|^2\right\rangle_\Omega\right] \end{aligned} \tag{6.20}$$

The minimum and maximum eigenvalues follow. Since $\left\langle\left|\hat{f}_{k,\theta}\right|^2\right\rangle >> \left\langle\left|\hat{h}_{k,\theta}\right|^2\right\rangle$

$$\begin{aligned} \lambda_{\min} &\approx -2\cos\left(\alpha_+ - \alpha_-\right)\frac{\left|a_\Omega^{+}\right|^2\left|a_\Omega^{-}\right|^2}{\left|a_\Omega^{+}\right|^2 + \left|a_\Omega^{-}\right|^2}\left\langle\left|\hat{h}_{k,\theta}\right|^2\right\rangle \\ \lambda_{\max} &\approx \left(\left|a_\Omega^{+}\right|^2 + \left|a_\Omega^{-}\right|^2\right)\left\langle\left|\hat{f}_{k,\theta}\right|^2\right\rangle \end{aligned} \tag{6.21}$$

Extraction of the stationary clutter image is performed either by forming the orthogonal complement of the data aligned with the maximum eigendirection or equivalently by projecting the data into the normalized minimum eigendirection. The latter is more straightforward and can be stated $\boldsymbol{x}_{\min}{}^{\dagger}\Gamma$, where

$$\Gamma(\kappa,\phi)=\begin{pmatrix}\gamma_{k,\theta}^{+}\\ \gamma_{k,\theta}^{-}\end{pmatrix};\ \chi_{\min}=\frac{\left|a_{\Omega}^{+}\right|}{\sqrt{\left|a_{\Omega}^{+}\right|^{2}+\left|a_{\Omega}^{-}\right|^{2}}}\begin{pmatrix}-\left\langle\gamma_{k,\theta}^{+}\gamma_{k,\theta}^{-*}\right\rangle_{\Omega}\Big/\left(\left\langle\left|\gamma_{k,\theta}^{+}\right|^{2}\right\rangle_{\Omega}-\lambda_{\min}\right)\\ 1\end{pmatrix} \tag{6.22}$$

It follows that:

$$\begin{aligned}\boldsymbol{\chi}_{\min}{}^{\dagger}\boldsymbol{\Gamma}&=\frac{\left|a_{\Omega}^{+}\right|}{\sqrt{\left|a_{\Omega}^{+}\right|^{2}+\left|a_{\Omega}^{-}\right|^{2}}}\begin{pmatrix}-a_{\Omega}^{-}/a_{\Omega}^{+} & 1\end{pmatrix}\begin{bmatrix}a_{\Omega}^{+}\left(\hat{f}_{k,\theta}+e^{i\alpha_{+}}\hat{h}_{k,\theta}\right)\\ a_{\Omega}^{-}\left(\hat{f}_{k,\theta}+e^{i\alpha_{-}}\hat{h}_{k,\theta}\right)\end{bmatrix}\\&=\frac{\left|a_{\Omega}^{+}\right|a_{\Omega}^{-}\left(e^{i\alpha_{-}}-e^{i\alpha_{+}}\right)}{\sqrt{\left|a_{\Omega}^{+}\right|^{2}+\left|a_{\Omega}^{-}\right|^{2}}}\hat{h}_{k,\theta}\end{aligned} \tag{6.23}$$

so the minimum eigenvector represents mover spectral amplitude. Performing an inverse Fourier transform of the image spectrum, a representation $h(x,y)$ of the ideal calibrated mover image $\hat{h}(x,y)$ is obtained. The faithfulness of $h(x,y)$ depends on the nature of the convolution kernel obtained by the Fourier transform of $\Omega$-dependent factor $|a_{\Omega}^{+}|a_{\Omega}^{-}/\sqrt{|a_{\Omega}^{+}|^{2}+|a_{\Omega}^{-}|^{2}}$ in (6.23). For this factor to have negligible influence, it must vary smoothly across the image spectrum, as indeed is what is to be expected. Furthermore, the validity of (6.23) depends on that $\Omega$ have been selected sufficiently small for the transfer characteristics (even though slow varying for $k,\theta$) to be nearly constant within $\Omega$. Since noise will upset the eigenvector estimates, so $\Omega$ cannot be selected arbitrarily small, however, as will be demonstrated.

To conveniently treat noise entering into the GMTI process, several assumptions are made. Since $a_{\Omega}^{\pm}$ are close to unity, the relation $||a_{\Omega}^{+}|^{2}-|a_{\Omega}^{-}|^{2}|$ $<<|a_{\Omega}^{+}|^{2}+|a_{\Omega}^{-}|^{2}$ holds. Equality between noise energy in the forward and aft antenna is assumed, that is, $\langle|n_{k,\theta}^{+}|^{2}\rangle_{\Omega}=\langle|n_{k,\theta}^{-}|^{2}\rangle_{\Omega}=W_{\Omega}$. Also, it is assumed that noise entering the antennas is independent, that is, $\langle|n_{k,\theta}^{+}{}^{*}n_{k,\theta}^{-}|^{2}\rangle_{\Omega}=0$. The SAR image signal to clutter ratio is the ratio with which the noise term in (6.13) fulfills the relation $\langle|\hat{f}_{k,\theta}|^{2}\rangle_{\Omega}<<W_{\Omega}$. The energy of the mover image component is generally much smaller, that is, $W_{\Omega}>>\langle|\hat{h}_{k,\theta}|^{2}\rangle_{\Omega}$. Adopting model

(6.13) with these assumptions, the modifications to the minimum eigenvalue (6.21) and thereby (6.22) follow as:

$$\lambda_{\min} \approx \frac{\left|a_\Omega^+\right|^2 + \left|a_\Omega^-\right|^2}{2} W_\Omega - 2\frac{\left|a_\Omega^+\right|^2 \left|a_\Omega^-\right|^2}{\left(\left|a_\Omega^+\right|^2 + \left|a_\Omega^-\right|^2\right)} \cos\left(\alpha_+ - \alpha_-\right)\left\langle\left|\hat{h}_{k,\theta}\right|^2\right\rangle$$
$$\Rightarrow \frac{\left\langle \gamma_{k,\theta}^+ \gamma_{k,\theta}^{-\,*} \right\rangle_\Omega}{\left\langle \left|\gamma_{k,\theta}^+\right|^2 \right\rangle_\Omega - \lambda_{\min}} = \frac{a_\Omega^+}{a_\Omega^{+\,*}} \frac{a_\Omega^+ a_\Omega^{-\,*}}{1 + \dfrac{\left|a_\Omega^+\right|^2 - \left|a_\Omega^-\right|^2}{2\left|a_\Omega^+\right|^2} \dfrac{W_\Omega}{\left\langle \left|\hat{f}_{k,\theta}\right|^2 \right\rangle_\Omega}} \tag{6.24}$$

From (6.21) and (6.24), $\lambda_{\max}/\lambda_{\min} = 2\langle|\hat{f}_{k,\theta}|^2\rangle_\Omega / W_\Omega$, that is, the eigenvalue quotient is twice the signal to noise ratio. This conclusion is, however, valid only if constant transfer conditions prevails for the segments $\Omega$. A check that these are sufficiently finely selected is thus that the eigenvalue quotient agrees with the expected signal-to-noise ratio. The test is not sensitive, however, since even a small amount of energy from stationary clutter leaking into the mover image may be sufficient to cause false alarms.

With noise imposed on (6.23), the following expression is obtained:

$$Q_\Omega = \frac{\left|a_\Omega^+\right|^2 - \left|a_\Omega^-\right|^2}{2\left|a_\Omega^+\right|^2} \frac{W_\Omega}{\left\langle \left|\hat{f}_{k,\theta}\right|^2 \right\rangle_\Omega} \Rightarrow$$
$$\chi_{\min}{}^\dagger \mathbf{\Gamma} \approx \frac{n_\Omega a_\Omega^-}{1 + Q_\Omega}\left[ n_{k,\theta}^- - n_{k,\theta}^+ + \left(e^{i\alpha_-} - e^{i\alpha_+}\right)\left(1 + \frac{Q_\Omega}{1 - e^{i(\alpha_+ - \alpha_-)}}\right)\hat{h}_{k,\theta} \right] \tag{6.25}$$

From (6.25), the impact of noise can be analyzed. Basically, the noise term $n_{k,\theta}^- - n_{k,\theta}^+$ sets the suppression limit of the stationary terrain. In the recent GMTI experiments with CARABAS III, which are accounted for below, signal-to-noise ratio was in the regime $\langle|\hat{f}_{k,\theta}|^2\rangle_\Omega / W_\Omega \approx 15$ to 20 dB. Since the noise term in (6.25) is $n_{k,\theta}^- - n_{k,\theta}^+$, the noise energy is 3 dB above $W_\Omega$. The suppression levels of up to 17 dB, which were obtained in the experiments, are thus in full accordance with expectations.

The quantity $Q_\Omega$ expresses the forward/aft antenna unbalance divided by signal-to-noise ratio. Hence, the lack of antenna balance upsets the eigenvalue estimates. Still no degradation of the mover image representation $h(x,y)$

will occur if $Q_\Omega$ is constant (or smoothly varying) across the image spectrum. Only when choosing $Q$ very small, $Q_\Omega$ will fluctuate strongly and rapidly, since the degree of averaging of $\langle|\hat{f}_{k,\theta}|^2\rangle_\Omega$ and $W_\Omega$ then becomes small. Defocusing of $h(x,y)$ will eventually lead to the total extinction of the mover SAR image. The image typically contains millions of pixels, meaning that the spectrum contains millions of points. Thereby, the spectrum might well contain a thousand subsets $\Omega$ without risking that this situation arises.

The covariance methods described were put to practical tests in GMTI trials conducted with CARABAS III in October 2016. The main objective was to confirm the GMTI modeling theory accounted for in Chapter 3, as well the covariance matrix approach just described. The anticipated detectability of slow and small targets, including humans walking in the open and undervegetation concealment, was tested. Measurements were performed with both horizontal and vertical polarizations. The RCS of humans modeled as cylinders erect on the ground surface was obtained from Chapter 2, and was further discussed in Chapter 3. RCS at CARABAS III frequencies is generally higher and certainly less dependent on incidence angle for horizontal polarization. Test outcome fully supported the predications that horizontal polarization provides better detectability, even though the targets remained traceable also at vertical polarization. In all, the outcome of the trials complied well with expectations.

A brief account of test results (restricted to horizontal polarization) is done with reference to Color Plates 8 to 11. Two test configurations were identified within the same general forested area (actually in the vicinity of the Gåra site shown in Color Plates 1 and 2). Color Plates 9 and 10 provide an example of tree cover with test persons walking back and forth along a straight narrow path in mainly spruce tree forest. Color Plate 11 exhibits an open configuration. Along a straight segment of a small lane two walkers, one person running, one biking person, and a personal car were moving back and forth. In this case, there was no screening from an overhead tree canopy, though clutter levels remained significant, with the lane surrounded by an abundance of young trees and bushes. The ground surface was comparatively rough and somewhat hilly. The walkers held a speed of 5 to 6 km/h, the runner a speed of 8 km/h, and the car and the bike speeds were in the interval 15 to 20 km/h. No strict synchronization of ground movements and radar overflights was imposed. Thus, the association between detections and actual movements relies on interpretations of the acquired imagery.

With the concealed configuration as an example, crucial spectral data properties are displayed in Color Plates 8 (left and right). To comprehend the charts, it should be understood that Color Plates 9 to 11 are map true with north pointing upward. The SAR path was selected at right angles to the heading of the movers, with the radar helicopter heading south for Color Plates 9 and 10 and ENE for Color Plate 11 (in Color Plates 9 and 10, the ground movements occurred along a forest path, and in Color Plate 11 along a lane, both traceable in the images). It is recalled from Chapter 3 that low-speed linear movement at right angles to the flight track yields full focusing of the mover; otherwise (but not necessary here), mover focusing can be regained by adjusting the platform speed assumed in the processing.

The charts in Color Plate 8 are related by Fourier transforms to Color Plate 9 and 10 and thus also map true. Color Plate 8 (left) exhibits the angle of the complex quotient $\gamma^{+}_{k,\theta}/\gamma^{-}_{k,\theta}$ of the forward and aft image spectra. The spectra have been restricted to the annular support of the main beam. Throughout the experiment, this support was set to range from 217.5 to 357.5 MHz and limited to a 60° sector (corresponding to the illumination/integration angle). As expected for radar data, the angle varies smoothly and consistently (apart from sporadic glitches of noise). The indication is therefore that data are dominated by the radar response and not noise. When collecting the common data for Color Plates 9 and 10, a strong easterly side wind prevailed, requiring the pilot to compensate by a yaw offset making the antenna beam tilt to a WSW direction in Color Plate 8. Color Plate 8 (right) exhibits the minimum to maximum eigenvalue quotient. The phase difference in Color Plate 8 (left) arises mainly due to the differences in interaction between the antennas and the airframe, depending on their different placements. The spectral subdivision must be sufficiently fine to mimic and compensate for these variations. A spectrum subdivision into 21 × 24 = 441 segments $\Omega$ seems appropriate. Increasing the number of spectral subdivisions was found to yield little.

As was established above, the eigenvalue quotient of Color Plate 8 (right) is twice the stationary-clutter-to-noise ratio. As stated, this ratio is in the regime in the regime 15 to 20 dB, so quotient values down to −18 dB are what can be expected at best. Bands of radio interference (the narrow circular bands in the figure) as well as the lower frequency end (where the transmissivity of the electrically small antennas is reduced) displays reduced cancellation depth, which is consistent with noise levels that are here higher. Hence, the chart in Color Plate 8 (right) is largely compliant with the expected noise variations across the spectrum.

The stationary and mover SAR images $f(x,y)$ and $h(x,y)$ are obtained by inverse Fourier transforms from eigendirection spectra $\boldsymbol{\chi}_{\max}{}^{\dagger}\Gamma$ and $\boldsymbol{\chi}_{\min}{}^{\dagger}\Gamma$, restricted to the frequency band and integration angle of Color Plate 8. The images in Color Plates 9 to 11 are all obtained by merging two SAR images with the stationary image tainted green and the mover image red. A constant color blending set up is exploited for all images. The image extension is 290 $\times$ 380 m$^2$, though covariance statistics is based on a 380 $\times$ 380 m$^2$ subsegment of the 1,000 $\times$ 1,000 m$^2$ CARABAS III standard image size. Covariance estimation is thus based on about 3.7 million 0.2m pixels.

Recall the analysis of motion nonlinearities in Chapter 3. It was concluded that with the type of wide band spectral support of Color Plate 8, the resulting mover SAR image may well display significant defocusing due to motion nonlinearities. The analysis in Chapter 3 concerned the requirement for the focused mover response to dominate clutter residuals. However, with this condition satisfied, further extending integration time and bandwidth, the mover image will still dominate clutter residuals while spreading over many resolution cells. It thereby exhibits a signature not only revealing the target but also the nonlinearities pertaining to its motion (Color Plate 11 exhibits several examples of such patterns). If available, this type information is obviously useful for the further discrimination of target type or just discriminating any mover from stationary clutter residuals, which may leak into $\boldsymbol{\chi}_{\min}{}^{\dagger}\Gamma$ as point-like responses (Color Plate 10 provides examples where this is the case). For the trials, the standoff range of the radar with respect to the image center was about 800m. With the 60° of integration angle and the helicopter moving at 25 m/s, the integration time for the images is around 30s. During integration time, the walkers thereby move for a distance of about 40m, while the car and bike move a distance of 120m to 160m.

While Color Plates 9 and 11 use the 441 $\Omega$ segments, Color Plate 10 shows the effect of poor equalization, using the entire spectral support as a single segment $\Omega$ for covariance estimation. The suppression measured as the energy $h(x,y)$ compared to the energy of $f(x,y)$ is $-16.0$ dB in Color Plate 9, while $-14.3$ dB in Color Plate 10. The energy wise dominant parts of the energy in $h(x,y)$ are the noise spread uniformly over the entire image as well as residuals of the stationary clutter entering due to the equalization errors. As is evident comparing Color Plates 9 and 10, these errors cause false alarms blending with, or even dominating, the true mover responses. Only Color Plate 9 allows unequivocal discrimination of the two human walkers. The arrows indicate their actual position on the forest path, with the offset from the path determining walking speed. As a straightforward application of the theory of Chapter 3, the offset of a mover, who heads at right angles to the SAR path, occurs at the angle of the tilt of the antenna beam, that is, the yaw offset due to strong side winds. Indications of movements just at the isolated tree in the open field may possibly be due to the wind. The tree and its immediate surroundings are located on a hill and very exposed. One of the walkers is noted to be well into the forest, the other not, without this situation causing any apparent difference in the strength of the responses.

From Chapter 3, it follows that a mover approaching the radar path at right angle still has a positive Doppler shift when the radar is abreast. Its representation in the SAR image will therefore be pushed in the direction of movement of the radar platform. Conversely, when the mover heads away from the radar path, its SAR image representation will be pushed to a position before the radar comes abreast. Since the helicopter in Color Plates 6 and 7 heads south, the walkers move into the forest.

Also, Color Plate 11 utilizes 421 $\Omega$ segments, by which suppression of $h(x,y)$ compared to $f(x,y)$ of −16.9 dB is obtained. With the entire spectral support for a single segment $\Omega$ (not illustrated), the suppression is −14.3 dB with a multitude of false alarms from stationary clutter entering into the image, just as Color Plate 10.

Several observations regarding the movers indicated in Color Plate 11 can be made. With the radar heading ENE, the targets to the east of the lane head north and those to the west head south. The bike and the car move at relatively high speed and can be identified in the SAR image by the correspondingly large displacements from the lane. The personal car was arranged to always lead in front of the bike. The image signature of the bike is evident, but that of the personal car is very weak, even though its RCS is very much larger than for any other of the targets. The one explanation is that at the onset of the radar registration, the car had nearly reached the end of its northerly run (where the lane bends in in the forest) and started to break. During initial light breaking, the response becomes defocused in azimuth and range, forming the just traceable residue in the image. Retarding harder, further contributions to $h(x,y)$ spread as noise. The weak but long tail from the otherwise well focused representation of the bike may possibly be due to the bike also breaking, though this happens at the end of the SAR registration. Walkers are seen well focused or just slightly defused in azimuth, indicating a good degree of linearity in their movement. The runner signature is far from focused, evidently containing image components displaced from the lane corresponding to up to twice the actual runner speed. A possible explanation is what is termed *micro-Doppler*, viz., that image these components represent just parts of the body moving at their individual speeds. For instance, legs move with up to twice the actual running speed.

With increased standoff range, a large integration angle corresponds to an extended integration time, during which the movement inevitably will include nonlinarites resulting in the mover response spreading over many resolution cells. In this case, detectability can be improved by splitting the imaging into a series of shorter integration intervals and obtaining the mover representation as a series of images, with the mover repeatedly appearing at displaced positions representing the actual motion of the mover (following the laws developed in Chapter 3). Thus, the mover becomes further discriminated from any false alarms by exhibiting a consistent pattern of motion.

The wideband covariance equalization of the SAR image spectrum readily generalizes to more than two channels. Rejection of stationary clutter is achieved by going to the subspace of the maximal eigendirection, much in line with what was done in the covariance-based scheme of nadir echo rejection of Chapter 4. The several dimensions of the remaining subspace allow bearing estimation of movers, resolving the velocity/location ambiguity of the present two channels of the DPCA system. Of course, for meter-wave SAR, this is only meaningful if sufficient separation between the channel phase centers is obtainable. For the low-speed ground movements considered here, misalignment is just a few meters, thus demanding very high angular resolution to be resolved. An antenna configuration enabling the required accuracy is perhaps possible with a bistatic arrangement utilizing two radar platforms, but would be no easy feat to realize.

Considering the literature on GMTI, the major part obviously concerns microwaves. Most work pertains to adaptive cancellation of clutter by STAP. Two distinct schools of performing GMTI can be discerned. The main bulk of publications represent radar data in the range Doppler domain to separate its stationary and nonstationary content. References [23] of Chapter 3, [13], and [14] are a few representatives of this school. The present approach, performing mover extraction based on SAR images, is an alternative addressed in [15–22] of Chapter 3 and [15–18]. Of course, generally for microwave GMTI coherent integration occurs for very short time (0.5s was cited in Chapter 3), making range and Doppler fully representative for the acquired data. SAR processing or operating on SAR images is then just a roundabout approach. In contrast, for low frequencies, prolonged integration time and SAR image formation are prerequisites for attaining the resolution necessary for mover detection. The publications [6.15] - [6.17] by Soumekh et al. come close to the methods developed by the team of the author. Reference [18] is another close example of utilizing meter-wave SAR images. The band used 220 to 450 MHz is similar to the CARABAS III high band. Data was obtained with the Sky SAR radar, offering one channel only, and so making coherent change detection necessary for clutter cancellation. The publication shows success, even though image equalization is not thoroughly carried out. To the experience of the author, and indeed as is tested in Color Plate 10, anything but very accurate equalization of channel transfer properties would result in an abundance of false alarms. In this sense, the present method also distinguishes itself from the Soumekh et al. publications, which do not adopt covariance equalization, nor do they address the adaption of equalization to the varying transfer characteristics of a wideband system.

There is also a set of publications relating to the Swedish LORA system (mentioned in Chapter 1) relying on the more conventional range Doppler approach to GMTI [19–21]. In the one experiment described, the target was significantly larger and faster (a truck) than the human walker targets, successfully detected with the CARABAS III high band, and the SAR-image-based methods here accounted for.

## 6.4 Polarimetric Subsurface Imaging

This last section exploits the covariance matrix method to further pursue the observations made in Section 2.5, regarding the similarity in the ground surface response for vertical and horizontal polarizations. The observations pertained to naked rough surfaces (i.e., bare ground), and were based on the SPM model developed in Chapter 2. Of course, they also applies to ground surface lightly overgrown (e.g. with grass), since such an overgrowth will not affect the meter-wave radar response. The findings of Chapter 2 can be summarized in that backscattering from the ground surface (as a function of frequency) is proportional to the Fourier transform of the ground height fluctuation (as a function of ground coordinates) with a proportionality constant depending on incidence angle and ground dielectric constant, as well as polarization. As was analyzed, backscattering from an underground target also significantly depends on these same parameters and, thus in particular, polarization. Polarization dependence is different for surface objects and the ground surface. The difference is well exhibited by formula (2.103), which provides the ratio of backscatter power from a nonpolarizing (e.g., sphere-like) subsurface target and the ground surface in the two cases of vertical and horizontal polarizations. The ratio is noteworthy demonstrating that at shallow depression angles, vertical surface backscattering tends to take precedence over horizontal compared to what is the case for backscattering from subsurface objects (which encounters the square of the vertically polarizing effect on refraction of a signal passing the ground surface). The ratio approaches its maximum value of 4 (i.e., 6 dB), as incidence angle tends to horizontal. It is reasonably close to this value already at practical (though shallow) angles of depression, say, at incidence angles larger than 70°.

Be as it may regarding polarization ratios, the strength of subsurface radar responses are typically weak and on par with the surface reflectivity fluctuations, even in the case of a bare ground surface. Linear subsurface structures like tunnels and buried cables can still (given the appropriate circumstances) be detectable from their shape, being objects of extended length. However, the response from small point-like buried targets, such as mines, will have no statistical significance, making detection by intensity tresholding not viable. The small amplitude changes that subsurface targets do give rise to will also be insufficient for the type of incoherent change detection discussed above. Coherent change detection may offer significantly higher sensitivity, but the geometric alignment and equalization issues are difficult to surmount. Hence, the observation made above regarding the polarimetric behavior of backscattering seems important in suggesting a possible solution. Just as for the GMTI method, conjunctly registered polarimetric measurements enable highly accurate geometric alignment and equalization. The GMTI method can to a high degree be reused in this new application.

To apply the covariance method, the detailed expressions (2.81) and (2.100) for vertically and horizontally polarized backscatters are not required, nor the exact expressions following from Fresnel's equations for subsurface targets. These have a meaning for SAR images that are absolutely calibrated, whereas the covariance methods relate only to quantities given up to proportionally, leaving calibration indeterminate. Also, the detailed expressions presume the conditions of the SPM model to be accurately fulfilled, which may perhaps not be the case. Presently, just consider as given the SAR images, that is, reflectivity density distributions, $\gamma_{V/H}(x,y)$ obtained for vertical and horizontal polarizations. For the SPM model, in particular, (2.81) and (2.100), $\gamma_{V/H}(x,y)$ commonly depend on the $x$ partial derivative of the ground surface height function (i.e., the ground surface slope), but with different factors of proportionality. Presently, this dependence will be substituted with a polarization independent surface object function $f(x,y)$, which determines either of $\gamma_{V/H}(x,y)$ by different factors of proportionality $\alpha_{V/H}$. It will not be necessary to consider the inner structure of this object function. For subsurface targets with no polarization affinity, the buried sphere scattering case may be considered a model. Then in line with (2.193), there is a polarization-independent target object function $h(x,y)$, which contributes to $\gamma_{V/H}(x,y)$, by the polarization-dependent constants $\beta_{V/H}$ arising from the refraction through the ground surface. In all, the following model of the scattering process is obtained:

$$\begin{bmatrix} \gamma_V(x,y) \\ \gamma_H(x,y) \end{bmatrix} = \begin{pmatrix} \alpha_V & \beta_V \\ \alpha_H & \beta_H \end{pmatrix} \begin{bmatrix} f(x,y) \\ h(x,y) \end{bmatrix} + \begin{bmatrix} n_V(x,y) \\ n_H(x,y) \end{bmatrix}$$
$$\Leftrightarrow \boldsymbol{\Gamma} = \boldsymbol{A}\begin{pmatrix} f \\ h \end{pmatrix} + \boldsymbol{n} \tag{6.26}$$

Here, $\boldsymbol{n}(x,y)$ is the additive noise across the SAR image. The model mimics and can be associated with the SPM scattering models. Indeed, (2.81) and (2.100) can then be applied and (6.26) assumes the form

$$\gamma_V(x,y) = -\frac{\sin^2\theta + \cos^2\tilde{\theta}}{\cos^2\left(\theta - \tilde{\theta}\right)} f(x,y) + \left[1 - \frac{\tan^2\left(\theta - \widehat{\theta}\right)}{\tan^2\left(\theta + \widehat{\theta}\right)}\right] h(x,y) + n_V(x,y)$$
$$\gamma_H(x,y) = f(x,y) + \left[1 - \frac{\sin^2\left(\theta - \widehat{\theta}\right)}{\sin^2\left(\theta + \widehat{\theta}\right)}\right] h(x,y) + n_H(x,y) \tag{6.27}$$

Note that at high incidence angles, the vertical and horizontal surface returns have opposite sign, and the subsurface responses are the same. As in Chapter 2, $\theta$ and $\tilde{\theta}$ are the angles of the incident and refracted signals, respectively, and are related by Snell's law. For complex index of refraction, also $\tilde{\theta}$ is complex.

Since $\alpha_{V/H}$ and $\beta_{V/H}$ all have the same magnitude and subsurface response is weak compared to the surface response $\langle|f|^2\rangle >> \langle|\hat{h}_{V/H}|^2\rangle$. Also assume $\langle fh^*\rangle = \langle fn^*_{V/H}\rangle = \langle hn^*_{V/H}\rangle = \langle n_H n^*_V\rangle = 0$. Thereby, although the context is different, the model (6.26) becomes similar to GMTI model (6.13) opening for a covariance method similar to that of the preceding section, but now applied for discriminating subsurface targets from the ground surface clutter.

Form the covariance matrix

$$\boldsymbol{M} = \begin{pmatrix} \langle \gamma_V^2 \rangle & \langle \gamma_V \gamma_H^* \rangle \\ \langle \gamma_H \gamma_V^* \rangle & \langle |\gamma_H|^2 \rangle \end{pmatrix} \tag{6.28}$$

Presently (with $\langle|n|^2\rangle = \langle|n_{V/H}|^2\rangle$)

$$\left\langle |\gamma_{V/H}|^2 \right\rangle = |\alpha_{V/H}|^2 \left\langle |f|^2 \right\rangle + |\beta_{V/H}|^2 \left\langle |h|^2 \right\rangle + \left\langle |n|^2 \right\rangle$$
$$\left\langle \gamma_V \gamma_H^* \right\rangle = \alpha_V \alpha_H^* \left\langle |f|^2 \right\rangle + \beta_V \beta_H^* \left\langle |h|^2 \right\rangle \tag{6.29}$$

Inserting (6.29) into the eigenvalue expression (6.19)

$$\lambda_{\min} = \frac{\left|\alpha_V\beta_H - \alpha_H\beta_V\right|^2}{\left(\left|\alpha_V\right|^2 + \left|\alpha_H\right|^2\right)}\left\langle |b|^2 \right\rangle + \left\langle |n|^2 \right\rangle$$
$$\lambda_{\max} = \frac{1}{2}\left(\left|\alpha_V\right|^2 + \left|\alpha_H\right|^2\right)\left\langle |f|^2 \right\rangle + \left\langle |n|^2 \right\rangle \tag{6.30}$$

Note when the vectors $(\alpha_V, \alpha_H)$ and $(\beta_V, \beta_H)$ are linearly dependent, $\alpha_V\beta_H - \alpha_H\beta_V = \det\boldsymbol{A} = 0$, whereby it follows that $\lambda_{\min} = \langle |n|^2 \rangle$, there is no longer any term in the minimum eigenvalue corresponding to the subsurface image.

In the same way as for GMTI, the subsurface image is obtained as

$$\boldsymbol{x}_{\min}{}^{\dagger}\boldsymbol{\Gamma} = \frac{\left(\alpha_V\beta_H - \alpha_H\beta_V\right)b + \alpha_V n_H - \alpha_H n_V}{\alpha_V\sqrt{\left|\alpha_V\right|^2 + \left|\alpha_H\right|^2}} \tag{6.31}$$

Again, the eigenvalue quotient is an indicator of the success in highlighting the subsurface image, with the subsurface image is reducing to zero when $(\alpha_V, \alpha_H)$ and $(\beta_V, \beta_H)$ become linearly dependent. From (6.31), noise remains constant implying that due to the degree of linera dependence, the subsurface image may be lost in the image noise. The associated signal factor (the reciprocal of the noise factor) is according to (6.31)

$$SF = \frac{\left|\alpha_V\beta_H - \alpha_H\beta_V\right|^2}{\left|\alpha_V\right|^2 + \left|\alpha_H\right|^2} \tag{6.32}$$

The signal factor is plotted for some typical ground media in Figure 6.2. For the typical soil constants given, it has a maximum at around 60° incidence angle, though it invariably remains significantly low implying significant amplitude loss in the process of canceling the surface return. The reason is that according to (6.27), the vectors $(\alpha_V, \alpha_H)$ and $(\beta_V, \beta_H)$ will never be at very large angles with each other. Hence, the signal loss is dictated by natural laws and cannot be circumvented by a suitable radar design. In contrast, by a proper GMTI antenna arrangement, mover data can be made essentially orthogonal to the stationary ground data in the GMTI covariance analysis. The signal loss in the GMTI stationary ground cancellation will therefore be small. Presently, the loss must be overcome having sufficient transmit power compared to radar range. A signal loss of 20 dB is balanced by a reduction of

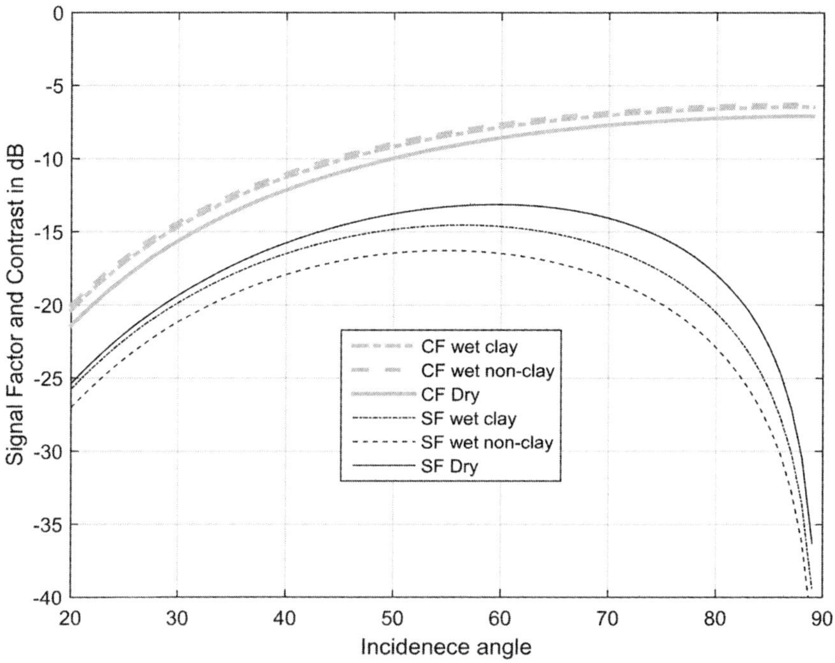


**Figure 6.2** Predicted contrast and noise factors for typical soil. The refraction values assumed are $\eta = 2.2 + 0.07i$ for dry soil, dry sand, and dry clay, $\eta = 4.4 + 0.24i$ for wet soil, and $\eta = 3.2 + 0.5i$ for wet clay.

radar range about five times, which will be the price for obtaining the capability of discriminating surface and subsurface returns.

Apart from the noise factor, another concern in the polarimetric surface cancelation is that prior to applying eigenvector decomposition of data, the initial contrast between surface and subsurface returns should be as favorable as possible. While the noise factor is determined by the strength of the subsurface residue compared to additive noise, the quotient in power ratio of vertical and horizontal responses for the surface and subsurface image determines the level of residual multiplicative noise. A quantity similar to the signal factor, but signifying the initial contrast between the surface and subsurface responses, may be defined as the contrast factor

$$CF = \left| 1 - \frac{\beta_V / \beta_H}{\alpha_V / \alpha_H} \right|^2 = \frac{\left| \alpha_V \beta_H - \alpha_H \beta_V \right|^2}{\left| \alpha_V \beta_H \right|^2} \tag{6.33}$$

The factor measures the degree to which the vectors ($\beta_V$,$\beta_H$) and ($\alpha_V$,$\alpha_H$) are differently directed. Thereby, it measures the robustness of the eigendirection separation with respect to multiplicative errors entering into the estimation of the covariance matrix. From (2.103), it is known that $1 \geq (\beta_V/\beta_H)/(\alpha_V/\alpha_H) \geq 0.5$, so $CF < 0.25$ is expected. The contrast factor was identified already in Section 2.5. As was analyzed there, *CF* improves monotonously for increasing incidence angle (as illustrated in Figure 6.2). In contrast, *SF* reaches a maximum around $\theta = 60°$ beyond which the subsurface response becomes heavily influenced by noise. At 60°, a deep suppression of the surface response is required and the covariance method must work very accurately. It would do so if the polarization response ratio for surface response would be truly constant, in accordance with the SPM theory. However, the theory is at best an approximation, and in analogous to the limitation of stationary ground suppression in the GMTI case, suppression will be limited by the approximation errors. The possible remedy is a large incidence angle yielding a sufficiently large contrast factor, to provide surface/subsurface separation also when the covariance matrix is poorly estimated. The cost doing so is a poor signal factor, implying a reduction of radar range.

A noteworthy observation is that while the Brewster angle is inherent in the model (6.30), it is certainly not the preferable incidence angle for minimizing surface clutter. To keep surface clutter low incidence angles significantly larger than the Brewster angle is preferable. Moreover, subsurface target information is mainly carried by the horizontally polarized data, not the vertically polarized. Though experimental subsurface data are limited, they support these conclusions clearly and unequivocally. The one account found published addressing the polarization dependence of subsurface SAR imaging is [23], which apart from the original publication on experiments in [22], constitute the meager list of references relevant for this topic.

Initial registrations of subsurface data have been made with CARABAS III. However, test conditions at the accessible test sites (i.e., locations in Sweden) have been far from ideal. An overriding issue has been the absence of penetrable, that is, essentially dry soil. Figure 6.2 states that for wet soil conditions, the noise factor increases sharply, severely reducing the possibilities for surface/subsurface separation, or for that matter verifying the scattering model developed here. In order to obtain at least some very limited set of data mimicking dry ground conditions, a site was artificially prepared in a 2013 test. Gravel quickly drains, so even after a short while of dry weather, the gravel retains the dielectric properties of dry sand, as was experimentally verified by network analyzer measurements (in contrast, sand was found to remain soaked for a long time). 65 tons of gravel was spread to form a shallow pile, into which subsurface targets were buried. To overall test area surrounding the pile consisted of flat farm fields of sand soil. The test was conducted late in the year, so the fields were all unvegetated, in all mimicking desert like terrain conditions.

The test results accounted for here are for the CARABAS III high-band 137.5 to 357.5 MHz, making simultaneous measurements with vertical and horizontal polarizations. Antennas were arranged with one high-band antenna vertical and the other horizontal at either side (see Figure 4.4.8 for the general CARABAS III antenna layout). To verify the dependence of polarization and incidence angle, a 40-L metal can test target was buried, with no part more shallow than 0.25m. A short standoff range of only 150m was selected to obtain sufficient signal-to-noise ratio, also in cases of a low signal factor. Operating at altitudes of 30m and 60m, the incidence angles of 77.5° and 65° were attained. Further test were made with no target buried into the gravel pile. It was established that the empty pile itself gave no distinguishable signature whatsoever in the SAR images. Another set of tests exploited a buried 1m corner reflector. However, such an object loses its unique property of a point scatterer when emerged into a lossy medium, making the results from this test difficult to interpret. For calibrated reflectivity measurements, a better option would have been a large metal sphere (for which RCS modeling of the sphere emerged in a lossy medium has been provided in Chapter 2).

Color Plates 12 to 17 illustrate the test results. The specific area imaged is a 50m-wide ground segment containing a 150m length of a dirt road. The road sides are rough and turfy, thus represented by the bright near horizontal pair of lines in the images. Below the road and to the left, there is pattern of lines. These are harrowing furrows. To the right, another field has the furrows at right angles and only the soil disturbances caused by the harrow turning are seen. The furrows are all quite shallow, less than 0.1m deep. The two fields are divided by a strip of turf. Above the road, a large flat expanse stretches out. The gravel pile and thus the buried target were located at the corner of the dividing turf strip and right-hand field.

Data covariance properties are well illustrated by the covariance statistics diagrams of Color Plates 12 and 15. These cross correlate the number of pixels falling into amplitude segments 2.5% of maximum image amplitude, for the horizontal and vertical polarization channels. As expected, the diagrams assume a generally elliptical shape stretching in the direction of the maximum covariance eigenvalue. Thereby, they also agree with the covariation of horizontal and vertical polarizations for the ground surface. Assuming dry ground condition with an index of refraction of $\eta = 2 + 0.07i$, directions calculated from $\tan^{-1}[\cos^2(\theta - \tilde{\theta})/(\sin^2\theta + \cos^2\tilde{\theta})$ have been inserted in the figures. The agreement with measured statistics is seen to be reasonable for 65° incidence and essentially perfect for 77.5°.

Color Plates 13 and 16 support the statement made above that visibility of subsurface responses is better in horizontal than vertical polarization. Moreover, the contrast between the subsurface and surface responses is more marked at 77.5° incidence than at 65°, a statement numerically verified in Table 6.2. The table compares the mean clutter level of any of the SAR images with the reflectivity of the subsurface target in each respective image. Evidently at 65°, the difference in contrast for the subsurface target between horizontal and vertical polarizations is 13.2 – 8.1 = 5.1 dB. At 77.5°, the difference is 18.4 – 11.3 = 7.1 dB. After eigendirection decomposition of surface and subsurface responses, the corresponding levels are 14.7 – 6.1 = 836 dB and 19 – 10.1 = 8.9 dB, respectively. These values are thus the suppression with which surface and subsurface targets have been categorized as belonging to respective category. Evidently, some advantage of exploiting the larger incidence angle has been found, though the advantage is limited, probably due to excessive additive noise.

Stepping back and considering the general problem of subsurface target detection, it should be stated that the present covariance method is not advocated as the universal way to detect subsurface targets. Generally, two somewhat opposite routes might be envisaged. The one discussed here is suited to subsurface target detection for rough naked or lightly vegetated terrain. It can perform this task by means of a single pass radar registration, achieving a fair amount of clutter suppression, but with the disadvantage of near grazing incidence. The resulting weak returns demand short-range operation, and radar angles beyond a few hundred meters are presumably impractical. The other situation is that ground clutter suppression for some reason is not a concern. It may be that surface clutter can be discriminated by change detection, which, for instance, is the case when surveying for newly laid mines. Another situation is that complementary sensors (e.g., optical) may be around

**Table 6.2**
Signal-to-Clutter Ratio for the Subsurface Target in the Horizontally and Vertically Polarized Surface and Subsurface SAR Images

| 65° Incidence | | | | 77.5° Incidence | | | |
|---|---|---|---|---|---|---|---|
| H-pol | V-pol | Max | Min | H-pol | V=pol | Max | Min |
| 13.2 | 8.1 dB | 6.1 dB | 14.7 dB | 18.4 dB | 11.3 dB | 10.1 dB | 19.0 dB |

for discriminating the surface response. There is also the extreme case of an entirely featureless ground surface, causing a very low response and thus no surface false alarms. In these cases, a better option than the polarimetric covariance method might be the exploitation of quite steep incidence angles, which according to Chapter 2 optimize the subsurface response compared to additive noise. Polarization would then no longer be the critical tool for detection; indeed, as incidence angles grow steeper, the polarimetric affinity of the radar returns become weak and are eventually lost.

## References

[1] Frölind, P. O., and L. M. H. Ulander, "Digital Elevation Map Generation Using VHF-Band SAR Data in Forested Areas", *IEEE Transactions on Geoscience and Remote Sensing*, Vol. 40, No. 8, August 2002.

[2] Rignot, E. J. M., and J. J. Van Zyl, "Change Detection Techniques for ERS-1 SAR Data", *IEEE Transactions on Geoscience and Remote Sensing*, Vol. 31, No. 4, July 1993.

[3] Ulander, L. M. H., et al., "Detection of Concealed Ground Targets in CARABAS SAR Images Using Change Detection", Proc. SPIE, 3721, pp. 243–252, April 1999.

[4] Novak, L. M., "Change detection for multi-polarization multi-pass SAR", Proc. SPIE, 5808, May 2005.

[5] Ulander, L. M. H., et al., "Change Detection for Low-Frequency SAR Ground Surveillance", *IEEE Proc. Radar Sonar Navigation*, Vol. 52, No. 6, December 2005.

[6] Lundberg, M., et al., "A Challenge Problem for Detection of Targets in Foliage", Proc. SPIE 6237, 2006.

[7] Ulander, L. M. H., and M. Lundberg, "Modeling of Change Detection in VHF- and UHF-band SAR", Proc. EUSAR, 2008.

[8] Ulander, L. M. H., et al., "Change detection of Vehicle-Sized Targets in Forest Concealment Using VHF- and UHF-band SAR", *IEEE AES Systems Magazine*, Vol. 26, No. 7, pp. 30–36, July 2011.

[9] Moser, G., and S. B. Serpico, "Generalized Minimum-Error Thresholding for Unsupervised Change Detection From SAR Amplitude Imagery", *IEEE Trans. on Geoscience and Remote Sensing*, Vol. 44, No. 10, October 2006

[10] Moser, G., and S. B. Serpico, "Unsupervised Change Detection From Multichannel SAR Data by Markovian Data Fusion", *IEEE Trans. on Geoscience and Remote Sensing*, Vol. 47, No. 7, July 2009.

[11] Hellsten, H., and R. Machado, "Bayesian Change Analysis for Finding Vehicle Size Targets In VHF Foliage Penetration SAR Data", *2015 International IEEE Radar Conference*, Oct. 27–30, 2015.

[12] Hellsten, H., et al., "Experimental Results on Change Detection Based on Bayes Probability Theorem", *IEEE IGARSS 2015*, July 26–31, 2015.

[13] Ender, J.H.G., "Space-Time Processing for Multichannel Synthetic Aperture Radar Data", *Electronics and Communications Engineering Journal*, February 1999.

[14] Greenewald, K., E. Zelnio, and A. Hero III, "Robust SAR STAP Via Kronecker Decomposition", *IEEE Trans. on Aerospace and Electronic Systems*, Vol. 52, No. 6, December 2016.

[15] Soumekh, M., "Signal Subspace Fusion of Uncalibrated Sensors with Application in SAR and Diagnostic Medicine", *IEEE Trans. on Image Processing*, Vol. 8, No. 1, January 1999.

[16] Majumder, U., et al., "Synthetic Aperture Radar Moving Target Indication Processing of Along-Track Monopulse Nonlinear Gotcha Data", *2009 IEEE Radar Conference*, Vol. 26, May 2009.

[17] Majumder, U., et al., "Spatially-Varying Calibration of Along-Track Monopulse Synthetic Aperture Radar Imagery for Ground Moving Target Indication and Tracking", *2010 IEEE Radar Conference*, May 10–14, 2010.

[18] Goldstein, J. S., et al., "Detection of Dismounts Using Synthetic Aperture Radar", *2010 IEEE Radar Conference*, May 10-14, 2010.

[19] Pettersson, M. I., "Detection of Moving Targets in Wideband SAR", *IEEE Trans. on Geoscience and Remote Sensing*, Vol. 40, No. 3, July 2004.

[20] Pettersson, M. I., L. M. H. Ulander, and H. Hellsten, "Simulations of Ground Moving Target Indication in an Ultrawideband and Wide-Beam SAR System", *SPIE Conference on Radar Processing, Technology and Applications IV*, Denver, Colorado, July 1999.

[21] Sjogren, T., et al., "Suppression of Clutter in Multichannel SAR GMTI", *IEEE Transactions on Geoscience and Remote Sensing*, Vol. 52, No. 7, 2013.

[22] Hellsten, H., S. Sahlin, and P. Dammert, "Polarimetric Subsurface SAR Imaging—Outcome of Theoretical Development and CARABAS III Tests", *International IEEE Radar Conference*, Oct. 13–17, 2014.

[23] Kalmykov, I. A., V. N. Tsymbal, and V. B. Yefimov, "Using Multifrequency Airborne Radar Complex MARS for Subsurface Remote Sensing", *Proc. European Microwave Conference*, 2005.

# About the Author

**Hans Hellsten**'s background is as a theoretical physics doctor. A ten-year period of research and graduate course lecturing provided close acquaintance with the main disciplines of physics. His involvement with meter-wave synthetic aperture radar commenced around 1984 at the Swedish Defense Research Agency FOI, which soon acquired significant national and international interest. As a consequence, a major Swedish foliage penetration meter-wave SAR research program was formed, running for almost a decade under the direction of the author. Among various achievements, the successful CARABAS I and II radar systems were produced.

In 2001 the author took the position of principal engineer for meter-wave SAR development in the Swedish radar industry. Transforming basic research into a product, the necessary development stages included system modeling to derive proper requirements for the basic radar components, such as transmitter, receiver, and antenna systems. The ambition was to obtain designs, which could be adapted into any aircraft, ranging from medium sized UAVs to helicopters and fighters. The various development stages were run in parallel with experimental verification, with the CARABAS III radar system built for this purpose.

A cohesive overall account of all the development was lacking. A book seemed to be the suitable way of documentation. The idea was well met by the author's employer, Saab. In late 2014, Artech House was contacted and agreed to the two-year project of composing the book.

Today, the author is the senior radar expert at Saab, and a professor in radar systems at Halmstad University. His technical achievements, including a large number of patents and many scientific publications, have earned him the Polhem Prize for technological innovation, the Gold Medal for outstanding engineering work from the Royal Swedish Academy of Engineering Sciences – IVA, and the Thulin Medal for aerospace technology achievements.

# Index

## Recent Titles in the Artech House Radar Series

Dr. Joseph R. Guerci, Series Editor

*Adaptive Antennas and Phased Arrays for Radar and Communications,* Alan J. Fenn

*Advanced Techniques for Digital Receivers*, Phillip E. Pace

*Advances in Direction-of-Arrival Estimation,* Sathish Chandran, editor

*Airborne Pulsed Doppler Radar, Second Edition*, Guy V. Morris and Linda Harkness, editors

*Basic Radar Analysis,* Mervin C. Budge, Jr. and Shawn R. German

*Bayesian Multiple Target Tracking, Second Edition* Lawrence D. Stone, Roy L. Streit, Thomas L. Corwin, and Kristine L Bell

*Beyond the Kalman Filter: Particle Filters for Tracking Applications,* Branko Ristic, Sanjeev Arulampalam, and Neil Gordon

*Cognitive Radar: The Knowledge-Aided Fully Adaptive Approach,* Joseph R. Guerci

*Computer Simulation of Aerial Target Radar Scattering, Recognition, Detection, and Tracking,* Yakov D. Shirman, editor

*Control Engineering in Development Projects,* Olis Rubin

*Design and Analysis of Modern Tracking Systems,* Samuel Blackman and Robert Popoli

*Detecting and Classifying Low Probability of Intercept Radar, Second Edition,* Phillip E. Pace

*Digital Techniques for Wideband Receivers, Second Edition,* James Tsui

*Electronic Intelligence: The Analysis of Radar Signals, Second Edition,* Richard G. Wiley

*Electronic Warfare in the Information Age*, D. Curtis Schleher

*Electronic Warfare Target Location Methods, Second Edition,* Richard A. Poisel

*ELINT: The Interception and Analysis of Radar Signals,* Richard G. Wiley

*EW 101: A First Course in Electronic Warfare,* David Adamy

*EW 102: A Second Course in Electronic Warfare,* David Adamy

*EW 103: Tactical Battlefield Communications Electronic Warfare,* David Adamy

*Fourier Transforms in Radar and Signal Processing, Second Edition,* David Brandwood

*Fundamentals of Electronic Warfare*, Sergei A. Vakin, Lev N. Shustov, and Robert H. Dunwell

*Fundamentals of Short-Range FM Radar,* Igor V. Komarov and Sergey M. Smolskiy

*Handbook of Computer Simulation in Radio Engineering, Communications, and Radar,* Sergey A. Leonov and Alexander I. Leonov

*High-Resolution Radar, Second Edition*, Donald R. Wehner

*Highly Integrated Low-Power Radars,* Sergio Saponara, Maria Greco, Egidio Ragonese, Giuseppe Palmisano, and Bruno Neri

*Introduction to Electronic Defense Systems, Second Edition,* Filippo Neri

*Introduction to Electronic Warfare,* D. Curtis Schleher

*Introduction to Electronic Warfare Modeling and Simulation,* David L. Adamy

*Introduction to RF Equipment and System Design*, Pekka Eskelinen

*Introduction to Modern EW Systems,* Andrea De Martino

*An Introduction to Passive Radar,* Hugh D. Griffiths and Christopher J. Baker

*Meter-Wave Synthetic Aperture Radar for Concealed Object Detection,* Hans Hellsten

*The Micro-Doppler Effect in Radar,* Victor C. Chen

*Microwave Radar: Imaging and Advanced Concepts,* Roger J. Sullivan

*Millimeter-Wave Radar Targets and Clutter,* Gennadiy P. Kulemin

*Modern Radar Systems, Second Edition,* Hamish Meikle

*Modern Radar System Analysis,* David K. Barton

*Modern Radar System Analysis Software and User's Manual, Version 3.0,* David K. Barton

*Monopulse Principles and Techniques, Second Edition*, Samuel M. Sherman and David K. Barton

*MTI and Pulsed Doppler Radar with MATLAB®, Second Edition*, D. Curtis Schleher

*Multitarget-Multisensor Tracking: Applications and Advances Volume III*, Yaakov Bar-Shalom and William Dale Blair, editors

*Precision FMCW Short-Range Radar for Industrial Applications*, Boris A. Atayants, Viacheslav M. Davydochkin, Victor V. Ezerskiy, Valery S. Parshin, and Sergey M. Smolskiy

*Principles of High-Resolution Radar,* August W. Rihaczek

*Principles of Radar and Sonar Signal Processing,* François Le Chevalier

*Radar Cross Section, Second Edition,* Eugene F. Knott, et al.

*Radar Equations for Modern Radar,* David K. Barton

*Radar Evaluation Handbook,* David K. Barton, et al.

*Radar Meteorology,* Henri Sauvageot

*Radar Reflectivity of Land and Sea, Third Edition*, Maurice W. Long

*Radar Resolution and Complex-Image Analysis*, August W. Rihaczek and Stephen J. Hershkowitz

*Radar RF Circuit Design,* Nickolas Kingsley and J. R. Guerci

*Radar Signal Processing and Adaptive Systems,* Ramon Nitzberg

*Radar System Analysis, Design, and Simulation,* Eyung W. Kang

*Radar System Analysis and Modeling,* David K. Barton

*Radar System Performance Modeling, Second Edition,* G. Richard Curry

*Radar Technology Encyclopedia*, David K. Barton and Sergey A. Leonov, editors

*Radio Wave Propagation Fundamentals,* Artem Saakian

*Range-Doppler Radar Imaging and Motion Compensation*, Jae Sok Son, et al.

*Robotic Navigation and Mapping with Radar,* Martin Adams, John Mullane, Ebi Jose, and Ba-Ngu Vo

*Signal Detection and Estimation, Second Edition*, Mourad Barkat

*Signal Processing in Noise Waveform Radar,* Krzysztof Kulpa

*Space-Time Adaptive Processing for Radar, Second Edition*, Joseph R. Guerci

*Special Design Topics in Digital Wideband Receivers,* James Tsui

*Theory and Practice of Radar Target Identification*, August W. Rihaczek and Stephen J. Hershkowitz

*Time-Frequency Signal Analysis with Applications,* Ljubiša Stanković, Miloš Daković, and Thayananthan Thayaparan

*Time-Frequency Transforms for Radar Imaging and Signal Analysis,* Victor C. Chen and Hao Ling

*Transmit Receive Modules for Radar and Communication Systems,* Rick Sturdivant and Mike Harris

For further information on these and other Artech House titles, including previously considered out-of-print books now available through our In-Print-Forever® (IPF®) program, contact:


Artech House
685 Canton Street
Norwood, MA 02062
Phone: 781-769-9750
Fax: 781-769-6334
e-mail: artech@artechhouse.com

Artech House
16 Sussex Street
London SW1V HRW UK
Phone: +44 (0)20 7596-8750
Fax: +44 (0)20 7630-0166
e-mail: artech-uk@artechhouse.com


Find us on the World Wide Web at: www.artechhouse.com